Shape Memory Alloys 2017

Special Issue Editor
Takuo Sakon

MDPI • Basel • Beijing • Wuhan • Barcelona • Belgrade

MDPI

Special Issue Editor
Takuo Sakon
Ryukoku University
Japan

Editorial Office
MDPI AG
St. Alban-Anlage 66
Basel, Switzerland

This edition is a reprint of the Special Issue published online in the open access journal *Metals* (ISSN 2075-4701) from 2017–2018 (available at: http://www.mdpi.com/journal/metals/special_issues/memory_alloys_2017).

For citation purposes, cite each article independently as indicated on the article page online and as indicated below:

Lastname, F.M.; Lastname, F.M. Article title. *Journal Name*. **Year**. *Article number, page range.*

First Edition 2018

ISBN 978-3-03842-773-5 (Pbk)
ISBN 978-3-03842-774-2 (PDF)

Table of Contents

About the Special Issue Editor

Takuo Sakon Doctor of Science, Professor of Faculty of Science and Technology, Ryukoku University. He was conferred a doctorate on for Physics department, Graduated School of Science, Tohoku University, Sendai, Japan in 1995. He served in Institute for Materials Research, Tohoku University (1995-2001) and Graduated School of Engineering and Resource Science, Akita University (2001-2012). His main research fields are condensed matter physics (experimental), magnetism of the Heusler type shape memory alloys in high magnetic fields, searching the magnetic and thermal functionality of Heusler type shape memory alloys. In an engineering field, development of the magnetic actuators by means of Shape memory alloys and other magnetic materials.

Preface to "Shape Memory Alloys 2017"

This Special Issue "Shape Memory Alloys 2017" is constructed articles reporting new and progressive research results, as well as reviews of particular classes of fundamental physics of the materials and their applications of shape memory alloys (SMAs). Through its 17 efficient articles, the reader will approach to researches related to SMAs with their peculiar magnetic, thermo-mechanical properties, superelasticity, plastic deformation and compression under pressure. These physical properties introduce a large number of applications as faster SMA actuators, application of medical devices, industrial joining parts,volts and magnetic/mechanical/thermal sensors. These articles are intended scientific researchers, professional engineers, students to obtain a better understanding in this field lately.

Takuo Sakon

Special Issue Editor

metals

MDPI

Editorial

Novel Research for Development of Shape Memory Alloys

Takuo Sakon

Department of Mechanical and Systems Engineering, Faculty of Science and Technology, Ryukoku University, Otsu, Shiga 520-2194, Japan; sakon@rins.ryukoku.ac.jp

Received: 7 February 2018; Accepted: 9 February 2018; Published: 11 February 2018

1. Introduction and Scope

Shape memory alloys have attracted much attention due to their attractive properties for applications as well as their basic aspects of deformation and transformation in structural and magnetic behavior. In 1951, the Au–Cd alloy was discovered [1]. After that, numberless shape memory alloys have been developed. A lot of applications of shape memory alloys were realized after the Ti–Ni alloy was found in 1963 and developed extensively in 1980s [2]. Recently, ferromagnetic shape memory alloys (FSMA) have been studied as candidates for highly functional materials [3]. Among FSMA, Ni_2MnGa is the most renowned [4]. The magnetostriction, or magnetic field induced strain (MFIS) is most featured phenomena. Few % MFIS were found for some Ni_2MnGa type single crystals. MFISs of 6.0% have been produced at room temperature in single crystals of $Ni_{49.8}Mn_{28.5}Ga_{21.7}$ (T_M = 318 K) by application of fields of order 400 kA/m (=0.50 T) under an opposing stress of order 1 MPa [5]. Fe–Pd and Fe–Pt alloys are also famous FSMA for MFIS. The strain is the result of field-induced twin boundary motion under an atmospheric pressure. A disordered Fe-31.2%Pd (at %) alloy (A1-type cubic) [6,7], and an ordered Fe_3Pt (L12-type cubic) ferromagnetic alloy [8,9], have attracted much interest due to the large MFISs. The alloys which was doped other elements are also studied. The magnetostriction studies on the premartensite phase of related Cr-substituted $Ni_2Mn_{1-x}Cr_xGa$ alloys were studied and robust 120 ppm magnetostriction was observed [10]. New alloys in the Ni–Mn–In, Ni–Mn–Sn, and Ni–Mn–Sb Heusler alloy systems that are expected to be ferromagnetic shape memory alloys have been studied [11]. A re-entrant magnetism was observed in some alloys [12,13]. These alloys are promising as a metamagnetic shape memory alloys with a magnetic field-induced shape memory effect and as magnetocaloric effect [14]. Consequently, these materials are finding use or are candidates as materials for sensors, actuators, magnetic refrigerator, etc. [15].

The dynamical functionality of the nickel–titanium (NiTi)-based alloy comes into the limelight. NiTi is an attractive alloy due to its unique functional properties, for example, shape memory effect (SMA), elastic deformation, super-elasticity, low stiffness, and damping characteristics [16,17]. In this special issue, physical properties are also shown for industrial objects, as joining between the different kind of metals, SMA bolts, and tubes. In our department in Ryukoku Univertsity, we are focused on robotics and applications (mechanical actuators). Therefore, our colleagues shed light on dynamic applications for SMAs.

2. Contributions

Xu et al. prepared $Cu_{71.5}Al_{17.5}Mn_{11}$ shape memory alloy by directional solidification by means of unique homemade equipment [18]. A large maximum recoverable strain of more than 11% was maintained due to the retained beneficial grain characteristics. Good superelastic behavior was observed.

Hu et al. investigated microstructural evolution of NiTi shape memory alloy (SMA) on the basis of heat treatment and severe plastic deformation (SPD) [19]. Consequently, SPD and subsequent aging contribute to enhancing the transformation temperature of martensite. The relation between softening of elastic constants and martensitic transformation has attracted considerable attention for many years and has been discussed [20].

Fukuda et al. investigated the relation between the softening of elastic constants and martensitic transformation in Fe_3Pt [21]. The softening in shear elastic coefficient $C' = (C_{11} - C_{12})/2$ is probably most strongly related to the formation of the face-centered-tetragonal with $c/a < 1$ (FCT1) martensite. This result implies that softening is most strongly related to the formation of the martensite.

Zhang et al. studied about physical mechanism for dynamic recrystallization of NiTi shape memory alloy subjected to local canning compression at various temperatures, 600, 700, and 800 °C [22]. In the case of 600 and 700 °C, continuous dynamic recrystallization and discontinuous dynamic recrystallization coexist in NiTi shape memory alloy. In the case of discontinuous dynamic recrystallization, the recrystallized grains are found to be nucleated at grain boundaries and even in grain interiors.

Jiang et al. studied about NiTi shape memory alloy (SMA) tube, which was coupled with mild steel cylinder in order to investigate deformation mechanisms of NiTi SMA tubes undergoing radial loading [23]. Microstructure was characterized by transmission electron microscope. When NiTi SMA tube is subjected to radial loading, strain induced martensite transformation is of great significance in the superelasticity of NiTi SMA, and reorientation and detwinning of twinned martensite lays a profound foundation for the shape memory effect of NiTi SMA.

Shiue et al. studied the infrared dissimilar joining $Ti_{50}Ni_{50}$ and 316L stainless steel using Cu foil in between Cusil-ABA and BAg-8 filler metals [24]. This study indicates great potential for industrial applications.

Jiang et al. studied the deformation behavior and microstructure evolution of NiTiCu SMA, which possesses martensite phase at room temperature based on a uniaxial compression test at the temperatures of 700~1000 °C [25]. Dislocations become the dominant substructures of martensite phase in NiTiCu SMA compressed at 700 °C. Martensite twins are dominant in NiTiCu SMA compressed at 800 and 900 °C. The microstructures resulting from dynamic recovery or dynamic recrystallization significantly influence the substructures in the martensite phase of NiTiCu SMA at room temperature.

Liang et al. studied NiTiFeNb and NiTiFeTa SMAs [26]. The microstructure, mechanical property, and phase transformation of NiTiFeNb and NiTiFeTa SMAs were investigated. As compared to NiTiFe SMA, quaternary NiTiFeNb and NiTiFeTa SMAs possess the higher strength, since solution strengthening plays a considerable role.

The processing map of $Ni_{47}Ti_{44}Nb_9$ (at %) shape memory alloy (SMA), which possesses B2 austenite phases and Nb phases at room temperature, is established by Wang et al. in order to optimize the hot working parameters [27].

Hu et al. investigated the texture evolution of NiTi shape memory alloy during uniaxial compression deformation at 673 K by combining crystal plasticity finite element method with electron back-scattered diffraction experiment and transmission electron microscope experiment [28]. Using the fitted material parameters, a crystal plasticity finite element method is used to predict texture evolution of NiTi shape memory alloy during uniaxial compression deformation. The simulation results agree well with the experimental ones. With the progression of plastic deformation, a crystallographic plane of NiTi shape memory alloy gradually rotates to be vertical to the loading direction, which lays the foundation for forming the <111> fiber texture.

Mitsui et al. observed field-induced reverse transformation in Co-doped Ni–Mn–In film by means of high field X-ray diffraction experiments under magnetic fields up to 5 T and temperature ranging from 293 to 473 K [29]. The reverse martensitic transformation induced by magnetic fields was directly observed in situ HF-XRD techniques.

Superelastic SMA bolts, which have a recentering capability upon unloading, are fabricated by Seo et al. as to solve drawbacks, and utilized by replacing conventional steel bolts in the partially restrained bolted T-stub connection [30].

Wu et al. investigated damping characteristics of inherent and intrinsic internal friction of Cu–Zn–Al Shape Memory Alloys [31]. The Cu-xZn-11Al SMAs are promising for practical high-damping applications under isothermal conditions because they possess good workability, low cost, acceptable mechanical properties, and the high damping capacities.

Adachi et al. made the phase diagram of $Ni_2MnGa_{1-x}Fe_x$ on the basis of the experimental results [32]. The magnetostructural transitions were observed at the Fe concentrations $x = 0.275$, 0.30, and 0.35. The obtained phase diagram was very similar to that for $Ni_2Mn_{1-x}Cu_xGa$ by Kataoka et al. [33]. The theoretical analysis of the phase diagram for $Ni_2Mn_{1-x}Cu_xGa$ showed that the biquadratic coupling term of the martensitic distortion and magnetization plays an important role in the interplay between the martensitic phase and ferromagnetic phase. Further theoretical investigation of the phase diagram for $Ni_2MnGa_{1-x}Fe_x$ is needed.

Umetsu et al. performed the specific heat measurements at low temperatures for $Ni_{50}Mn_{50-x}In_x$ alloys to determine their Debye temperatures θ_D and electronic specific heat coefficients γ, and investigated the change in the density of states during the martensitic phase transformation [34]. The γ was slightly larger in the parent phase, in good agreement with the reported density of states around the Fermi energy obtained by the first-principle calculations. The martensite (M) phase ($x \leq 15$), θ_D decreases linearly and γ increases with increasing Indium content.

Anelastic properties of Ti–Ni-based alloys are widely explored, both for microstructural characterization of the alloys and for their application as high-damping materials [35]. Sapozhnikov et al. investigated the linear and non-linear internal friction, effective Young's modulus, and amplitude-dependent modulus defect of a $Ti_{50}Ni_{46.1}Fe_{3.9}$ alloy after different heat treatments, affecting hydrogen content [36]. They found that the internal friction maximum is associated with a competition of two different temperature-dependent processes affecting the hydrogen concentration in the core regions of twin boundaries.

Carl et al. investigated the SMA with high martensitic transformation temperature [37]. Small changes in at % Ni have a dramatic effect on the transformation temperatures of the NiTi-20 at % Zr system, even more so than in binary NiTi. The transformation temperature can be tuned for a given application through aging treatments. Ni–Ti-based shape memory alloys (SMAs) have now become an important technological material for a wide array of applications not only for specifically medical devices [38] but actuators above room temperature [39].

As introduced here, many fundamental research results are shown in 17 articles. There are many SMA materials in which there is possibility of the application for mechanical and/or magnetic actuators, springs, bolts, magnetic refrigeration, sensors, etc.

I hope the results of the research in this special issue contribute to further development of SMAs.

Conflicts of Interest: The author declares no conflict of interest.

References

1. Chang, L.C.; Read, T.A. Plastic Deformation and Diffusionless Phase Changes in Metals-the Gold-Cadmium Beta Phase. *Trans. AIME* **1951**, *189*, 47–52. [CrossRef]
2. Buehler, W.J.; Gilfrich, J.V.; Wiley, R.C. Effect of Low-Temperature Phase Changes on the Mechanical Properties of Alloys near Composition TiNi. *J. Appl. Phys.* **1963**, *34*, 1475–1477. [CrossRef]
3. Chernenko, V.A.; Besseghini, S. Ferromagnetic shape memory alloys: Scientific and applied aspects. *Sens. Actuators A Phys.* **2008**, *142*, 542–548. [CrossRef]
4. Ullakko, K.; Huang, J.K.; Kantner, C.; O'Handley, R.C.; Kokorin, V.V. Large magnetic-field-induced strains in Ni_2MnGa single crystals. *Appl. Phys. Lett.* **1996**, *69*, 1966–1968. [CrossRef]
5. Murray, S.J.; Marioni, M.; Allen, S.M.; O'Handley, R.C. 6% magnetic-field-induced strain by twin-boundary motion in ferromagnetic Ni–Mn–Ga. *Appl. Phys. Lett.* **2000**, *77*, 886–888. [CrossRef]
6. Kakeshita, T.; Fukuda, T.; Sakamoto, T.; Takeuchi, T.; Kindo, K.; Endo, S.; Kishino, K. Martensitic transformation in shape memory alloys under magnetic field and hydrostatic pressure. *Mater. Trans.* **2002**, *43*, 887–892. [CrossRef]

7. Sakamoto, T.; Fukuda, T.; Kakeshita, T.; Takeuchi, T.; Kishino, K. Magnetic field-induced strain in iron-based ferromagnetic shape memory alloys. *J. Appl. Phys.* **2003**, *93*, 8647–8649. [CrossRef]
8. Kakeshita, T.; Fukuda, T. Conversion of variants by magnetic field in Iron-based ferromagnetic shape memory alloys. *Mater. Sci. Forum* **2003**, *426–432*, 2309–2314. [CrossRef]
9. Fukuda, T.; Sakomoto, T.; Inoue, T.; Kakeshita, T.; Kishio, K. Influence of magnetic field direction on recoverable strain due to rearrangement of variants in Fe3Pt. *Trans. Mater. Res. Soc. Jpn.* **2004**, *29*, 3059–3060.
10. Sakon, T.; Fujimoto, N.; Kanomata, T.; Adachi, Y. Magnetostriction of $Ni_2Mn_{1-x}Cr_xGa$ Heusler Alloys. *Metals* **2017**, *7*, 410. [CrossRef]
11. Oikawa, K.; Ito, W.; Imano, Y.; Sutou, Y.; Kainuma, R.; Ishida, K.; Okamoto, S.; Kitakami, O.; Kanomata, T. Effect of magnetic field on martensite transformation of $Ni_{46}Mn_{41}In_{13}$ Heusler alloy. *Appl. Phys. Lett.* **2006**, *88*, 122507. [CrossRef]
12. Kainuma, R.; Imano, Y.; Ito, W.; Sutou, Y.; Morino, H.; Okamoto, S.; Kitakami, O.; Oikawa, K.; Fujita, A.; Kanomata, T.; et al. Magnetic-field-induced shape recovery by reverse phase transformation. *Nature* **2006**, *439*, 957–960. [CrossRef] [PubMed]
13. Sakon, T.; Sasaki, K.; Numakura, D.; Abe, M.; Nojiri, H.; Adachi, Y.; Kanomata, T. Magnetic field-induced transition in Co Doped $Ni_{11}Co_9Mn_{21.5}Ga_{18.5}$ Heusler Alloy. *Mater. Trans.* **2013**, *54*, 9–13. [CrossRef]
14. Sakon, T.; Kitaoka, T.; Tanaka, K.; Nakagawa, K.; Nojiri, H.; Adachi, Y.; Kanomata, T. Vadim Glebovsky. Magnetocaloric and magnetic properties of meta-magnetic Heusler alloy $Ni_{41}Co_9Mn_{31.5}Ga_{18.5}$. In *Progress in Metallic Alloys*; InTech: Rijeka, Croatia, 2016; pp. 265–287.Ni41Co9Mn31.5Ga18.5. In *Progress in Metallic Alloys*; InTech: Rijeka, Croatia, 2016; pp. 265–287.
15. Niemann, R.; Hahn, S.; Diestel, A.; Backen, A.; Schultz, L.; Nielsch, K.; Wagner, M.F.-X.; Fähler, S. Reducing the nucleation barrier in magnetocaloric Heusler alloys by nanoindentation. *Appl. Phys. Letter. Mater.* **2016**, *4*, 064101. [CrossRef]
16. Otsuka, K.; Ren, X. Physical metallurgy of Ti–Ni-based shape memory alloys. *Prog. Mater. Sci.* **2005**, *50*, 511–678. [CrossRef]
17. Elahinia, M.; Moghaddam, N.S.; Andani, M.T.; Amerinatanzi, A.; Bimber, B.A.; Hamilton, R.F. Fabrication of NiTi through additive manufacturing: A review. *Prog. Mater. Sci.* **2016**, *83*, 630–663. [CrossRef]
18. Xu, S.; Huang, H.; Xie, J.; Kimura, Y.; Xu, X.; Omori, T.; Kainuma, R. Dynamic Recovery and Superelasticity of Columnar-Grained Cu–Al–Mn Shape Memory Alloy. *Metals* **2017**, *7*, 141. [CrossRef]
19. Hu, L.; Jiang, S.; Zhang, Y. Role of Severe Plastic Deformation in Suppressing Formation of R Phase and Ni_4Ti_3 Precipitate of NiTi Shape Memory Alloy. *Metals* **2017**, *7*, 145. [CrossRef]
20. Dai, L.; Cullen, J.; Wuttig, M. Intermartensitic transformation in a NiMnGa alloy. *J. Appl. Phys.* **2004**, *95*, 6957–6959. [CrossRef]
21. Fukuda, T.; Kakeshita, T. Lattice Softening in Fe3Pt Exhibiting Three Types of Martensitic Transformations. *Metals* **2017**, *7*, 156. [CrossRef]
22. Zhang, Y.; Jiang, S.; Hu, S. Investigation of Dynamic Recrystallization of NiTi Shape Memory Alloy Subjected to Local Canning Compression. *Metals* **2017**, *7*, 208. [CrossRef]
23. Jiang, S.; Junbo, Y.; Hu, L.; Zhang, Y. Investigation on Deformation Mechanisms of NiTi Shape Memory Alloy Tube under Radial Loading. *Metals* **2017**, *7*, 268. [CrossRef]
24. Shiue, R.K.; Wu, S.K.; Yang, S.H.; Liu, C.K. Infrared Dissimilar Joining of $Ti_{50}Ni_{50}$ and 316L Stainless Steel with Copper Barrier Layer in between Two Silver-Based Fillers. *Metals* **2017**, *7*, 276. [CrossRef]
25. Jiang, S.; Sun, D.; Zhang, Y.; Hu, L. Deformation Behavior and Microstructure Evolution of NiTiCu Shape Memory Alloy Subjected to Plastic Deformation at High Temperatures. *Metals* **2017**, *7*, 294. [CrossRef]
26. Liang, Y.; Jiang, S.; Zhang, Y.; Yu, J. Microstructure, Mechanical Property, and Phase Transformation of Quaternary NiTiFeNb and NiTiFeTa Shape Memory Alloys. *Metals* **2017**, *7*, 309. [CrossRef]
27. Wang, Y.; Jiang, S.; Zhang, Y. Processing Map of NiTiNb Shape Memory Alloy Subjected to Plastic Deformation at High Temperatures. *Metals* **2017**, *7*, 328. [CrossRef]
28. Hu, L.; Jiang, S.; Zhang, Y. A Combined Experimental-Numerical Approach for Investigating Texture Evolution of NiTi Shape Memory Alloy under Uniaxial Compression. *Metals* **2017**, *7*, 356.
29. Mitsui, Y.; Koyama, K.; Ohtsuka, M.; Umetsu, R.Y.; Kainuma, R.; Watanabe, K. High Field X-ray Diffraction Study for $Ni_{46.4}Mn_{38.8}In_{12.8}Co_{2.0}$ Metamagnetic Shape Memory Film. *Metals* **2017**, *7*, 364. [CrossRef]
30. Seo, J.; Hu, J.W.; Kim, K.H. Analytical Investigation of the Cyclic Behavior of Smart Recentering T-Stub Components with Superelastic SMA Bolts. *Metals* **2017**, *7*, 386. [CrossRef]

31. Wu, S.K.; Chan, W.J.; Chang, S.H. Damping Characteristics of Inherent and Intrinsic Internal Friction of Cu-Zn-Al Shape Memory Alloys. *Metals* **2017**, *7*, 397. [CrossRef]
32. Adachi, Y.; Ogi, Y.; Kobayashi, N.; Hayasaka, Y.; Kanomata, T.; Umetsu, R.Y.; Xu, X.; Kainuma, R. Temperature Dependences of the Electrical Resistivity on the Heusler Alloy System $Ni_2MnGa_{1-x}Fe_x$. *Metals* **2017**, *7*, 413. [CrossRef]
33. Kataoka, M.; Endo, K.; Kudo, N.; Kanomata, T.; Nishihara, H.; Shishido, T.; Umetsu, R.Y.; Nagasako, M.; Kainuma, R. Martensitic transition, ferromagnetic transition, and their interplay in the shape memory alloys $Ni_2Mn_{1-x}Cu_xGa$. *Phys. Rev. B* **2010**, *82*, 214423. [CrossRef]
34. Umetsu, R.Y.; Xu, X.; Ito, W.; Kainuma, R. Evidence of Change in the Density of States during the Martensitic Phase Transformation of Ni-Mn-In Metamagnetic Shape Memory Alloys. *Metals* **2017**, *7*, 414. [CrossRef]
35. Blanter, M.S.; Golovin, I.S.; Neuhäuser, H.; Sinning, H.-R. *Internal Friction in Metallic Materials; A Handbook*, Springer. Berlin/Heidelberg, Germany, 2007.
36. Sapozhnikov, K.; Torrens-Serra, J.; Cesari, E.; Humbeeck, J.V.; Kustov, S. Effect of Hydrogen on the Elastic and Anelastic Properties of the R Phase in $Ti_{50}Ni_{46.1}Fe_{3.9}$ Alloy. *Metals* **2017**, *7*, 493. [CrossRef]
37. Carl, M.; Smith, J.D.; Doren, B.V.; Young, M.L. Effect of Ni-Content on the Transformation Temperatures in NiTi-20 at % Zr High Temperature Shape Memory Alloys. *Metals* **2017**, *7*, 511. [CrossRef]
38. Pelton, A.; Duerig, T.; Stöckel, D. A guide to shape memory and superelasticity in nitinol medical devices. *Minim. Invasive Ther. Allied Technol.* **2004**, *13*, 218–221. [CrossRef] [PubMed]
39. Benafan, O.; Brown, J.; Calkins, F.; Kumar, P.; Stebner, A.; Turner, T.; Vaidyanathan, R.; Webster, J.; Young, M. Shape memory alloy actuator design: Casmart collaborative best practices and case studies. *Int. J. Mech. Mater. Des.* **2014**, *10*, 1–42. [CrossRef]

metals

MDPI

Article

Dynamic Recovery and Superelasticity of Columnar-Grained Cu–Al–Mn Shape Memory Alloy

Sheng Xu [1,2], Haiyou Huang [1], Jianxin Xie [1,*], Yuta Kimura [2], Xiao Xu [2], Toshihiro Omori [2] and Ryosuke Kainuma [2,*]

[1]   Key Laboratory for Advanced Materials Processing of the Ministry of Education, University of Science and Technology Beijing, Beijing 100083, China; xu.sheng@hotmail.com (S.X.); huanghy@mater.ustb.edu.cn (H.H.)

[2]   Department of Materials Science, Graduate School of Engineering, Tohoku University, Sendai 980-8579, Japan; yuta.kimura.s8@dc.tohoku.ac.jp (Y.K.); xu@material.tohoku.ac.jp (X.X.); omori@material.tohoku.ac.jp (T.O.)

*   Correspondence: jxxie@mater.ustb.edu.cn (J.X.); kainuma@material.tohoku.ac.jp (R.K.); Tel.: +86-010-6233-2254 (J.X.); +81-022-7957-323 (R.K.)

Academic Editor: Takuo Sakon
Received: 23 March 2017; Accepted: 11 April 2017; Published: 15 April 2017

Abstract: The columnar-grained $Cu_{71.5}Al_{17.5}Mn_{11}$ shape memory alloy was treated by single-pass hot rolling at 900 °C with a thickness reduction of 67.3% followed by immediate water quenching. Dynamic recovery other than discontinuous dynamic recrystallization occurred during the treatment process, bringing about retained columnar grains with <001> textures, as well as dislocations introduced into the parent matrix. As a result, a large maximum recoverable strain of more than 11% was maintained due to the retained beneficial grain characteristics. The critical stress for inducing martensitic transformation and stress hysteresis were enhanced mainly due to the existence of dislocations.

Keywords: shape memory alloy; Cu–Al–Mn; columnar grain; hot deformation; dynamic recovery; martensitic transformation; superelasticity

1. Introduction

Shape memory alloys (SMAs) undergo reversible martensitic phase transformation in response to changes in temperature or applied stress, and hence have unique properties known as the shape memory effect and superelasticity [1,2]. Despite the availability of a commercial Ni–Ti alloy, Cu-based SMAs are the most attractive for practical applications owing to their low cost and good shape memory properties, as well as advantages with regard to electrical and thermal conductivities [1,3].

The shape memory properties of Cu-based SMAs are significantly influenced by their microstructure. Specifically, large grain size and <001> textures are preferred for enhanced superelasticity [4,5]. Recently, directional solidification has been demonstrated to be an effective way for the fabrication of high-performance Cu-based SMAs due to the formation of columnar grains with <001> textures [6–8]. Directionally solidified Cu-based SMAs can be used directly. Sometimes, however, the solidified Cu-based SMAs must undergo plastic processing for different product shapes required for various applications. In this case, two problems are encountered. First, Cu-based SMAs such as Cu–Al–Ni and Cu–Zn–Al are generally too brittle to be sufficiently cold-worked due to the high degree of order in the parent phase and the high elastic anisotropy [1]; thus currently, they can only be used in a solidified state. Second, Cu–Al–Mn SMAs is more ductile than Cu–Al–Ni and Cu–Zn–Al due to a lower degree of order in the parent phase. Therefore, conventional thermomechanical treatment, i.e., cold working and annealing, can be applied to Cu–Al–Mn for plastic processing [3,4,9]. However, the finally obtained {112} <110> recrystallization texture is not the most favorable texture for obtaining ideal superelastic behavior [4,9]. In a word, it is still difficult to plastically process the solidified Cu-based SMAs without the loss of good superelastic behavior.

When a solidified alloy is subjected to hot deformation, dynamic recovery and discontinuous dynamic recrystallization are involved, the former one usually occurring prior to the latter one [10]. However, it has been found that the dynamic recovery other than discontinuous dynamic recrystallization is preferred to occur for columnar-grained Cu-based alloys during hot deformation, especially when the deformation is perpendicular to the solidification direction (SD) [11,12]. Therefore, in this study, we conducted hot deformation for plastic processing of a columnar-grained Cu–Al–Mn SMA, expecting good superelastic behavior by retention of the beneficial grain characteristics via dynamic recovery.

2. Materials and Methods

A $Cu_{71.5}Al_{17.5}Mn_{11}$ alloy ingot with a columnar-grained microstructure was prepared by directional solidification using homemade equipment. The fabrication process is schematically shown in Figure 1a, and the details can be found in a previous study [7]. The volume of the fabricated ingot was around 150 mm × 80 mm × 40 mm. Dog-bone shaped tensile samples with a gauge size of 20 mm × 4 mm × 1 mm were cut out using wire electro-discharge machining with their longitudinal direction paralleling to SD. These samples were solution-treated (ST) at 800 °C for 5 min followed by quenching in water to obtain a single β_1 phase (L2$_1$ structure) [3], hereafter notated as the as-ST samples. Plates with a size of 50 mm × 15 mm × 5.5 mm were also cut out from the ingot for hot rolling with their longitudinal direction paralleling to SD. These plates were first heated to 900 °C with a holding time of 5 min and subsequently hot-rolled with a twin roller by single pass followed by immediate quenching in water. The diameter of the roller was 200 mm and the rolling speed was set to be 2.0 m/min. The final thickness of the plates after rolling was 1.8 mm with a reduction ratio of 67.3%. The above thermomechanical process, hereafter notated as SHRQ treatment, is illustrated in Figure 1b. Dog-bone shaped tensile samples with a gauge area of 20 mm × 4 mm were then cut out from the hot-rolled plates with their longitudinal direction parallel the rolling direction (RD), hereafter notated as the SHRQ-treated samples.

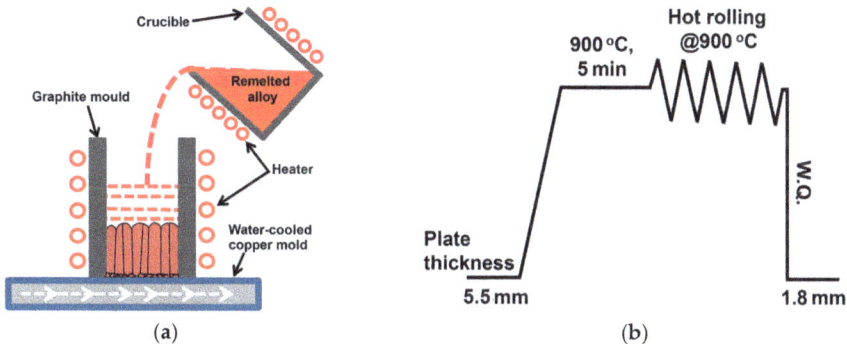

Figure 1. (**a**) Schematic illustration of the directional solidification process. (**b**) Overview of the single-pass hot rolling followed by immediate water quenching (SHRQ) treatment process.

The superelastic response was examined by a cyclic loading–unloading tensile test with increasing applied strain at room temperature with a strain rate of 4.2×10^{-4} s^{-1}. A non-contact video extensometer was used to measure the strain. The crystallographic orientations of the as-ST and SHRQ-treated samples were examined by electron backscattered diffraction (EBSD) employing a field-emission scanning electron microscope (SEM) operated at 20 kV. EBSD samples were finally polished using a colloidal silica suspension with particle size of 0.04 μm. The martensitic transformation temperatures were determined by the differential scanning calorimetry (DSC) with a heating/cooling rate of 10 °C/min. The crystal structure was analyzed by X-ray diffraction (XRD) method employing Cu Kα radiation at 40 kV with a 2θ angle ranging from 20° to 80°. The microstructure was also observed by using transmission electron microscopy (TEM) operated at 200 kV at room temperature, and the TEM samples were prepared by twin jetting with an electrolyte consisting of 250 mL of H_3PO_4, 50 mL of HCl, 250 mL of C_2H_5OH, 500 mL of distilled water, and 5 g of carbamide.

3. Results and Discussion

Figure 2a,b show the cyclic tensile loading–unloading curves of the as-ST and SHRQ-treated samples at room temperature, respectively. For the as-ST sample, the maximum recoverable strain reaches more than 14%, the critical stress σ_c is 135 MPa, stress hysteresis between forward and reverse transformation is 73 MPa at an applied strain of 13%, where the stress hysteresis is defined as the stress difference between the first loading and unloading curves at half of the applied strain.

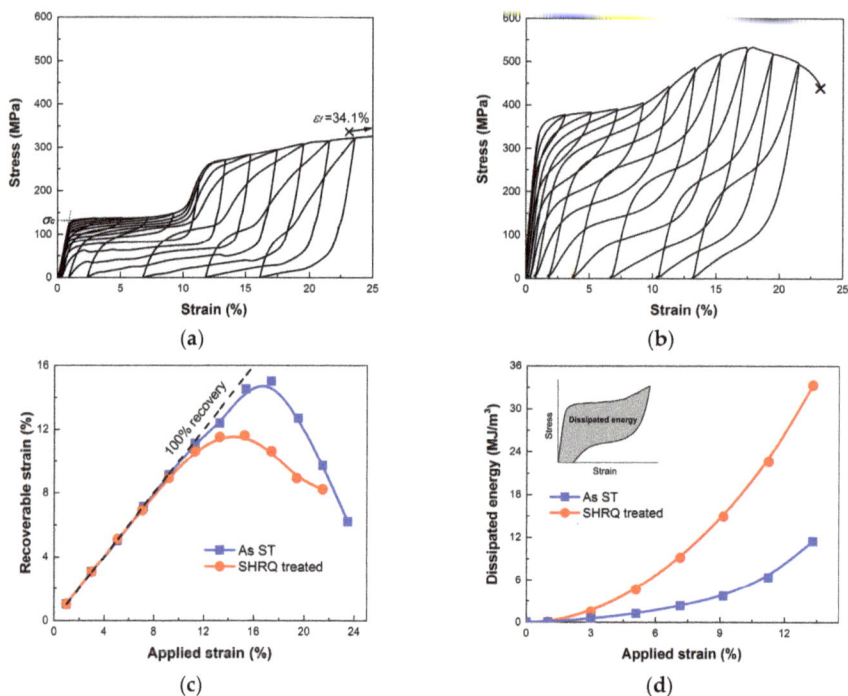

Figure 2. Cyclic tensile stress–strain curves obtained in the (**a**) as-solution-treated (as-ST) sample and (**b**) SHRQ-treated sample at room temperature. (**c**) Recoverable strain as a function of applied tensile strain. (**d**) Mechanical energy dissipated by one superelastic cycle as a function of applied tensile strain for both samples.

For the SHRQ-treated sample, its maximum recoverable strain is maintained at higher than 11%, while the critical stress σ_c is 368 MPa and the stress hysteresis is 254 MPa at an applied strain of 13%, which are 2.7 and 3.5 times as high as those of the as-ST sample, respectively. There is also an increase in apparent Young's modulus after SHRQ treatment. Summarized from Figure 2a,b, the recoverable strain of both samples is plotted against the applied strain in Figure 2c, where the dashed line represents perfect shape recovery. For an applied strain up to 11%, both samples display almost perfect shape recovery. With a further increase in applied strain, the recovery strain decreases earlier in the SHRQ-treated sample than that of the as-ST sample. Nevertheless, the SHRQ-treated sample still exhibits much better shape recovery properties than that of the ones treated by conventional thermomechanical treatment, in which the maximum recoverable stain is less than 7% [9].

Based on the above results, we can infer that, by using SHRQ treatment, a deformation reduction of 67.3% can be achieved. Meanwhile, the large recoverable strain remained. Moreover, the critical stress σ_c and the stress hysteresis were enhanced, which is potentially important for practical applications of the Cu–Al–Mn SMAs. For example, when SMAs are used as damping vibration components, high critical stress σ_c and stress hysteresis, together with a large recoverable strain, are usually required for a strong damping capacity ΔW, which can be characterized by the enclosed area of the superelastic curves [13,14], as illustrated in the inset of Figure 2d. We made a comparison of damping capacity ΔW in one superelastic cycle at every strain level between the as-ST and SHRQ-treated samples, as shown in Figure 2d. With the increase in applied strain, the ΔW of both samples increases gradually; when the applied strain is 11%, the ΔW of SHRQ-treated sample reaches 22.6 MJ/m^3, which is 3.5 times higher than 6.4 MJ/m^3 in the as-ST sample, when the applied strain is up to 13%, the ΔW of the SHRQ-treated sample reaches 33.3 MJ/m^3, which is 2.9 times higher than the ΔW of the as-ST sample.

As described previously, the superelastic strain of the Cu–Al–Mn SMAs has a strong dependence on microstructure [4,5]. The strong <001> texture and straight grain morphology are the key factors making columnar-grained Cu–Al–Mn alloy show a high recoverable strain [6]. Figure 3a, Figure 4a show the EBSD map and corresponding inverse pole figure of the as-ST sample. It shows a strong <001> texture and straight grain boundaries along the SD, which are typical characteristics of columnar-grained microstructure, and the grain orientations perpendicular to the SD are distributed randomly between <001> and <101> [6]. After SHRQ treatment, even though there was a little loss of orientation uniformity within each separated grain, the <001> texture along the RD was retained, and no recrystallized grain was observed, with the grains in a columnar morphology, as shown in Figure 3b, Figure 4b. The theoretically calculated maximum superelastic strain caused by reversible $\beta_1/6$ M martensitic transformation with stressing along different grain orientations is drawn as contour lines in the inverse pole figure in Figure 3c, Figure 4b [9]. It can be seen that the corresponding theoretical superelastic strain is basically larger than 9% in the SHRQ-treated sample. Therefore, the maintaining of a high recoverable strain in the SHRQ-treated sample is ascribed to the retained <001> texture and columnar grain morphology.

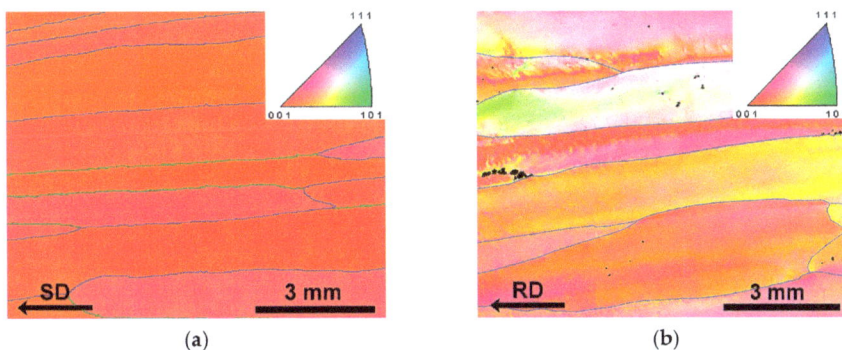

(a)   (b)

Figure 3. Quasi-colored orientation mapping of the (**a**) as-ST sample and the (**b**) SHRQ-treated sample, where the blue line represents a high-angle grain boundary and the green line represents a low-angle grain boundary lower than 15°.

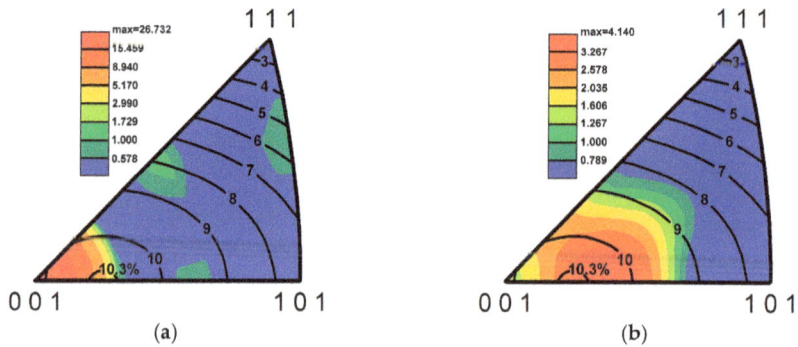

Figure 4. Inverse pole figures with theoretical superelastic strain contour lines of the (**a**) as-ST sample and (**b**) SHRQ-treated sample [9].

It should also be noted that the texture shows a small orientation deviation away from <001> after SHRQ treatment, which probably causes the increase in apparent Young's modulus because the elastic anisotropy is very high in Cu-based SMAs and the Young's modulus along <001> is usually the minimum one [1]. Moreover, the lower intensity of texture after SHRQ treatment should make grain constraint from neighboring grains increase, which may result in an additional increase of Young's modulus. It is also known that the Cu–Al–Mn SMAs display the lowest Tailor factor for the <001> orientation [4]. Therefore, the deviation and weakening of texture for the SHRQ-treated sample should also yield an increase in critical stress σ_c as well as stress hysteresis due to the grain constraint effect [4]. However, even compared with an equiaxed polycrystalline counterpart [9], the critical stress σ_c and stress hysteresis are much higher in the SHRQ-treated sample, which should be caused by other factors.

To further clarify the reasons for the enhanced critical stress σ_c and stress hysteresis, we observed the microstructure of the SHRQ-treated sample by TEM, with Figure 5a showing the TEM bright-field image and corresponding selected area electron diffraction pattern (SAEDP), respectively. It was found that many dislocation distribute in the alloy matrix. The corresponding SAEDP demonstrates that the matrix is the parent phase with a $L2_1$ crystallographic structure. Moreover, no martensite was observed, which is also verified by the XRD pattern, as shown in Figure 5b. It can be inferred then that SHRQ treatment did not introduce residual stress-induced martensite but dislocations into the parent phase. These pre-existing dislocations may have played a key role for the enhanced critical stress σ_c and stress hysteresis in the SHRQ-treated sample, as discussed later.

The thermoelastic martensite in Cu-based SMAs is induced by lowering the temperature or applying stress. The relationship of these two factors for inducing martensitic transformation can be explained by the Clausius–Clapeyron equation [15]:

$$\frac{d\sigma_c}{dT} = -\frac{\Delta S}{\varepsilon \cdot V_m{}'} \tag{1}$$

where ΔS is the molar entropy change between the parent phase and martensite, ε is the transformation strain, and V_m is the molar volume. In a thermo-induced martensitic transformation, the lower martensitic transformation starting temperature M_s indicates a more stable parent phase, which corresponds to a higher critical stress σ_c in the stress-induced martensitic transformation. It is well known that, for a given chemical composition, the depression of thermo-induced martensitic transformation can be induced by imperfections (dislocations, second-phase particles, precipitates, etc.) [16]. We conducted DSC measurement to determine the martensitic transformation temperatures of both samples, as shown in Figure 5c. The martensitic transformation starting and finishing points M_s, M_f, and the reverse transformation starting and finishing points A_s, and A_f for the as-ST sample are -55.9 °C, -67.6 °C, -51.6 °C, and -39.6 °C, respectively.

Using $d\sigma_c/dT = 1.85$ MPa/°C for a columnar-grained Cu–Al–Mn SMA [17], the critical stress σ_c at room temperature is calculated to be around 140 MPa for the as-ST sample, which is consistent with the experimental value of 135 MPa. However, no martensitic transformation is observed for the SHRQ-treated sample, even cooled down to 80 °C, indicating that the parent phase was stabilized due to the existence of dislocations strongly hindering the nucleation of martensite, and therefore the M_s was greatly depressed. In addition, it is also known that a lower degree of order in the parent phase results in a decrease in M_s in the Cu–Al–Mn SMA [18]. By comparing the ratio of the intensities (I) between (200) and (400) peaks for both samples using the XRD patterns in Figure 5b, i.e., $I_{(200)}/I_{(400)}$, a lower degree of order for the SHRQ-treated sample can be verified due to a lower value of $I_{(200)}/I_{(400)}$, which may also give a decrease in M_s and subsequent increase in critical stress σ_c at room temperature.

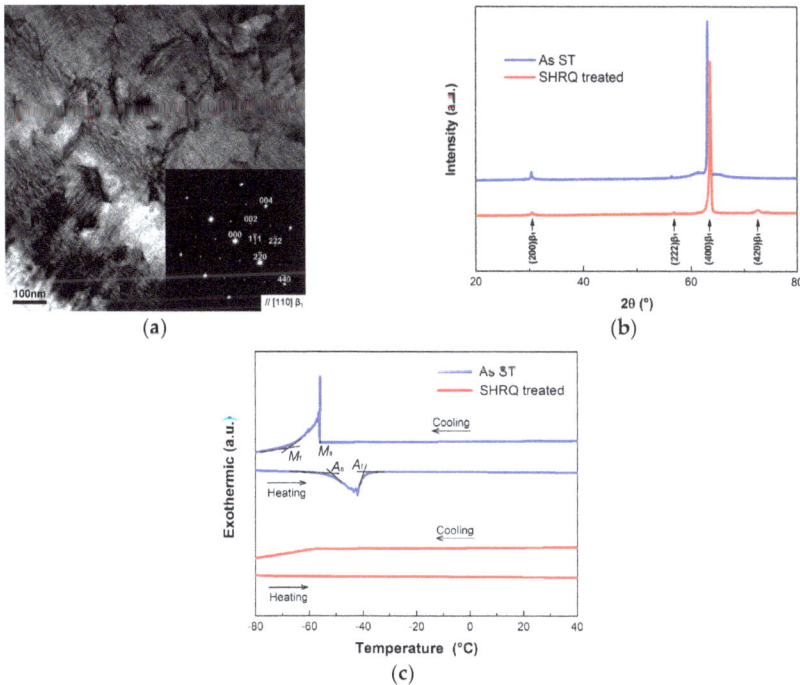

Figure 5. (a) TEM bright-field image of the SHRQ-treated sample showing dislocations distributed in the matrix, the inset shows the selected area electron diffraction pattern of the matrix. (b) XRD patterns and (c) DSC curves of the as-ST sample and SHRQ-treated sample.

For SMAs, the stress hysteresis is generally affected by friction against migration of parent/martensite interfaces and dislocation generation. [19,20]. The intrinsic interface friction is usually dominated by the lattice incompatibility between transforming phases, which is mainly related to the chemical composition. In our case, the SHRQ treatment did not change the chemical composition of the alloy, so the increase in stress hysteresis in the SHRQ-treated sample should be attributed to the dislocations introduced into the parent phase accordingly. The pre-existing dislocations in the parent phase have already been demonstrated to improve stress hysteresis by hindering the growth of martensite by impeding the motion of parent phase/martensite interfaces [21]. Therefore, it is apparent that the dislocations introduced into the non-martensite parent phase by SHRQ treatment significantly enhanced the stress hysteresis.

4. Conclusions

In conclusion, with single-pass hot rolling followed by immediate water quenching treatment, a plastic reduction of 67.3% was achieved in the columnar-grained $Cu_{71.5}Al_{17.5}Mn_{11}$ shape memory alloy. Meanwhile, a large maximum recoverable strain of more than 11%, as well as enhanced critical stress σ_c and stress hysteresis were obtained. The retained columnar grains with <001> textures due to dynamic recovery during treatment were believed to be the reasons for the maintenance of a high recoverable strain. The dislocations introduced into the parent matrix served mainly for enhancing the critical stress σ_c and stress hysteresis. This study should be useful in further improvement in the plastically processing the solidified brittle Cu-based shape memory alloys while maintaining good superelastic behavior at the same time.

Acknowledgments: This work was financially supported by the National Natural Science Foundation of China (Grant No. 51574027), the National Key Research and Development Program of China (Grant No. 2016YFB0700505), and the Discipline Innovative Engineering Plan of China (111 Project) (Grant No. B17003).

Author Contributions: Sheng Xu, Haiyou Huang, and Xiao Xu conceived and designed the experiments; Sheng Xu performed the experiments, Sheng Xu and Haiyou Huang analyzed the data, and wrote the paper; Yuta Kimura conducted the TEM experiments; Jianxin Xie, Xiao Xu, Toshihiro Omori and Ryosuke Kainuma supported the paper writing. All authors have participated in the discussions of the results.

Conflicts of Interest: The authors declare no conflict of interest.

References

1. Otsuka, O.; Wayman, C.M. *Shape Memory Materials*, 1st ed.; Cambridge University Press: Cambridge, UK, 1998; pp. 97–116.
2. Juan, J.S.; Nó, M.L.; Schuh, C.A. Nanoscale shape-memory alloys for ultrahigh mechanical damping. *Nat. Nanotechnol.* **2009**, *4*, 415–419. [CrossRef] [PubMed]
3. Sutou, Y.; Omori, T.; Kainuma, R.; Ishida, K. Ductile Cu–Al–Mn shape memory alloys: General properties and applications. *Mater. Sci. Technol.* **2008**, *24*, 896–901. [CrossRef]
4. Sutou, Y.; Omori, T.; Yamauchi, K.; Ono, N.; Kainuma, R.; Ishida, K. Effect of grain size and texture on pseudoelasticity in Cu–Al–Mn-based shape memory wire. *Acta Mater.* **2005**, *53*, 4121–4133. [CrossRef]
5. Sutou, Y.; Omori, T.; Kainuma, R.; Ishida, K. Grain size dependence of pseudoelasticity in polycrystalline Cu–Al–Mn-based shape memory sheets. *Acta Mater.* **2013**, *61*, 3842–3850. [CrossRef]
6. Liu, J.L.; Huang, H.Y.; Xie, J.X. The roles of grain orientation and grain boundary characteristics in the enhanced superelasticity of $Cu_{71.8}Al_{17.8}Mn_{10.4}$ shape memory alloys. *Mater. Des.* **2014**, *64*, 427–433. [CrossRef]
7. Liu, J.L.; Huang, H.Y.; Xie, J.X. Superelastic anisotropy characteristics of columnar-grained Cu–Al–Mn shape memory alloys and its potential applications. *Mater. Des.* **2015**, *85*, 211–220. [CrossRef]
8. Fu, H.D.; Song, S.; Zhuo, L.; Zhang, Z.H.; Xie, J.X. Enhanced mechanical properties of polycrystalline Cu–Al–Ni alloy through grain boundary orientation and composition control. *Mater. Sci. Eng. A* **2016**, *650*, 218–224. [CrossRef]
9. Sutou, Y.; Omori, T.; Kainuma, R.; Ono, N.; Ishida, K. Enhancement of superelasticity in Cu–Al–Mn-Ni shape-memory alloys by texture control. *Metall. Mater. Trans. A* **2002**, *22*, 2814–2824. [CrossRef]
10. Humphreys, F.J.; Hatherly, M. *Recrystallization and Related Annealing Phenomena*, 2nd ed.; Elsevier: Oxford, UK, 2004; pp. 415–437.
11. Yu, J.; Liu, X.; Xie, J. Study on dynamic recrystallization of a Cu-based alloy BFe10-1-1 with continuous columnar grains. *Acta Metall. Sin.* **2011**, *47*, 482–488.
12. Liu, Y.K.; Huang, H.Y.; Xie, J.X. Effect of compression direction on the dynamic recrystallization behavior of continuous columnar-grained $CuNi_{10}Fe_1Mn$ alloy. *Int. J. Miner. Mater. Metall.* **2015**, *28*, 851–859. [CrossRef]
13. Ma, J.; Karaman, I. Expanding the repertoire of shape memory alloy. *Science* **2010**, *327*, 1468–1469. [CrossRef] [PubMed]
14. Tanaka, Y.; Himuro, Y.; Kainuma, R.; Sutou, Y.; Ishida, K. Ferrous polycrystalline shape-memory alloy showing huge superelasticity. *Science* **2010**, *327*, 1488–1490. [CrossRef] [PubMed]

15. Wollants, P.; Bonte, M.D.; Roos, J.R. Thermodynamic analysis of the stress-induced martensitic transformation in a single crystal. *Z. Metallk.* **1979**, *70*, 113–117.

16. Sutou, Y.; Koeda, N.; Omori, T.; Kainuma, R.; Ishida, K. Effect of aging on bainitic and thermally induced martensitic transformations in ductile Cu–Al–Mn-based shape memory alloys. *Acta Mater.* **2009**, *57*, 5748–5758. [CrossRef]

17. Xu, S.; Huang, H.Y.; Xie, J.; Takekawa, S.; Xu, X.; Omori, T.; Kainuma, R. Giant elastocaloric effect covering wide temperature range in columnar-grained $Cu_{71.5}Al_{17.5}Mn_{11}$ shape memory alloy. *APL Mater.* **2016**, *4*, 106106. [CrossRef]

18. Kainuma, R.; Takahashi, S.; Ishida, K. Ductile shape memory alloys of the Cu–Al–Mn system. *J. Phys. IV* **1995**, *5*, 961–966. [CrossRef]

19. Ortín, J.; Planes, A. Thermodynamic analysis of thermal measurements in thermoelastic martensitic transformations. *Acta Metall.* **1988**, *36*, 1873–1889. [CrossRef]

20. Karaca, H.E.; Acar, E.; Basaran, B.; Noebe, R.D.; Chumlyakov, Y.I. Superelastic response and damping capacity of ultrahigh-strength [111]-oriented NiTiHfPd single crystals. *Scri. Mater.* **2012**, *67*, 447–450. [CrossRef]

21. Romero, R.; Lovey, F.C.; Ahlers, M. The effect of β phase plastic deformation on martensitic transformation in Cu–Zn and Cu–Zn–Al single crystals. *Scri. Metall.* **1990**, *24*, 285–289. [CrossRef]

metals

MDPI

Article

Role of Severe Plastic Deformation in Suppressing Formation of R Phase and Ni₄Ti₃ Precipitate of NiTi Shape Memory Alloy

Li Hu [1,2], Shuyong Jiang [1,*] and Yanqiu Zhang [1]

[1] College of Mechanical and Electrical Engineering, Harbin Engineering University, Harbin 150001, China; heu_huli@126.com (L.H.); zhangyanqiu0924@sina.com (Y.Z.)

[2] College of Materials Science and Chemical Engineering, Harbin Engineering University, Harbin 150001, China

* Correspondence: jiangshuyong@hrbeu.edu.cn; Tel.: +86-451-8251-9710

Academic Editor: Takuo Sakon

Received: 22 March 2017; Accepted: 17 April 2017; Published: 19 April 2017

Abstract: Microstructural evolution of NiTi shape memory alloy (SMA) with a nominal composition of $Ni_{50.9}Ti_{49.1}$ (at %) is investigated on the basis of heat treatment and severe plastic deformation (SPD). As for as-rolled NiTi SMA samples subjected to aging, plenty of R phases appear in the austenite matrix. In terms of as-rolled NiTi SMA samples undergoing solution treatment and aging, Ni_4Ti_3 precipitates arise in the austenite matrix. In the case of as-rolled NiTi SMA samples subjected to SPD and aging, martensitic twins are observed in the matrix of NiTi SMA. With respect to as-rolled NiTi SMA samples subjected to solution treatment, SPD, and aging, neither R phases nor Ni_4Ti_3 precipitates are observed in the matrix of NiTi SMA. The dislocation networks play an important role in the formation of the R phase. SPD leads to amorphization of NiTi SMA, and in the case of annealing, amorphous NiTi SMA samples are subjected to crystallization. This contributes to suppressing the occurrence of R phase and Ni_4Ti_3 precipitate in NiTi SMA.

Keywords: shape memory alloy; NiTi alloy; severe plastic deformation; microstructure

1. Introduction

NiTi shape memory alloys (SMAs) have been widely used in aviation, medical, dental, and automotive fields because of their excellent abrasion resistance, good functional properties, and high mechanical properties [1,2]. Furthermore, plastic deformation and heat treatment have a significant influence on the microstructures of SMAs, which consequently affect their operation performance [3–5]. In general, solution treatment leads to one-step transformation of NiTi SMAs from austenite (B2 structure) to martensite (B19′ structure). Solution treatment along with aging results in the precipitation of Ni_4Ti_3 phase, which further contributes to the occurrence of two-stage transformation, three-stage transformation, or even four-stage transformation of NiTi SMAs [6–9]. It has been proposed that the existence of Ni_4Ti_3 precipitates contribute to the emergence of the R phase, which plays a significant role in multiple-stage transformation of SMAs. Severe plastic deformation (SPD) methods, such as high pressure torsion (HPT) [10,11], cold drawing [12,13], cold rolling [14], surface mechanical attrition treatment (SMAT) [15], and local canning compression [16], can lead to amorphization of NiTi SMAs at lower temperatures. In addition, amorphous NiTi SMA induced by means of SPD is able to be subjected to crystallization in the case of proper heat treatment, where even nanocrystalline NiTi SMA can be produced. However, almost no literature has reported the influence of SPD and subsequent heat treatment on the formation of Ni_4Ti_3 precipitates and the R phase.

Based on local canning compression, comprehensive influence of heat treatment, as well as SPD on microstructural evolution of NiTi SMA was investigated in the present work. The influence of

SPD and subsequent heat treatment on the formation of Ni_4Ti_3 precipitate and R phase, in particular, have been emphasized.

2. Materials and Methods

A NiTi SMA bar, which possesses the nominal composition of $Ni_{50.9}Ti_{49.1}$ (at %), was manufactured by virtue of vacuum induction melting followed by hot rolling. In order to obtain the transformation temperatures of the NiTi SMA bar, differential scanning calorimetry (DSC) test was conducted on a DSC-204 type equipment (Netzsch Group, Freistaat Bayern, Germany). During the DSC test, both the heating rate and cooling rate were set as 10 °C/min and the temperatures ranged from -100 to 100 °C. Consequently, the transformation temperatures of NiTi SMA bar are determined as: $Ms = -27.2$ °C, $Mf = -41.7$ °C, $As = -17.3$ °C, $Af = -4.1$ °C. The as-rolled NiTi SMA bar was cut into halves, where one half was subjected to a solution treatment at 850 °C for 2 h and subsequently was quenched into ice water, and the other half was used for a contrast sample. Subsequently, the samples with the a diameter of 4 mm and the height of 6 mm were removed from the as-rolled and solution treated NiTi SMA samples by virtue of electro-discharge machining (EDM) (DK7725, Jiangsu Dongqing CNC Machine Tool Co., Ltd., Taizhou, China), respectively. Afterward, these NiTi SMA samples were inserted into low carbon steel cans (Baosteel, Shanghai, China) which possess an inner diameter of 4 mm, outer diameter of 10 mm, and height of 3 mm. The locally canned NiTi SMA samples were compressed between the two anvils of the INSTRON 5500R equipment (Instron Corporation, Norwood, MA, USA) by 75% in height at the strain rate of 0.05 s^{-1} and at room temperature, as shown in Figure 1. It can be noted that the NiTi SMA samples are under a three-dimensional compressive stress state due to the constraint of the steel cans. After the compression, the NiTi SMA samples were removed from the steel cans, respectively. Two samples with the diameter of 4 mm and the height of 6 mm were cut from the as-rolled NiTi SMA bar and the solution-treated NiTi SMA bar, respectively. Finally, the two samples along with a compressed sample were aged for 2 h at 600 °C and then they were cooled to room temperature in the ambient atmosphere. All the NiTi SMA samples that were subjected to heat treatment were vacuum-sealed in quartz tubes separately.

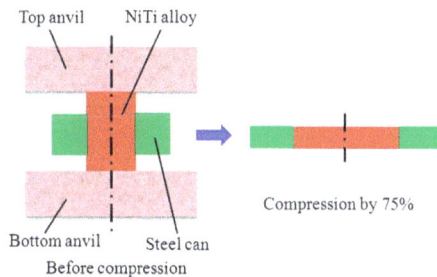

Figure 1. Schematic diagram of NiTi SMA sample subjected to local canning compression.

Microstructural evolution of the NiTi SMA samples was characterized using transmission electron microscopy (TEM). Foils used for TEM characterization were ground to 70 μm by virtue of mechanical method and subsequently were thinned by means of twin-jet polishing in an electrolyte with the composition of 6% $HClO_4$, 34% $C_4H_{10}O$ and 60% CH_3OH (volume fraction). TEM observations were implemented on a FEI TECNAI G2 F30 type microscope (FEI Corporation, Hillsboro, OR, USA), which possesses a side-entry and double-tilt specimen stage with an angular range of $\pm40°$.

3. Results

3.1. Microstructures of As-Rolled and Solution-Treated NiTi SMA Samples

The microstructures of as-rolled NiTi SMA sample are shown in Figure 2. It can be found from Figure 2 that plenty of dislocations are distributed in the B2 austenite matrix. It can be deduced that plenty of dislocations are induced by plastic deformation. However, in the case of solution treatment, plenty of dislocations have disappeared in the B2 austenite matrix, as shown in Figure 3.

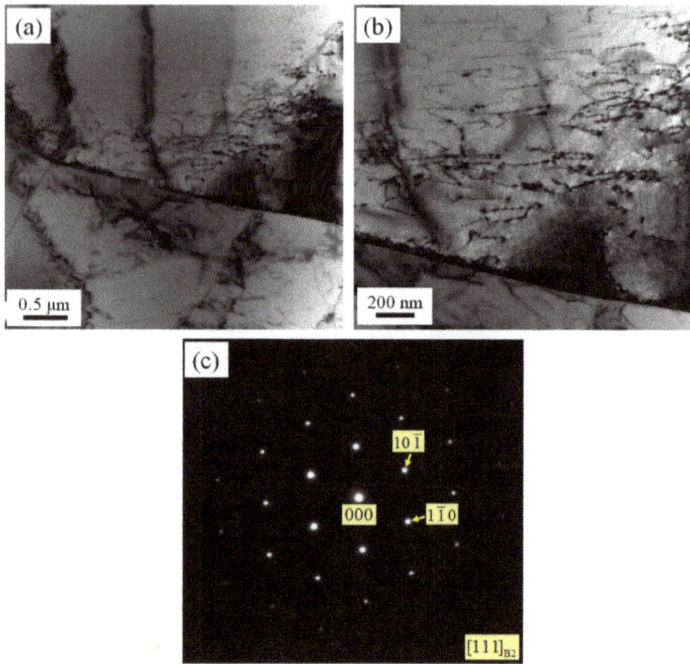

Figure 2. TEM micrographs of as-rolled NiTi SMA sample: (**a**) Bright field image (low magnification); (**b**) Bright field image (high magnification); (**c**) Diffraction pattern of (**b**).

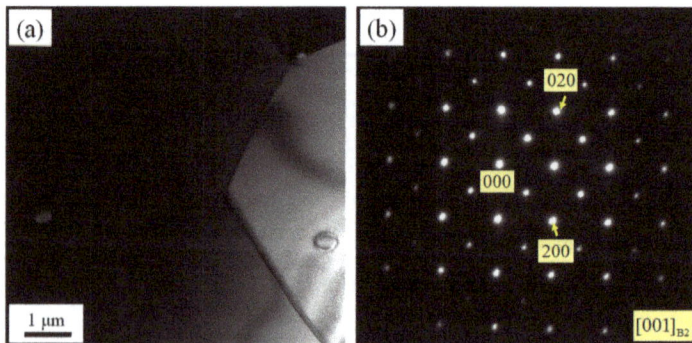

Figure 3. TEM micrographs of solution-treated NiTi SMA sample: (**a**) Bright field image; (**b**) Diffraction pattern of (**a**).

3.2. Microstructures of NiTi SMA Samples Subjected to SPD

In the case of local canning compression, the as-rolled NiTi SMA sample is subjected to SPD so that a mixture of amorphous and nanocrystalline phases appears, where a small amount of nanocrystalline phase is distributed in the dominant amorphous matrix, as shown in Figure 4. In the same manner, solution-treated NiTi SMA samples exhibit a mixture of amorphous and nanocrystalline phases after suffering from SPD as well, but the amorphization seems to be completed more thoroughly, as shown in Figure 5.

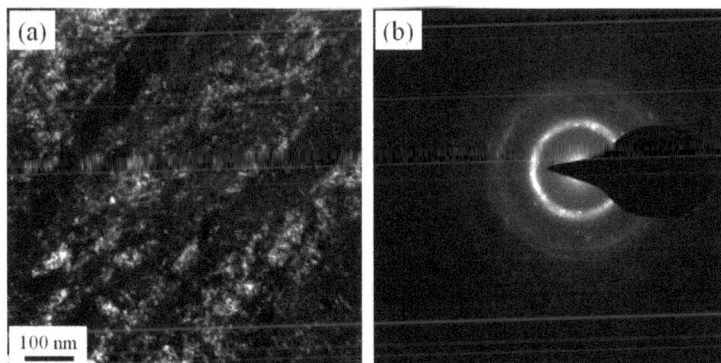

Figure 4. TEM micrographs of as-rolled NiTi SMA sample subjected to SPD: (**a**) Dark field image; (**b**) Diffraction pattern of (**a**).

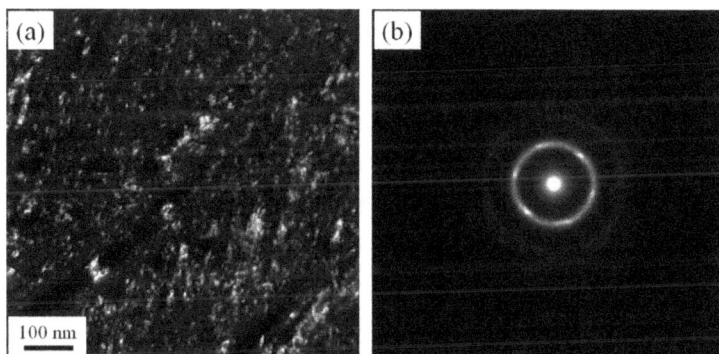

Figure 5. TEM micrographs of solution-treated NiTi SMA sample subjected to SPD: (**a**) Dark field image; (**b**) Diffraction pattern of (**a**).

3.3. Microstructures of As-Rolled NiTi SMA Sample Subjected to Aging

Figure 6 shows TEM micrographs of as-rolled NiTi SMA sample subjected to aging. It can be observed from Figure 6 that R phase appears in the B2 austenite matrix of NiTi SMA. In addition, there is a certain orientation relationship between R phase and B2 austenite. It is generally accepted that the appearance of the R phase is attributed to the existence of Ni_4Ti_3 precipitate. However, it seems that no obvious diffraction pattern with respect to Ni_4Ti_3 precipitate is captured in the present work.

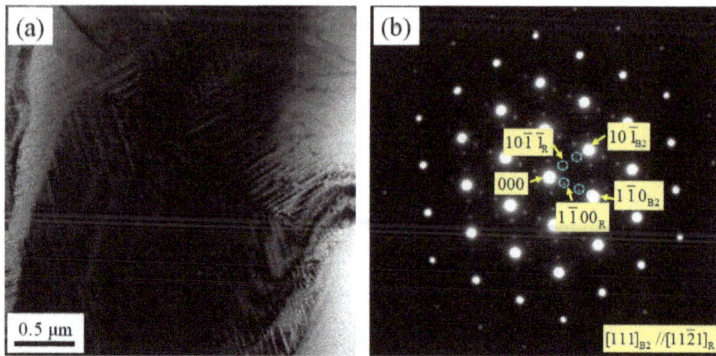

Figure 6. TEM micrographs of an as-rolled NiTi SMA sample subjected to aging: (**a**) Bright field image; (**b**) Diffraction pattern of (**a**).

3.4. Microstructures of NiTi SMA Sample Subjected to Solution Treatment and Aging

Figure 7 shows TEM micrographs of NiTi SMA sample subjected to solution treatment and aging. It can be found from Figure 7 that the Ni_4Ti_3 precipitates exhibit the inhomogeneous distribution in the matrix of NiTi SMA sample and they are characterized by a typical lenticular shape [9]. In addition, it can be observed from Figure 6 that the Ni_4Ti_3 precipitates arise in the grain interior as well as in the grain boundary pinned by TiC phase and they are obviously coarser in the grain interior than in the grain boundary.

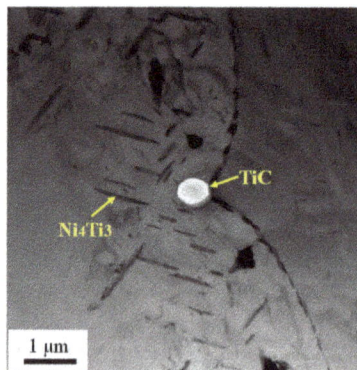

Figure 7. TEM micrographs of a solution-treated NiTi SMA sample subjected to aging.

3.5. Microstructures of As-Rolled NiTi SMA Sample Subjected to SPD and Aging

Figure 8 shows the microstructures of an as-rolled NiTi SMA sample subjected to SPD and subsequent aging. It can be observed from Figure 8 that the microstructure of NiTi SMA sample has a feature of martensite morphology. Furthermore, it can be found that martensite twins are distributed in the matrix. The phenomenon has been validated by the previous study [17]. In addition, neither R phase nor Ni_4Ti_3 precipitate appear in the matrix of NiTi SMA.

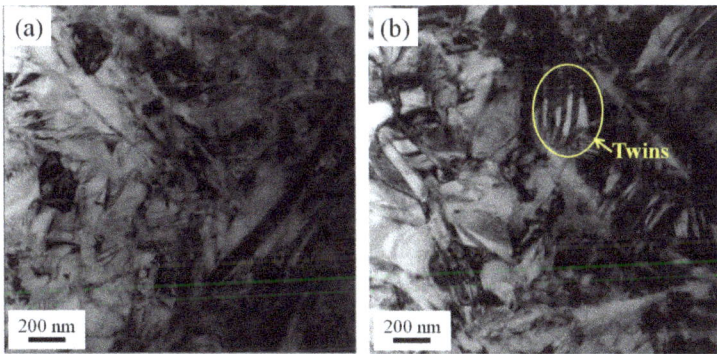

Figure 8. TEM micrographs of an as-rolled NiTi SMA sample subjected to SPD and subsequent aging: (**a**) Bright field image showing martensite morphology; (**b**) Bright field image showing the existence of martensite twins.

3.6. Microstructures of Solution-Treated NiTi SMA Sample Subjected to SPD and Aging

Figure 9 TEM micrographs of solution-treated NiTi SMA sample subjected to SPD and subsequent aging. It can be found from Figure 9 that, as for solution-treated NiTi SMA samples subjected to SPD and subsequent aging, B2 austenite phase and B19' martensite phase coexist in the matrix of NiTi SMA. In particular, neither R phase nor Ni_4Ti_3 precipitate appear in the matrix of NiTi SMA as well.

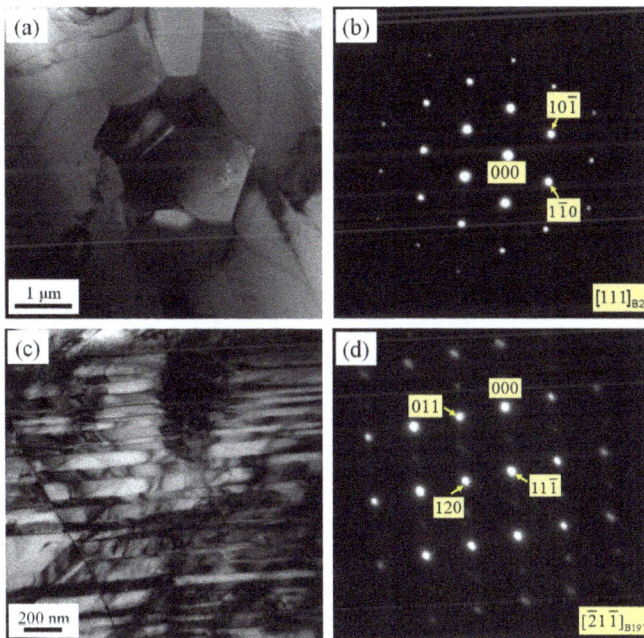

Figure 9. TEM micrographs of a solution-treated NiTi SMA sample subjected to SPD and subsequent aging: (**a**) Bright field image showing B2 austenite phase; (**b**) Diffraction pattern of (**a**); (**c**) Bright field image showing B19' martensite phase; (**d**) Diffraction pattern of (**c**).

4. Discussion

In general, Ni_4Ti_3 precipitates are nucleated and grow preferentially at the grain boundary on the basis of the theory for phase transformation kinetics. The nucleation rate of the Ni_4Ti_3 precipitates at the grain boundary is greater than that in the grain interior because nucleation barrier of the former is lower than that of the latter. Furthermore, the grain boundary energy plays an important role in accelerating the nucleation of Ni_4Ti_3 precipitates at the grain boundary. However, in the present study, the curvature of the grain boundary pinned occurs due to the pinning effect of the secondary phase TiC at the grain boundary. As a consequence, the grain boundary energy is reduced as compared to the original grain boundary, which suppresses the nucleation and growth of the Ni_4Ti_3 precipitates to a certain degree. However, whether the grain boundary is pinned or not, the nucleation rate of these Ni_4Ti_3 precipitates at the grain boundary is always greater than that in the grain interior. Firstly, the nucleation and growth of Ni_4Ti_3 precipitates at the grain boundary results in the Ni concentration at grain boundary being lowered as compared to the grain interior, which causes the Ni atoms in the interior of grain to diffuse towards the grain boundary. However, coherent interface is unable to be formed between the Ni_4Ti_3 precipitates at grain boundary and the B2 matrix, so the Ni_4Ti_3 phases at the grain boundary are hard to grow towards the grain interior due to the impediment of grain boundary. With the proceeding of aging, the Ni_4Ti_3 precipitates are nucleated and grow along the coherent interface between them and the B2 matrix in the grain interior. In addition, because the Ni_4Ti_3 phases in the grain interior can absorb Ni atoms from the B2 austenite matrix around them, they grow to be larger and larger, and consequently the smaller Ni_4Ti_3 precipitates are gradually merged by the larger Ni_4Ti_3 precipitates.

It is generally accepted that the existence of the Ni_4Ti_3 phases is able to suppress the transformation of martensite, which consequently promotes the occurrence of R phase. Because R phase plays an important role in multi-stage phase transformation of NiTi SMAs, the occurrence of two-stage phase transformation (B2-R-B19′) is regarded as a reasonable phenomenon in Ni-rich NiTi SMAs subjected to aging. However, abnormal three-stage transformation often appears in the aged Ni-rich NiTi SMAs. Bataillard et al. [18] suggested that the small-scale stress inhomogeneity in the B2 austenite matrix around Ni_4Ti_3 precipitates is responsible for the emergence of three-stage transformation during cooling, where the three-stage transformation includes an one-stage transformation from B2 austenite to R phase and a two-stage transformation from R phase to B19′ martensite, which corresponds to the high stress region near Ni_4Ti_3 precipitate and the low stress region away from Ni_4Ti_3 precipitate, respectively. Khalil Allafi et al. [19] suggested that the three-stage transformation during cooling is induced by the small-scale chemical composition inhomogeneity in the B2 austenite matrix and between the Ni_4Ti_3 precipitates, which leads to one B2-R transformation as well as two R-B19′ transformations. Furthermore, the two R-B19′ transformations occur in a low Ni region near Ni_4Ti_3 precipitate or at high Ni region away from Ni_4Ti_3 precipitate. Dlouhý et al. [20] gave microstructural evidence that R phase is nucleated at the interface between the B2 austenite matrix and Ni_4Ti_3 precipitate and B19′ martensite is nucleated at the interface between R phase and Ni_4Ti_3 precipitate. According to the aforementioned literatures, it can be proposed that the occurrence of R phase is mainly attributed to the inhomogeneous stress filed or the heterogeneous chemical composition, which results from the existence of Ni_4Ti_3 precipitates. However, in the present study, it seems that the dislocation networks lay the foundation for the formation of R phase. Under the condition of aging for 2 h at 600 °C, the Ni_4Ti_3 precipitates are easier to occur in the solution-treated NiTi SMA as compared to NiTi SMA which contains plenty of dislocations due to previous processing history.

When NiTi SMA is subjected to local canning compression, SPD results in amorphization of NiTi SMA. Crystallization of the amorphous NiTi SMA takes place in the course of subsequent aging for 2 h at 600 °C. As a matter of fact, SPD results in a mixture containing the amorphous phase and the retained nanocrystalline phase. The crystallization mechanism for the amorphous phase accompanied by the retained nanocrystalline phase can be summarized as the following procedures. Firstly, the retained nanocrystalline phase preferentially acts as the nucleus under the action of thermal driving force.

In addition, a high density of dislocations is able to be induced in the retained nanocrystalline phase, where the stored energy is enhanced. As a consequence, the release of the stored energy makes a contribution to the formation of the crystal nucleus. With the proceeding of aging time, new crystals are nucleated in the amorphous matrix because the stored energy aroused by the lattice defects of the amorphous phase is relaxed, and simultaneously the retained nanocrystals continue to grow up under the thermal driving force. Once all the grains impinge each other, complete crystallization occurs in the NiTi SMA sample suffering from SPD. Consequently, the amorphous phase is preferentially crystallized in the studied NiTi SMA, which contributes to suppressing the formation of Ni_4Ti_3 precipitates because the crystallized NiTi SMA is unable to meet the requirements for the precipitation of Ni_4Ti_3 phase. On the one hand, the energy has been consumed by the crystallization process and consequently the crystallized NiTi SMA is unable to meet the requirements for the precipitation of Ni_4Ti_3 phase in terms of energy. On the other hand, the crystallized NiTi SMA does not meet the requirements for the precipitation of Ni_4Ti_3 phase in terms of chemical composition. It can be concluded that SPD plays a significant role in suppressing the formation of R phase and Ni_4Ti_3 precipitate of NiTi SMA.

5. Conclusions

(1) In the case of as-rolled NiTi SMA sample subjected to aging for 2 h at 600 °C, R phase appears in the B2 austenite matrix. The as-rolled NiTi SMA sample contains plenty of dislocation networks aroused by previous processing history. The dislocation networks are responsible for the formation of the R phase.

(2) Solution treatment for 2 h at 850 °C can result in the annihilation of the dislocations in the as-rolled NiTi SMA sample. As a consequence, the Ni_4Ti_3 precipitates more easily arise in a solution-treated NiTi SMA sample when aged for 2 h at 600 °C.

(3) SPD is capable of inducing amorphization of a NiTi SMA sample at room temperature. Amorphous NiTi SMA is able to be subjected to crystallization when aged for 2 h at 600 °C. Consequently, SPD and subsequent aging contribute to enhancing the transformation temperature of martensite. In addition, SPD plays an important role in suppressing the occurrence of R phase and Ni_4Ti_3 precipitate.

Acknowledgments: The work was financially supported by National Natural Science Foundation of China (Nos. 51475101, 51305091 and 51305092).

Author Contributions: Li Hu performed the experiments and wrote the manuscript; Shuyong Jiang supervised the research; Yanqiu Zhang performed TEM analysis.

Conflicts of Interest: The authors declare no conflict of interest.

References

1. Otsuka, K.; Ren, X. Physical metallurgy of Ti-Ni-based shape memory alloys. *Prog. Mater. Sci.* **2005**, *50*, 511–678. [CrossRef]

2. Delobelle, V.; Chagnon, G.; Favier, D.; Alonso, T. Study of electropulse heat treatment of cold worked NiTi wire: From uniform to localised tensile behaviour. *J. Mater. Process. Technol.* **2016**, *227*, 244–250. [CrossRef]

3. Jiang, S.; Zhao, Y.; Zhang, Y.; Li, H.; Liang, Y. Effect of solution treatment and aging on microstructural evolution and mechanical behavior of NiTi shape memory alloy. *Trans. Nonferr. Met. Soc.* **2013**, *23*, 3658–3667. [CrossRef]

4. Tadayyon, G.; Mazinani, M.; Guo, Y.; Zebarjad, S.M.; Tofail, S.A.; Biggs, M.J. The effect of annealing on the mechanical properties and microstructural evolution of Ti-rich NiTi shape memory alloy. *Mater. Sci. Eng. A* **2016**, *662*, 564–577. [CrossRef]

5. Tadayyon, G.; Mazinani, M.; Guo, Y.; Zebarjad, S.M.; Tofail, S.A.; Biggs, M.J. Study of the microstructure evolution of heat treated Ti-rich NiTi shape memory alloy. *Mater. Charact.* **2016**, *112*, 11–19. [CrossRef]

6. Kuang, C.-H.; Chien, C.; Wu, S.-K. Multistage martensitic transformation in high temperature aged Ti 48 Ni 52 shape memory alloy. *Intermetallics* **2015**, *67*, 12–18. [CrossRef]

7. Wang, X.; Verlinden, B.; Van Humbeeck, J. Effect of post-deformation annealing on the R-phase transformation temperatures in NiTi shape memory alloys. *Intermetallics* **2015**, *62*, 43–49. [CrossRef]

8. Zhao, C.; Zhao, S.; Jin, Y.; Meng, X. Nanoscale interwoven structure of B2 and R-phase in Ni-Ti alloy film. *Vacuum* **2016**, *129*, 45–48. [CrossRef]

9. Jiang, S.; Zhang, Y.; Zhao, Y.; Liu, S.; Li, H.; Zhao, C.-Z. Influence of Ni_4Ti_3 precipitates on phase transformation of NiTi shape memory alloy. *Trans. Nonferr. Met. Soc.* **2015**, *25*, 4063–4071. [CrossRef]

10. Resnina, N.; Belyaev, S.; Zeldovich, V.; Pilyugin, V.; Frolova, N.; Glazova, D. Variations in martensitic transformation parameters due to grains evolution during post-deformation heating of Ti-50.2 at.% Ni alloy amorphized by HPT. *Thermochim. Acta* **2016**, *627*, 20–30. [CrossRef]

11. Shahmir, H.; Nili-Ahmadabadi, M.; Huang, Y.; Jung, J.M.; Kim, H.S.; Langdon, T.G. Shape memory effect in nanocrystalline NiTi alloy processed by high-pressure torsion. *Mater. Sci. Eng. A* **2015**, *626*, 203–206. [CrossRef]

12. Yu, C.; Aoun, B.; Cui, L.; Liu, Y.; Yang, H.; Jiang, X.; Cai, S.; Jiang, D.; Liu, Z.; Brown, D.E. Synchrotron high energy X-ray diffraction study of microstructure evolution of severely cold drawn NiTi wire during annealing. *Acta Mater.* **2016**, *115*, 35–44. [CrossRef]

13. Shi, X.; Guo, F.; Zhang, J.; Ding, H.; Cui, L. Grain size effect on stress hysteresis of nanocrystalline NiTi alloys. *J. Alloys Compd.* **2016**, *688*, 62–68. [CrossRef]

14. Li, Y.; Li, J.; Liu, M.; Ren, Y.; Chen, F.; Yao, G.; Mei, Q. Evolution of microstructure and property of NiTi alloy induced by cold rolling. *J. Alloys Compd.* **2015**, *653*, 156–161. [CrossRef]

15. Ke, C.; Cao, S.; Zhang, X. Phase field modeling of Ni-concentration distribution behavior around Ni_4Ti_3 precipitates in NiTi alloys. *Comp. Mater. Sci.* **2015**, *105*, 55–65. [CrossRef]

16. Jiang, S.; Hu, L.; Zhang, Y.; Liang, Y. Nanocrystallization and amorphization of NiTi shape memory alloy under severe plastic deformation based on local canning compression. *J. Non-Cryst. Solids* **2013**, *367*, 23–29. [CrossRef]

17. Jiang, S.; Ming, T.; Zhao, Y.; Li, H.; Zhang, Y.; Liang, Y. Crystallization of amorphous NiTi shape memory alloy fabricated by severe plastic deformation. *Trans. Nonferr. Met. Soc.* **2014**, *24*, 1758–1765. [CrossRef]

18. Bataillard, L.; Bidaux, J.-E.; Gotthardt, R. Interaction between microstructure and multiple-step transformation in binary NiTi alloys using in-situ transmission electron microscopy observations. *Philos. Mag. A* **1998**, *78*, 327–344. [CrossRef]

19. Allafi, J.K.; Ren, X.; Eggeler, G. The mechanism of multistage martensitic transformations in aged Ni-rich NiTi shape memory alloys. *Acta Mater.* **2002**, *50*, 793–803. [CrossRef]

20. Dlouhý, A.; Bojda, O.; Somsen, C.; Eggeler, G. Conventional and in-situ transmission electron microscopy investigations into multistage martensitic transformations in Ni-rich NiTi shape memory alloys. *Mater. Sci. Eng. A* **2008**, *481*, 409–413. [CrossRef]

metals

MDPI

Article

Lattice Softening in Fe₃Pt Exhibiting Three Types of Martensitic Transformations

Takashi Fukuda * and Tomoyuki Kakeshita

Department of Materials Science and Engineering, Graduate School of Engineering, Osaka University,
2-1, Yamada-oka, Suita, Osaka 565-0871, Japan; kakeshita@mat.eng.osaka-u.ac.jp
* Correspondence: fukuda@mat.eng.osaka-u.ac.jp; Tel.: +81 6-6879-7483

Academic Editor: Volodymyr A. Chernenko
Received: 24 March 2017; Accepted: 26 April 2017; Published: 27 April 2017

Abstract: We have investigated the relation between the softening of elastic constants and martensitic transformation in Fe₃Pt, which exhibits various kinds of martensitic transformation depending on its long-range order parameter S. The martensite phases of the examined alloys are BCT ($S = 0.57$), FCT1 ($S = 0.75$, $c/a < 1$) and FCT2 ($S = 0.88$, $c/a > 1$). The elastic constants C' and C_{44} of these alloys decrease almost linearly with decreasing temperature. Although the temperature coefficient of C' decreases as S increases, C' at the transformation temperature is the smallest in the alloy with $S = 0.75$, which transforms to FCT1. This result implies that softening is most strongly related to the formation of the FCT1 martensite with tetragonality $c/a < 1$ among the three martensites.

Keywords: elastic constants; band Jahn–Teller effect; disorder–order transformation

1. Introduction

The relation between softening of elastic constants and martensitic transformation has attracted considerable attention for many years and has been discussed by many researchers [1–10]. In some alloys exhibiting martensitic transformation, softening of elastic constants $C' = (C_{11} - C_{12})/2$ and large elastic anisotropy (C_{44}/C') was observed in the parent phase, but the significance of the softening is largely different between the alloys. For example, the value of C' near the transformation start temperature is approximately 0.01 GPa in In–27Tl (at %) alloy [3], 1 GPa in Au–30Cu–47Zn (at %) alloy [2], 5 GPa in Fe–30Pd (at %) alloy [7], 8 GPa in Cu–14Al–4Ni (at %) alloy [10], and 14 GPa in Ti–50.8Ni (at %) [8] and Al–63.2Ni (at %) alloys [4]. Because of such a large distribution of C' at the M_s temperature, the influence of softening of C' on martensitic transformation is expected to be significantly different between these alloys. Martensitic transformation in some alloys is probably strongly related to the softening of C', while that in others is weakly related despite the fact that the softening appears before the transformation.

In some alloys, several kinds of martensite phases appear by slightly changing the composition or long-range order parameter (degree of order). We consider that a study on elastic softening in such alloys will help us understand the relation between softening and martensitic transformation. Iron–platinum alloys are one such alloy system. An iron–platinum alloy with Pt content of 25 at % (Fe₃Pt) exhibits disorder–order transformation from the A1-type disordered structure to the L1₂-type ordered structure. Depending on the degree of order S of the L1₂-type structure, Fe₃Pt exhibits various kinds of martensitic transformations. (Here, S is the Bragg–Williams long-range order parameter [11]). The disordered alloy transforms to the BCC (body-centered-cubic) martensite like iron–nickel alloys. As the degree of order increases, the structure of the martensite phase changes to BCT (body-centered-tetragonal) and then to FCT (face-centered-tetragonal) with $c/a < 1$ [12–15]. Recently, another type of FCT martensite with $c/a > 1$ and also an orthorhombic martensite were found in highly ordered Fe₃Pt [16,17]. We call the former FCT ($c/a < 1$) martensite FCT1 and the latter FCT

($c/a > 1$) martensite FCT2 in the following. Incidentally, the Strukturbericht symbol for the BCT, FCT1 and FCT2 martensites are L6$_0$-type, but we use the traditional names in this paper for convenience.

As mentioned above, at least five kinds of martensite phases appear in Fe$_3$Pt depending on degree of order S. Among the five, three (BCT, FCT1 and FCT2) have tetragonal structures. When the BCT martensite is formed, the tetragonality c/a changes drastically from 1 to 0.79 at the transformation start temperature. This means that the L1$_2$-BCT transformation is an obvious first order. On the other hand, the jump in c/a at the transformation temperature is undetectable by conventional X-ray techniques for the FCT1 and the FCT2 martensites; c/a changes gradually in the cooling process from 1 to 0.94 when the FCT1 martensite is formed and from 1 to 1.005 when the FCT2 martensite is formed [16]. This means that the L1$_2$-FCT1 and the L1$_2$-FCT2 transformations are very weak first order. Because of the difference in transformation behavior, we may expect different softening behavior in elastic constants of the L1$_2$-type parent phase.

Several reports have been made on elastic constants of Fe–Pt alloys with Pt content near 25%. Huash [18] examined the elastic constants of a disordered Fe–28Pt (at %) alloy. Although the alloy does not show a martensitic transformation, it exhibits significant softening in elastic constants C' and C_{44} below its Curie temperature. Influence of degree of order S on elastic constants was examined by Ling and Owen [19] using an Fe–25Pt alloy. According to their report, the softening becomes less significant as S increases. They discussed the behavior from the view point of Invar effect. At the time of their report, the various types of martensites in Fe–25Pt alloys were not identified; therefore, the relation between the softening and martensitic transformations was not discussed there. The relation between elastic constant and martensitic transformation was discussed by Kawald et al. [20] in an Fe–25Pt (at %) with $S = 0.6$. They observed that C' approaches nearly zero in Fe–25Pt alloy, but the structure of the martensite phase and the transformation start temperature was not identified in the report. Owen [21] discussed the relation between softening and martensitic transformations in Fe–Pt alloys, and suggested that the softening affects the growth of the martensite phase. However, the structure of the martensite phase was not mentioned and it seems that only the BCC martensite was considered.

The elastic properties of Fe–Pt alloys were also examined by constructing phonon dispersion curves [22–24]. Tajima et al. [22] reported that [110] TA$_1$ mode of an ordered Fe–27.8Pt (at %) alloy exhibits significant softening near the Γ-point (center of the Brillourin zone) with decreasing temperature. Kästner et al. [23] found that softening at the Γ-point of the ordered Fe–28Pt is less significant compared with the disordered Fe–28Pt alloy. However, the alloys they examined do not show a martensitic transformation; therefore, the relation between the softening and martensitic transformations are not discussed there.

As mentioned above, although there exist several reports on the elastic properties of Fe–Pt alloys, the relation between the softening of elastic constant and martensitic transformations is not clear. The present study is motivated to make progress on the interpretation of the relation between the softening in elastic constant and martensitic transformation using several Fe–25Pt alloys, which transform to three types of tetragonal martensite phases (BCT, FCT1 and FCT2) depending on degree of order.

2. Materials and Methods

An ingot of Fe–25.0Pt (at %) was prepared by melting an iron bar (99.99 mass %) and a platinum plate (99.95 mass %) in an arc melting furnace (DIAVAC, Chiba, Japan) under an argon gas atmosphere. By using the ingot, a boule (single crystal) was grown by a floating zone method in an argon gas atmosphere with a growth rate of 3 mm/h. Three parallelepiped specimens (Specimen-A, Specimen-B, Specimen-C) with all faces parallel to {100} were cut from the boule and subjected to homogenization heat treatment at 1373 K for 1 h. Then, three kinds of ordering heat treatment were applied to obtain different degrees of order S of the L1$_2$-type structure.

Specimen-A ($3.07 \times 3.06 \times 3.03$ mm^3) was cooled from 1373 to 973 K with a cooling rate of 1 K/min, and then kept at 973 K for 10 h followed by quenching into iced water. Specimen-B ($3.03 \times 3.01 \times 3.00$ mm^3) was cooled from 1373 to 923 K with a cooling rate of about 1 K/min, and then kept at 923 K for 100 h followed by quenching into iced water. Specimen-C ($3.28 \times 3.23 \times 3.10$ mm^3) was cooled from 1373 to 1103 K with a cooling rate of about 1 K/min, and then cooled to 773 K with a cooling rate of 10 K/day followed by furnace cooling to room temperature. The degree of order S (L1$_2$-type) of the three specimens was determined by comparing the intensity of 100 and 200 reflections obtained by X-ray diffraction. Details of the method were described elsewhere [25], and the obtained value of S was 0.57, 0.75 and 0.88 with an error of approximately ±0.05 for Specimens-A, -B and -C, respectively; in the following, we refer to the three specimens as S57, S75 and S88 in order to clearly indicate the degree of order.

All three specimens exhibit ferromagnetic transition, and the Curie temperature, T_c, increases with increasing degree of order (S) and the value is 353, 394 and 441 K for S57, S75 and S88, respectively [15]. The specimen S57 ($S = 0.57$) transforms to the BCT martensite at 145 K, S75 transforms to the FCT1 martensite with $c/a < 1$ at 85 K, and S88 transforms to the FCT2 martensite with $c/a > 1$ at 60 K [15].

The elastic constants of the three specimens were measured by a rectangular parallelepiped resonance (RPR) method [26,27]. More than 40 resonance peaks in the frequency range of 200 and 800 Hz were used to optimize the elastic constants. During the measurements, a magnetic field of 1 T was applied to the [111] direction of the specimen to avoid the movement of magnetic domains. The experimental setup is shown in Figure 1a, and an example of the spectra obtained for the S88 specimen at 300 K is shown in Figure 1b with indexes of modes.

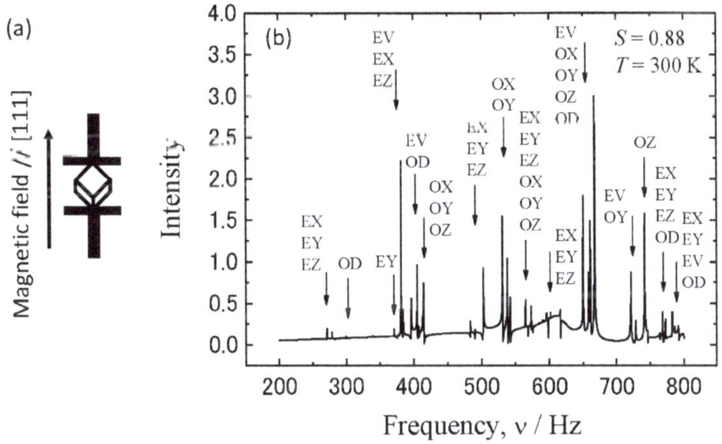

Figure 1. Schematic illustration showing the setting of the specimen for rectangular parallelepiped resonance (RPR) measurements under a magnetic field (**a**), and an example of the spectra of S88 at 300 K (**b**).

3. Results and Discussion

Figure 2 shows the temperature dependence of elastic constants $C_L = (C_{11} + C_{12} + 2C_{44})/2$, C_{44} and $C' = (C_{11} - C_{12})/2$. The errors of data are estimated to be the size of the marks. The temperature dependence of C_L is small, and C_L increases as the degree of order S increases. The value of C_{44} decreases almost linearly in the examined temperature range for all the specimens. The value of C' decreases almost linearly for all the specimens, but it slightly deviates from the linear relation below about 200 K for S75 and S57. This deviation could be due to the formation of tweed microstructure reported previously [12]. The bulk modulus B, which is given by $B = (C_{11} + 2C_{12})/3 = C_L - C_{44} - C'/3$, increases as temperature decreases; the value of B is between 85 and 100 GPa for S57, between 105 and 125 GPa for S75, and between 125 and 150 GPa for S88 in the examined temperature range.

Since C_{44} and C' decrease almost linearly in a wide temperature range, we fitted C_{44} and C' by linear functions of T, and the results are shown in Figure 2 by dashed lines. On each line, Curie temperature is shown by a double arrow. We notice that the value of C_{44} and C' at T_c on the extrapolated line is nearly the same for all the specimens: $C_{44}(T_c)$ ~100 GPa and $C'(T_c)$ ~30 GPa. This result supports a previous discussion that the softening of C' and C_{44} is caused by the band structure of the ferromagnetic phase [27–29].

The martensitic transformation temperature is also shown by a single arrow on each line in Figure 2. The extrapolated value of C' at M_s is 1.7 GPa for S57, 0.6 GPa for S75 and 6.9 GPa for S88. The value of C' is the smallest when the martensite phase is FCT1 with $c/a < 1$. The present result suggests that the softening in C' is most strongly related to the formation of FCT1 martensite with $c/a < 1$ compared with other martensites.

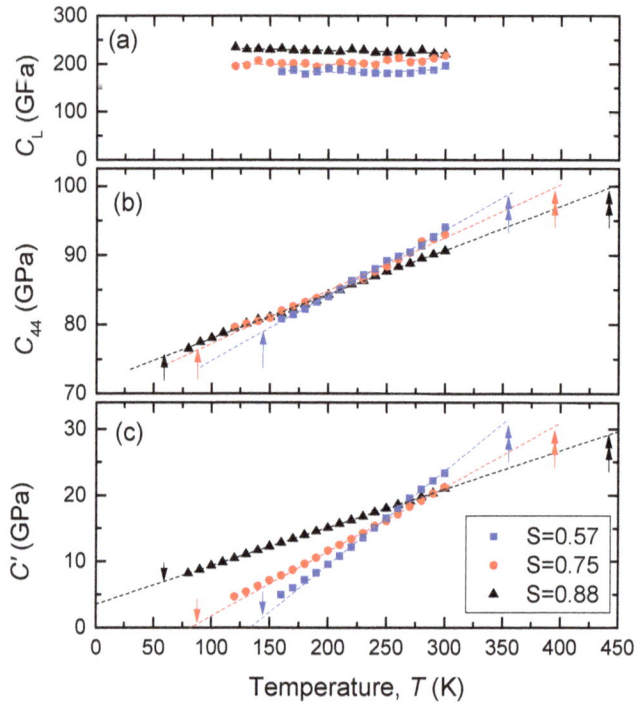

Figure 2. Elastic constants C_L (a), C_{44} (b) and C' (c) of Fe$_3$Pt with different degrees of order S. Single arrows indicate the martensitic transformation start temperature. Double arrows indicate the Curie temperature.

The anomalies in elastic anisotropy $A = C_{44}/C'$ are frequently discussed in alloys exhibiting thermoelastic martensitic transformations. Figure 3 shows the elastic anisotropy evaluated from Figure 2. The vertical lines indicate the martensitic transformation temperature. If we simply extend the experimental data, the approximate value of A at M_s temperature is expected to be 10 for S88, 30 for S75 and 20 for S57. The elastic anisotropy at M_s is the largest for S75 which transforms to FCT1 with $c/a < 1$. This again suggests that softening in C' is most strongly related to the formation of FCT1 martensite. Presumably, the band Jahn–Teller effect is the main reason for the softening of C' as previously reported [30]. Incidentally, the solid curve in Figure 3 is the calculated anisotropy using the linear relation shown by the dotted lines in Figure 2. If the linear relations were satisfied to M_s, the anisotropy at M_s would be 50 in S57 and 280 in S75.

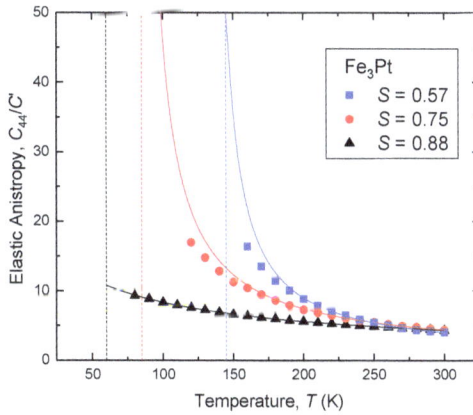

Figure 3. Temperature dependence of the elastic anisotropy of Fe$_3$Pt with different degrees of order S. The vertical lines indicate the martensitic transformation temperature.

4. Conclusions

The three alloys of Fe$_3$Pt (S = 0.57, 0.75, 0.88) all exhibit softening of elastic constant C'. Although the temperature coefficient of C' is largest for S = 0.57, which transforms to the BCT martensite, the value of C' at M_s is the smallest for S = 0.75, which transforms to the FCT1 martensite. In addition, the elastic anisotropy is the largest for S = 0.75. The softening in C' is probably most strongly related to the formation of the FCT1 martensite.

Acknowledgments: We appreciate technical advices by Katsushi Tanaka (Kobe University) on the measurements and analysis of PRP method. We also appreciate Masataka Yamamoto, Sayaka Sekida and Motoyoshi Yasui, who were students of Osaka University, for their support on sample preparation, measurements and analysis on the raw data. The present work was supported by JSPS KAKENHI Grant Number JP26289230 and JP25249086.

Author Contributions: Takashi Fukuda designed the experiments and prepared the manuscript; Tomoyuki Kakeshita supervised the research.

Conflicts of Interest: The authors declare no conflict of interest.

References

1. Nakanishi, N. Elastic constants as they relate to lattice properties and martensite formation. *Prog. Mater. Sci.* **1980**, *24*, 143–265. [CrossRef]
2. Murakami, Y. Lattice softening, phase stability and elastic anomaly of the β-Au-Cu-Zn alloys. *J. Phys. Soc. Jpn.* **1972**, *33*, 1350–1360. [CrossRef]
3. Gunton, D.J.; Sanuders, G.A. The elastic behavior of In-Tl alloys in the vicinity of the martensitic tansformation. *Solid State Commun.* **1974**, *14*, 865–868. [CrossRef]
4. Enami, K.; Hasunuma, J.; Nagasawa, A.; Nenno, S. Elastic softening and electron-diffraction anomalies prior to the martensitic transformation in Ni-Al β$_1$ alloy. *Scr. Metall.* **1976**, *10*, 879–884. [CrossRef]
5. Nagasawa, A.; Ueda, Y. Elastic Soft Mode Preceeding Martensitic Phase Transition in β Phase Alloys. *J. Phys. Soc. Jpn.* **1978**, *45*, 1249–1252. [CrossRef]
6. Verlinden, B.; Delay, L. On the elastic constants and Ms-temperatures in β-Hume-Rothery alloys. *Acta Metall.* **1988**, *36*, 1771–1779. [CrossRef]
7. Muto, S.; Oshima, R.; Fujita, F.F. Elastic softening and elastic strain energy consideration in the fcc-fct transformation of FePd alloys. *Acta Metall. Mater.* **1990**, *38*, 685–694. [CrossRef]
8. Ren, X.; Miura, N.; Zhang, N.; Otsuka, K.; Tanaka, K.; Koiwa, M.; Suzuki, T.; Chumlyakov, Yu.I.; Asai, M. A comparative study of elastic constants of Ti-Ni based alloys prior to martensitic transformation. *Mater. Sci. Eng. A* **2001**, *312*, 196–206. [CrossRef]

9   Sedlák, P.; Seiner, H.; Landa, M.; Novák, V.; Šittner, P.; Mañosa, L.I. Elastic constants of bcc austenite and 2H orthorhombic martensite in CuAlNi shape memory alloy. *Acta Mater.* **2005**, *53*, 3643–3661. [CrossRef]

10.   Graczykowski, G.; Mielcarek, S.; Breczewski, T.; No, M.L.; San-Juan, J. Martensitic phase transition in Cu-14%Al-4%Ni shape memory alloys studied by Brillouin light scattering. *Smart Mater. Struct.* **2013**, *22*, 085027. [CrossRef]

11.   Warren, B.E. X-ray studies of order-disorder. In *X-ray Diffraction*; Dover: New York, NY, USA, 1990.

12.   Muto, S.; Oshima, R.; Fujita, F.E. Electron microscopy study on martensitic transformations in Fe–Pt alloys: General features of internal structure. *Metall. Trans. A* **1988**, *19*, 2723–2731. [CrossRef]

13.   Dunne, D.P.; Wayman, C.M. The effect of austenite ordering on the martensite transformation in Fe–Pt alloys near the composition Fe_3Pt: I. Morphology and transformation characteristics. *Metall. Trans.* **1973**, *4*, 137–152. [CrossRef]

14.   Tadaki, T.; Shimizu, K. High tetragonality of thermoelastic Fe_3Pt martensite and small volume change during the transformation. *Scr. Metall.* **1975**, *9*, 771–776. [CrossRef]

15.   Umemoto, M.; Wayman, C.M. The effect of austenite ordering on the transformation temperature, transformation hysteresis, and thermoelastic behavior in Fe–Pt alloys. *Metall. Trans. A* **1978**, *9*, 891–897. [CrossRef]

16.   Yamamoto, M.; Sekida, S.; Fukuda, T.; Kakeshita, T.; Takahashi, K.; Koyama, K.; Nojiri, H. A new type of FCT martensite in single-crystalline Fe_3Pt Invar alloy. *J. Alloys Compd.* **2011**, *509*, 8530–8533. [CrossRef]

17.   Yamamoto, M.; Fukuda, T.; Kakeshita, T.; Takahashi, K. Orthorhombic martensite formed in $L1_2$-type Fe_3Pt Invar alloys. *J. Alloys Compd.* **2013**, *577*, S503–S506. [CrossRef]

18.   Hausch, G. Elastic constants of Fe–Pt alloys. I. Single crystalline elastic constants of $Fe_{72}Pt_{28}$. *J. Phys. Soc. Jpn.* **1974**, *37*, 819–823. [CrossRef]

19.   Ling, H.C.; Owen, W.S. The magneto-elastic properties of the invar alloy, Fe_3Pt. *Acta Metall.* **1983**, *31*, 1343–1352. [CrossRef]

20.   Kawald, U.; Zemke, W.; Bach, H.; Pelzl, J.; Saunders, G.A. Elastic constants and martensitic transformations in FePt and FeNiPt Invar alloys. *Physica B* **1989**, *161*, 72–74. [CrossRef]

21.   Owen, W. The influence of lattice softening of the parent phase on the martensitic transformation in FeNi and FePt alloys. *Mater. Sci. Eng. A* **1990**, *127*, 197–204. [CrossRef]

22.   Tajima, K.; Endoh, Y.; Ishikawa, Y. Acoustic-phonon softening in the invar alloy Fe_3Pt. *Phys. Rev. Lett.* **1976**, *37*, 519–522. [CrossRef]

23.   Kästner, J.; Petry, W.; Shapiro, S.M.; Zheludev, A.; Neuhaus, J.; Wassermann, E.F.; Bach, H. Influence of atomic order on $TA_1[110]$ phonon softening and displacive phase transition in $Fe_{72}Pt_{28}$ Invar alloys. *Eur. Phys. J. B* **1999**, *10*, 641–648. [CrossRef]

24.   Noda, Y.; Endoh, Y. Lattice dynamics in ferromagnetic Invar alloys, Fe_3Pt and $Fe_{65}Ni_{35}$. *J. Phys. Soc. Jpn.* **1988**, *57*, 4225–4231. [CrossRef]

25.   Fukuda, T.; Yamamoto, M.; Yamaguchi, T.; Kakeshita, T. Magnetocrystalline anisotropy and magnetic field-induced strain of three martensites in Fe_3Pt ferromagnetic shape memory alloys. *Acta Mater.* **2014**, *62*, 182–187. [CrossRef]

26.   Migliori, A.; Sarrao, J.L.; Visscher, W.M.; Bell, T.M.; Lei, M.; Fisk, Z.; Leisure, R.G. Resonant ultrasound spectroscopic techniques for measurement of the elastic moduli of solids. *Physica B* **1993**, *183*, 1–24. [CrossRef]

27.   Tanaka, K.; Okamoto, K.; Inui, H.; Minonishi, Y.; Yamaguchi, M.; Koiwa, M. Elastic constants and their temperature dependence for the intermetallic compound Ti_3Al. *Philos. Mag. A* **1996**, *73*, 1475–1488. [CrossRef]

28.   Gruner, M.E.; Adeagbo, W.A.; Zayak, A.T.; Hucht, A.; Entel, P. Lattice dynamics and structural stability of ordered Fe_3Ni, Fe_3Pt and Fe_3Pt alloys using density functional theory. *Phys. Rev. B* **2010**, *81*, 064109. [CrossRef]

29.   Sternik, M.; Couet, S.; Łażewski, J.; Parlinski, P.T.; Vantomme, A.; Temst, K.; Piekarz, P. Dynamical properties of ordered Fe-Pt alloys. *J. Alloys Compd.* **2015**, *651*, 528–536. [CrossRef]

30.   Yamamoto, T.; Yamamoto, M.; Fukuda, T.; Kakeshita, T.; Akai, H. An interpretation of martensitic transformation in $L1_2$-type Fe_3Pt from its electronic structure. *Mater. Trans.* **2010**, *51*, 896–898. [CrossRef]

metals

MDPI

Article

Investigation of Dynamic Recrystallization of NiTi Shape Memory Alloy Subjected to Local Canning Compression

Yanqiu Zhang [1], Shuyong Jiang [1,*] and Li Hu [2]

[1] College of Mechanical and Electrical Engineering, Harbin Engineering University, Harbin 150001, China; zhangyq@hrbeu.edu.cn
[2] College of Materials Science and Chemical Engineering, Harbin Engineering University, Harbin 150001, China; hcu_huli@126.com
* Correspondence: jiangshuyong@hrbeu.edu.cn; Tel.: +86-451-82519710

Academic Editor: Takuo Sakon
Received: 28 April 2017; Accepted: 1 June 2017; Published: 6 June 2017

Abstract: Physical mechanism for dynamic recrystallization of NiTi shape memory alloy subjected to local canning compression at various temperatures, 600, 700 and 800 °C, was investigated via electron backscattered diffraction experiments and transmission electron microscopy observations. With increasing deformation temperature, fractions of recrystallized grains and substructures increase, whereas fraction of deformed grains decreases. In the case of 600 and 700 °C, continuous dynamic recrystallization and discontinuous dynamic recrystallization coexist in NiTi shape memory alloy. In the case of discontinuous dynamic recrystallization, the recrystallized grains are found to be nucleated at grain boundaries and even in grain interior. The pile-up of statistically stored dislocation lays the foundation for the nucleation of the recrystallized grains during discontinuous dynamic recrystallization of NiTi shape memory alloy. Geometrically necessary dislocation plays as an important role in the formation of new recrystallized grains during continuous dynamic recrystallization of NiTi shape memory alloy.

Keywords: shape memory alloy; NiTi alloy; plastic deformation; dynamic recrystallization; microstructure

---

1. Introduction

NiTi-based shape memory alloys (SMAs) have been widely used in the engineering fields because they possess shape memory effect as well as superelasticity [1,2]. It is well known that plastic deformation at high temperatures is of great importance in manufacturing the products of NiTi-based SMAs [3–5]. In particular, dynamic recrystallization (DRX) frequently occurs during plastic deformation of NiTi-based SMAs at high temperatures [6–9]. It can be generally accepted that DRX has an influence on the microstructures of metallic alloys, which further have an effect on the mechanical properties [10,11]. As a consequence, it is important to investigate the DRX mechanisms of metallic alloys subjected to plastic deformation at high temperatures. In recent years, many researchers have devoted themselves to investigating the DRX of various metals subjected to uniaxial compression at high temperatures [12–17]. However, only a few literatures related to the DRX of NiTi-based SMAs have been reported. Yin et al. studied the mechanisms of DRX in the 50Ti-47Ni-3Fe SMA by means of uniaxial compression tests at the temperatures ranging from 750 to 1050 °C and at the strain rates ranging from 0.01 to 10 s^{-1}, where continuous dynamic recrystallization (CDRX) was found [6]. Basu et al. investigated the DRX in an Ni-Ti-Fe SMA subjected to uniaxial compression at the temperatures of 750, 850 and 950 °C, respectively, and they found that the DRX is able to suppress calorimetric signatures of phase transformations from austenite to martensite [7]. Mirzadeh and Parsa investigated hot compression behavior of 50.5 at % Ni–49.5 at % Ti SMA via flow stress curves at the temperatures ranging from 700 to 1000 °C and at the strain rate of 0.1 s^{-1}, where the typical single-peak DRX behavior was observed [9].

All the aforementioned investigations related to the DRX of NiTi-based SMAs were performed in the case of uniaxial compression at the temperatures above 700 °C. In addition, none of them focused on the mechanism of DRX in NiTi-based SMAs. In fact, the DRX may occur in the NiTi-based SMA which is deformed at a lower temperature of 600 °C [8]. It is well known that the deformation temperature is able to influence the mechanism of DRX. Furthermore, local canning compression may result in a different DRX behavior since it is able to provide a three-dimensional compression stress for the deformed materials. As a consequence, in the present work, a binary NiTi SMA was subjected to local canning compression at three temperatures (600, 700 and 800 °C) so as to investigate the physical mechanism of DRX.

2. Materials and Methods

2.1. Local Canning Compression

Commercially hot-rolled binary NiTi SMA bar with the composition of $Ni_{50.9}Ti_{49.1}$ (at %) and the diameter of 12 mm was used as raw material. The NiTi SMA bar was provided by Xi'an saite metal materials development Co., Ltd., Xi'an, China. The NiTi SMA samples, whose diameter and height are 4 and 6 mm, respectively, were removed from the as-rolled NiTi SMA bar along the axial direction via an electro-discharge machining (EDM) (DK7725, Jiangsu Dongqing CNC Machine Tool Co., Ltd., Taizhou, China). Subsequently, they were locally canned by the cans made of low carbon steel. Therein, the inner diameter, outer diameter and the height of the cans are 4, 10 and 3 mm, respectively. Afterward, the local canning compression experiments were carried out on the INSTRON 5500 equipment (Instron Corporation, Norwood, MA, USA) at the strain rate of 0.05 s^{-1} and at 600, 700 and 800 °C, respectively. Furthermore, all the canned NiTi SMA samples were compressed by 75% in height. Subsequently, all the compressed NiTi SMA samples were quenched into water at room temperature. Finally, the compressed NiTi SMA samples were removed from the steel cans.

2.2. Materials Characterization

The samples for electron backscattered diffraction (EBSD) measurements were prepared based on the cross section which is parallel to axial direction of the as-rolled NiTi SMA bar as well as the compressed NiTi SMA samples. The NiTi SMA samples for EBSD observation were mechanically polished and subsequently electropolished in a solution, which is composed of 30% HNO_3 and 70% CH_3OH by volume fraction, at −30 °C. EBSD experiments were performed on the NiTi SMA samples via Zeiss ULTRA plus scanning electron microscope (SEM) (University of South Carolina, Columbia, SC, USA), which is equipped with Oxford Instruments AZtec integrated energy-dispersive spectroscopy (EDS) and EBSD system. Microstructures of the compressed NiTi SMA samples were characterized via transmission electron microscopy (TEM). Foils for TEM observation were mechanically ground to 70 μm and then thinned by twin-jet polishing in an electrolyte which is composed of 6% $HClO_4$, 34% $C_4H_{10}O$ and 60% CH_3OH by volume fraction. TEM observations were carried out on a FEI TECNAI G2 F30 microscope (FEI Corporation, Hillsboro, OR, USA), which possesses a side-entry and double-tilt specimen stage with angular range of ±40° at an accelerating voltage of 300 kV.

3. Results

3.1. EBSD Analysis of Microstructural Evolution

Based on EBSD experiment, the microstructure of the as-rolled NiTi SMA bar was characterized, and the corresponding results were shown in Figure 1. It can be seen from Figure 1 that the grains of the as-rolled NiTi SMA bar exhibit a certain preferential orientation, where <110> texture is dominant according to orientation distribution function (ODF). In addition, it can be found that the misorientation of the grains basically follows a normal distribution except that the grains with the misorientation of 1–2° exhibit a higher frequency, as shown in Figure 1c. In addition, it can be noted from Figure 1d that the size of grains mainly ranges from 10 to 60 μm.

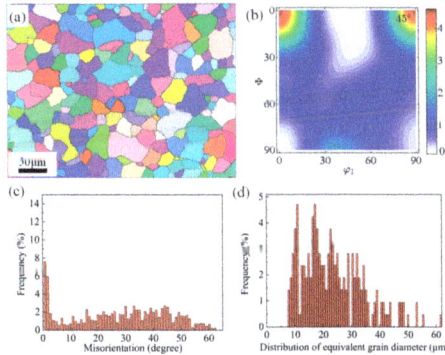

Figure 1. Microstructure characterization of as-rolled NiTi shape memory alloy (SMA) based on electron backscattered diffraction (EBSD) experiment: (**a**) distribution of grains with various orientations; (**b**) orientation distribution function (ODF); (**c**) misorientation angle distribution; (**d**) distribution of equivalent grain diameter.

Figure 2 shows the EBSD results of NiTi SMA subjected to canning compression at various temperatures, 600, 700 and 800 °C. It can be observed from Figure 2 that as compared to the as-rolled NiTi SMA, the grains with low angle grain boundaries are dominant in the compressed NiTi SMA. In particular, <110> texture of the as-rolled NiTi SMA is transformed into <111> texture of the compressed NiTi SMA as a result of plastic deformation. Furthermore, the intensity of <111> texture increases with increasing deformation temperature. The aforementioned results indicate that DRX seems to have a significant influence on the texture evolution of NiTi SMA.

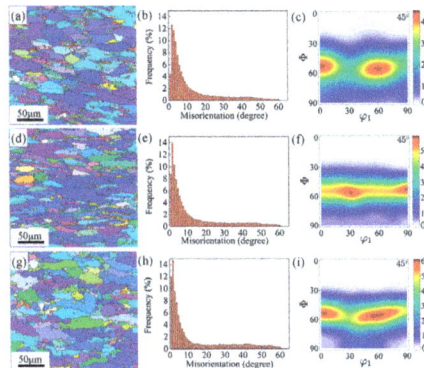

Figure 2. EBSD analysis of NiTi SMA subjected to local canning compression at various temperatures: (**a**) orientation map at 600 °C; (**b**) misorientation distribution at 600 °C; (**c**) ODF at 600 °C; (**d**) Orientation map at 700 °C; (**e**) misorientation distribution at 700 °C; (**f**) ODF at 700 °C; (**g**) orientation map at 800 °C; (**h**) misorientation distribution at 800 °C; (**i**) ODF at 800 °C.

For the purpose of further understanding DRX mechanisms of NiTi SMA, microstructures of NiTi SMA subjected to local canning compression at the various temperatures were characterized by means of EBSD experiment, as shown in Figure 3. The fractions of the recrystallized grains, the substructures and the deformed grains are captured in NiTi SMA subjected to local canning compression at 600, 700 and 800 °C. It can be noted that with increasing deformation temperature, the fractions of the recrystallized grains and the substructures increase, whereas the fraction of the deformed grains decreases. In fact, the three structures do not show such a considerable discrepancy between 600 and 700 °C. However, as compared to 600 and 700 °C, the three structures present an apparent discrepancy in NiTi SMA subjected to local canning compression at 800 °C. The aforementioned phenomena are attributed to the fact that dynamic recovery (DRV) and DRX are dominant in NiTi SMA subjected to local canning compression at 600 and 700 °C, while only DRX is dominant in NiTi SMA subjected to local canning compression at 800 °C.

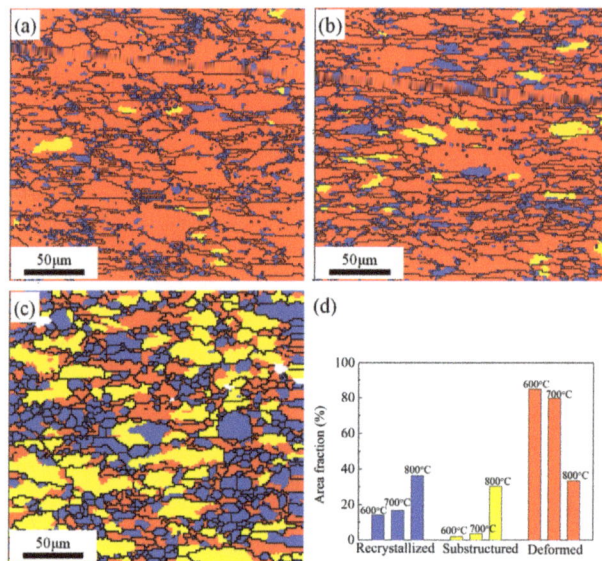

Figure 3. EBSD analysis of dynamic recrystallized microstructures in NiTi SMA subjected to local canning compression at various temperatures: (**a**) DRX map at 600 °C; (**b**) DRX map at 700 °C; (**c**) DRX map at 800 °C; (**d**) Area fraction distribution of various microstructures. (The grains with the misorientation equal to or larger than 10° was defined as the recrystallized structures, and the grains with the misorientation ranging from 3 to 10° were regarded as the substructured structures, while the grains with the misorientation equal to or lower than 3° was identified as the deformed structures.)

Figure 4 shows a distribution of geometrically necessary dislocation (GND) density of NiTi SMA samples based on EBSD experiments. It can be observed that as compared to the as-rolled NiTi SMA sample, GND density shows little difference in the NiTi SMA samples subjected to local canning compression at 600, 700 and 800 °C. In addition, the deformation temperatures have little influence on GND density of NiTi SMA samples. Furthermore, as is expected, GND is mainly distributed at the grain boundaries.

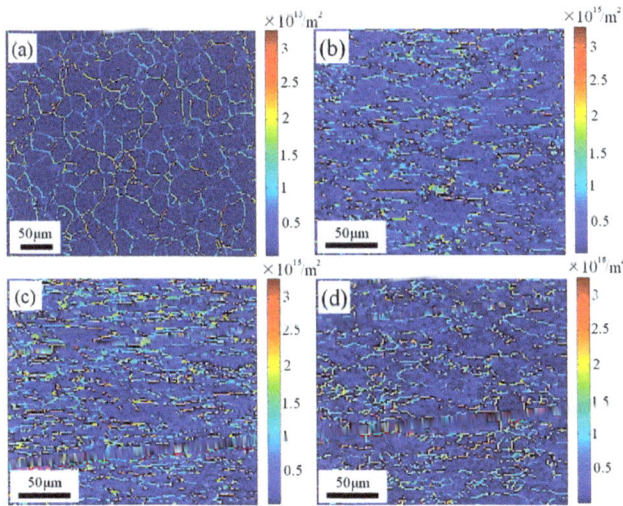

Figure 4. Distribution of geometrically necessary dislocation (GND) density in NiTi SMA subjected to local canning compression at various temperatures: (**a**) As-rolled; (**b**) 600 °C; (**c**) 700 °C; (**d**) 800 °C.

3.2. TEM Observation of Microstructural Evolution

TEM micrographs of NiTi SMA subjected to local canning compression at 600, 700 and 800 °C are shown in Figures 5–7, respectively, so that the corresponding mechanisms of DRX are further clarified. It can be found from Figure 5a that in the case of 600 °C, a high density of dislocations is distributed in the matrix of NiTi SMA with B2 austenite phase. In addition, the subgrain with low angle grain boundary can be observed in Figure 5c, where plenty of dislocations are distributed in the subgrain interior. In particular, the recrystallized grains can be observed in the grain interior in Figure 5d. However, as for NiTi SMA subjected to local canning compression at 700 °C, dislocation density is obviously reduced, as shown in Figure 6a. The recrystallized grain can be observed in the grain interior, as shown in Figure 6c, where plenty of dislocations are distributed in the recrystallized grain. Moreover, a triple-junction grain boundary can be observed in Figure 6d, where a few dislocations are distributed near the grain boundary. In particular, no recrystallized grain can be seen in the grain interior. However, when NiTi SMA is subjected to local canning compression at 800 °C, dislocation density is substantially reduced, as shown in Figure 7a. It is very difficult to capture the whole recrystallized grain as well as the subgrain by means of TEM observation. It can be observed that the parallel dislocation array is distributed near the grain boundary due to the pile-up of dislocations at the grain boundary.

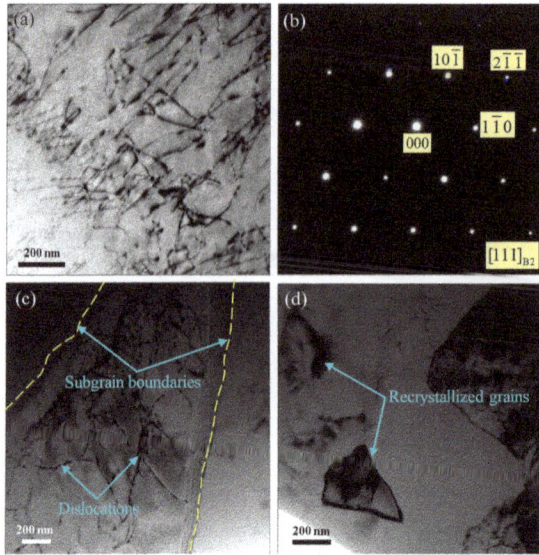

Figure 5. TEM micrographs of NiTi SMA subjected to local canning compression at 600 °C: (**a**) bright field image showing a high density of dislocations; (**b**) diffraction patterns of (**a**) showing the existence of B2 austenite; (**c**) bright field image showing the existence of subgrain boundary; (**d**) bright field image showing the existence of recrystallized grains.

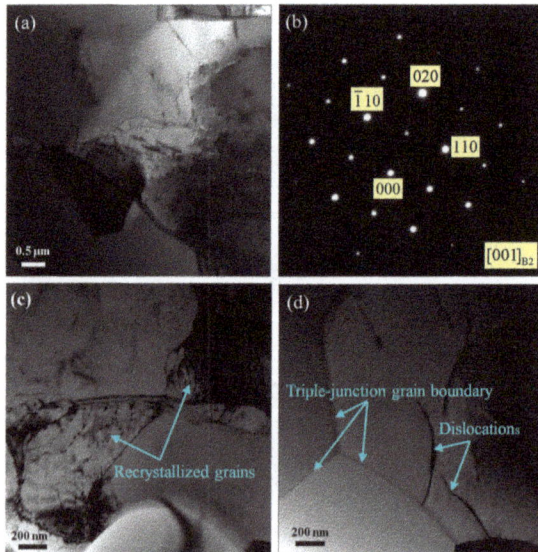

Figure 6. TEM micrographs of NiTi SMA subjected to local canning compression at 700 °C: (**a**) Bright field image; (**b**) diffraction patterns of (**a**) showing the existence of B2 austenite; (**c**) bright field image showing the formation of recrystallized grains at the grain boundary; (**d**) bright field image showing the existence of triple-junction grain boundary.

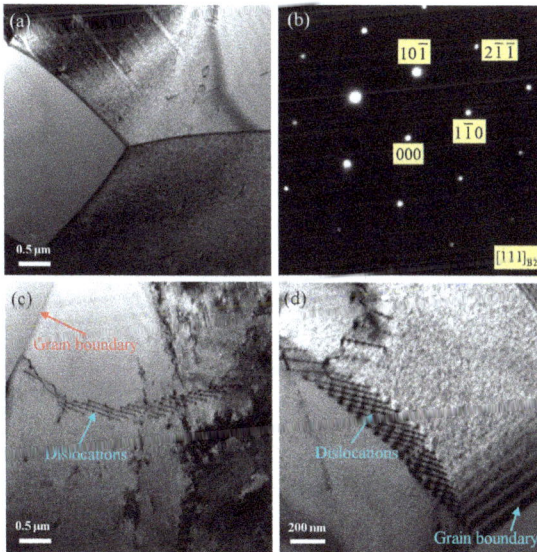

Figure 7. TEM micrographs of NiTi SMA subjected to local canning compression at 800 °C: (a) bright field image showing the existence of triple-junction grain boundary; (b) diffraction patterns of (a) showing the existence of B2 austenite, (c) bright field image showing the pile-up of dislocations at the grain boundary; (d) bright field image showing the existence of substructures due to parallel dislocations.

4. Discussion

It is well known that DRX frequently occurs in the metal materials subjected to hot working. Consequently, the new grains, which are completely different from the original ones, arise in the initial microstructures. Furthermore, the factors, which have an influence on DRX, deal with deformation temperature, plastic strain, strain rate and nature of metal materials. So far as the latter is concerned, stacking fault energy is related closely to DRX of metal materials. In particular, it is generally accepted that DRX can be divided into discontinuous dynamic recrystallization (DDRX) and CDRX. In the case of DDRX, the formation of the recrystallized grains deals with the nucleation and the growth of new grains. In other words, DDRX has a characteristic of repeated nucleation and finite growth. However, in terms of CDRX, subgrain structures with low angle grain boundary are induced at the primary stage of plastic deformation and they are progressively transformed into the new grains with high angle grain boundaries after experiencing large plastic strain. In general, DDRX takes place in the metal materials with low to medium stacking fault energy at high temperatures more than $0.5T_m$, where T_m is the melting point of the metal materials. The melting point of the NiTi SMA used in the present study is determined as 1300 °C or so. CDRX is found to occur in all the metal materials regardless of their stacking fault energy when the deformation temperature is above $0.5T_m$ and the metal materials are subjected to severe plastic deformation. However, at high temperatures above $0.5T_m$, CDRX are frequently observed in the metal materials with high stacking fault energy [10,11].

It is well known that dislocation density plays an important role in DRX of metals. In the case of DDRX, in particular, dislocation density shall become a driving force that facilitates the nucleation of new recrystallized grains during plastic deformation. In general, the new recrystallized grains are free of dislocations during DDRX. However, in the case of CDRX, the dislocations are progressively absorbed during transformation from low angle grain boundaries to high angle grain boundaries. Furthermore, CDRX frequently occurs along with DRV. Consequently, the dislocations can be frequently observed in the recrystallized grains based on CDRX. According to the aforementioned analysis, it can be deduced that in the present study, DDRX and CDRX coexist during plastic deformation of NiTi SMA at 600 and 700 °C. It has been reported in the literatures that stacking fault energy is not a rigorous factor distinguishing between DDRX and CDRX since DDRX is observed in the metals with high stacking fault energy [18], while CDRX is observed in the metals with low stacking fault energy [19]. It can be generally accepted that DDRX frequently occurs in NiTi-based SMA [6–8]. However, recent literature has shown that CDRX is captured in NiTiFe SMA [9]. However, it seems that DDRX is dominant in NiTi SMA subjected to plastic deformation at 800 °C, while it is unclear that whether CDRX is able to arise since the experimental evidence is insufficient in the present work.

It can be generally accepted that during DDRX of NiTi SMA, there is a competition between hardening mechanism due to dislocation storage and softening mechanism due to dislocation elimination. The competition mechanism is described as follows. When NiTi SMA is subjected to plastic deformation, the dislocation density increases with increasing plastic strain, and consequently the increase of the dislocation density leads to the occurrence of hardening mechanism. Simultaneously, the increasing dislocation density provides a driving force for the nucleation of the dynamic recrystallized grains, and thus the formation of the dynamic recrystallized grains result in the decrease of the dislocation density [11,20]. In the meantime, the DRV process is relatively slow, so the dislocation density is able to achieve a sufficiently high value to promote the nucleation of new recrystallized grains during plastic deformation. In general, the new recrystallized grains are nucleated preferentially at the grain boundaries, especially at the triple-junction of grain boundaries. The phenomenon is attributed to the fact that the dislocation density exhibits a higher value at the grain boundaries due to the pile-up of dislocations. However, according to the aforementioned results, it seems that the recrystallized grains are able to be nucleated in the grain interior. In general, NiTi SMA is in a three-dimension compressive stress state when it is subjected to local canning compression. This shall lead to the occurrence of high plastic strain in the local region of the grain interior. As a consequence, local high plastic strain results in local high dislocation density, which provides a driving force for the nucleation of the recrystallized grains in the grain interior. Of course, the corresponding experiment evidence shall be performed in the future.

In order to better give an insight into the distinction between DDRX and CDRX and to deeply understand the role of the dislocation density in DRX of NiTi SMA, the dislocation can be divided into statistically stored dislocation (SSD) and GND. In general, SSD is inherently responsible for plastic flow based on crystallographic slip on the distinct slip planes. However, GND results from the inhomogeneous plastic deformation, which is responsible for accommodating a given strain gradient and preserving crystallographic lattice compatibility. When NiTi SMA is subjected to plastic deformation at high temperatures, the slip systems are activated and consequently plenty of SSDs pile up towards the grain boundary. A high density of dislocations at the grain boundary lays the foundation for the nucleation of the recrystallized grains during DDRX of NiTi SMA. SSDs can tangle with each other even in the grain interior, and thus a high density of dislocations is formed in the grain interior. As a consequence, it is possible for the recrystallized grains to be nucleated in the grain interior. It can be generally accepted that there is not a rigorous dividing line between SSD and GND. In the case of CDRX, SSDs can be transformed into GNDs, which shall constitute the subgrain boundary. With the progression of plastic deformation, the subgrain boundary progressively absorbs GNDs and finally the new recrystallized grains are formed. It seems that GND plays a more important role in the formation of the new recrystallized grains during CDRX.

5. Conclusions

(1) In the case of 600 and 700 °C, DDRX and CDRX coexist in NiTi SMA. In the case of DDRX, the recrystallized grains are found to be nucleated at the grain boundary and even in the grain interior. The pile-up of SSDs lays the foundation for the nucleation of the recrystallized grains during DDRX of NiTi SMA. GND plays a significant role in the formation of new recrystallized grains during CDRX

of NiTi SMA. DDRX of NiTi SMA is attributed to a competition between hardening mechanism due to dislocation storage and softening mechanism due to dislocation elimination. In the case of CDRX, subgrain structures with low angle grain boundary are induced at the primary stage of plastic deformation and they are progressively transformed into the new grains with high angle grain boundaries after being subjected to large plastic strain. Furthermore, the dislocations are progressively absorbed during transformation from low angle grain boundaries to high angle grain boundaries.

(2) With increasing deformation temperature, fractions of the recrystallized grains and the substructures increase, whereas the fraction of the deformed grains decreases. Dynamic recrystallization has a significant influence on the texture evolution of NiTi SMA. In particular, <110> texture of the as-rolled NiTi SMA is transformed into <111> texture of the compressed NiTi SMA as a result of plastic deformation. Furthermore, the intensity of <111> texture increases with increasing deformation temperature. It can be deduced that local canning compression causes NiTi SMA to be in a three-dimensional compressive stress state, which contributes to the occurrence of DRX. Furthermore, under the action of a three-dimensional compressive stress, the grains of NiTi SMA are preferentially oriented along with plastic deformation. As a consequence, <111> texture is formed under the simultaneous action of DRX and compression deformation. The formation mechanism of <111> texture shall be further investigated in the future.

Acknowledgments: The work was financially supported by National Natural Science Foundation of China (Nos. 51475101 and 51305091).

Author Contributions: Yanqiu Zhang performed EBSD analysis and wrote the manuscript; Shuyong Jiang supervised the research; Li Hu performed experiments and TEM analysis.

Conflicts of Interest: The authors declare no conflict of interest.

References

1. Tadayyon, G.; Mazinani, M.; Guo, Y.; Zebarjad, S.M.; Tofail, S.A.M.; Biggs, M.J. The effect of annealing on the mechanical properties and microstructural evolution of Ti-rich NiTi shape memory alloy. *Mater. Sci. Eng. A* **2016**, *662*, 564–577. [CrossRef]
2. Shamsolhodaei, A.; Zarei-Hanzaki, A.; Ghambari, M.; Moemeni, S. The high temperature flow behavior modeling of NiTi shape memory alloy employing phenomenological and physical based constitutive models: A comparative study. *Intermetallics* **2014**, *53*, 140–149. [CrossRef]
3. Yeoma, J.T.; Kima, J.H.; Hong, J.K.; Kim, S.W.; Park, C.H.; Nam, T.H.; Lee, K.Y. Hot forging design of as-cast NiTi shape memory alloy. *Mater. Res. Bull.* **2014**, *58*, 234–238. [CrossRef]
4. Etaati, A.; Dehghani, K. A study on hot deformation behavior of Ni-42.5Ti-7.5Cu alloy. *Mater. Chem. Phys.* **2013**, *140*, 208–215. [CrossRef]
5. Morakabati, M.; Aboutalebi, M.; Kheirandish, S.; Karimi Taheri, A.; Abbasi, S.M. Hot tensile properties and microstructural evolution of as cast NiTi and NiTiCu shape memory alloys. *Mater. Des.* **2011**, *32*, 406–413.
6. Mirzadeh, H.; Parsa, M.H. Hot deformation and dynamic recrystallization of NiTi intermetallic compound. *J. Alloys Compd.* **2014**, *614*, 56–59. [CrossRef]
7. Basu, R.; Jain, L.; Maji, B.; Krishnan, M. Dynamic recrystallization in a Ni–Ti–Fe shape memory alloy: Effects on austenite–martensite phase transformation. *J. Alloys Compd.* **2015**, *639*, 94–101. [CrossRef]
8. Jiang, S.; Zhang, Y.; Zhao, Y. Dynamic recovery and dynamic recrystallization of NiTi shape memory alloy under hot compression deformation. *Trans. Nonferrous Met. Soc. China* **2013**, *23*, 140–147. [CrossRef]
9. Yin, X.Q.; Park, C.H.; Li, Y.F.; Ye, W.J.; Zuo, Y.T.; Lee, S.W.; Yeom, J.T.; Mi, X.J. Mechanism of continuous dynamic recrystallization in a 50Ti–47Ni-3Fe shape memory alloy during hot compressive deformation. *J. Alloys Compd.* **2017**, *693*, 426–431. [CrossRef]
10. Huang, K.; Logé, R.E. A review of dynamic recrystallization phenomena in metallic materials. *Mater. Des.* **2016**, *111*, 548–574. [CrossRef]
11. Sakai, T.; Belyakov, A.; Kaibyshev, R.; Miura, H.; Jonas, J.J. Dynamic and post-dynamic recrystallization under hot, cold and severe plastic deformation conditions. *Prog. Mater. Sci.* **2014**, *60*, 130–207. [CrossRef]
12. Xu, T.C.; Peng, X.D.; Qin, J.; Chen, Y.F.; Yang, Y.; Wei, G.B. Dynamic recrystallization behavior of Mg–Li–Al–Nd duplex alloy during hot compression. *J. Alloys Compd.* **2015**, *639*, 79–88. [CrossRef]

13. Liang, H.; Guo, H.; Ning, Y.; Peng, X.; Qin, C.; Shi, Z.; Nan, Y. Dynamic recrystallization behavior of Ti–5Al–5Mo–5V–1Cr–1Fe alloy. *Mater. Des.* **2014**, *63*, 798–804. [CrossRef]
14. Zhang, Z.; Yang, X.; Xiao, Z.; Jun, W.; Zhang, D.; Liu, C.; Sakai, T. Dynamic recrystallization behaviors of a Mg-4Y-2Nd-0.2Zn-0.5Zr alloy and the resultant mechanical properties after hot compression. *Mater. Des.* **2016**, *97*, 25–32. [CrossRef]
15. Li, Y.; Koizumi, Y.; Chiba, A. Dynamic recrystallization in biomedical Co-29Cr-6Mo-0.16N alloy with low stacking fault energy. *Mater. Sci. Eng. A* **2016**, *668*, 86–96. [CrossRef]
16. Wu, Y.; Zhang, X.; Deng, Y.; Tang, C.; Yang, L.; Zong, Y. Dynamic recrystallization mechanisms during hot compression of Mg–Gd–Y–Nd–Zr alloy. *Trans. Nonferr. Met. Soc. China* **2015**, *25*, 1831–1839. [CrossRef]
17. Zhang, B.; Liu, C.; Zhou, J.; Tao, C. Dynamic recrystallization of single-crystal nickel-based superalloy. *Trans. Nonferr. Met. Soc. China* **2014**, *24*, 1744–1749. [CrossRef]
18. Yamagata, H.; Ohuchida, Y.; Saito, N.; Otsuka, M. Nucleation of new grains during discontinuous dynamic recrystallization of 99.998 mass% Aluminum at 453 K. *Scr. Mater.* **2001**, *45*, 1055–1061. [CrossRef]
19. Yanushkevich, Z.; Belyakov, A.; Kaibyshev, R. Microstructural evolution of a 304-type austenitic stainless steel during rolling at temperatures of 773–1273 K. *Acta Mater.* **2015**, *82*, 244–254. [CrossRef]
20. Ouyang, D.L.; Fu, M.W.; Lu, S.Q. Study on the dynamic recrystallization behavior of Ti-alloy Ti–10V–2Fe–3V in β processing via experiment and simulation. *Mater. Sci. Eng. A* **2014**, *619*, 26–34. [CrossRef]

metals MDPI

Article

Investigation on Deformation Mechanisms of NiTi Shape Memory Alloy Tube under Radial Loading

Shuyong Jiang [1,*], Junbo Yu [2], Li Hu [2] and Yanqiu Zhang [1]

[1]   College of Mechanical and Electrical Engineering, Harbin Engineering University, Harbin 150001, China; zhangyq@hrbeu.edu.cn

[2]   College of Materials Science and Chemical Engineering, Harbin Engineering University, Harbin 150001, China; yujunbo@hrbeu.edu.cn (J.Y.); heu_huli@126.com (L.H.)

*   Correspondence: jiangshuyong@hrbeu.edu.cn; Tel.: +86-451-8251-9710

Received: 9 June 2017; Accepted: 11 July 2017; Published: 13 July 2017

Abstract: NiTi shape memory alloy (SMA) tube was coupled with mild steel cylinder in order to investigate deformation mechanisms of NiTi SMA tubes undergoing radial loading. NiTi SMA tubes of interest deal with two kinds of nominal compositions; namely, Ni-50 at % Ti and Ni-49.1 at % Ti, where at room temperature, B19′ martensite is dominant in the former, and B2 austenite is complete in the latter. The mechanics of the NiTi SMA tube during radial loading were analyzed based on elastic mechanics and plastic yield theory, where effective stress and effective strain are determined as two important variables that investigate deformation mechanisms of the NiTi SMA tube during radial loading. As for the NiTi SMA tube with austenite structure, stress-induced martensite (SIM) transformation as well as plastic deformation of SIM occur with the continuous increase of effective stress. As for NiTi SMA tube which possesses martensite structure, reorientation and detwinning of twinned martensite as well as plastic deformation of reoriented and detwinned martensite occur with the continuous increase in the effective stress. Plastic deformation for dislocation slip has a negative impact on superelasticity and shape memory effect of NiTi SMA tube.

Keywords: shape memory alloy; NiTi alloy; NiTi tube; radial loading; deformation mechanism

---

1. Introduction

NiTi shape memory alloy (SMA) has been substantially employed in the domain of engineering since it possesses perfect shape memory effect (SME) and excellent superelasticity [1]. Based on SME, NiTi SMA can remember the shape in the austenite phase when it is heated to austenite finish temperature (A_f) after undergoing a certain extent of deformation in the martensite state. In terms of superelasticity, NiTi SMA is characterized by nonlinear recoverable deformation behavior when the temperature is above A_f. Superelasticity stems from stress-induced martensite (SIM) transformation by exerting force and spontaneous reverse martensite transformation after removing force.

NiTi SMA tube—which possesses perfect SME, good corrosion resistance, high plateau stress, ultimate tensile strength, high fatigue life, and excellent superelasticity at/around body temperature—has been the best candidate for biomedical stents and pipe coupling [2–4]. More and more researchers have been engaged in plastic working of NiTi SMA tube, such as drawing [5–7], extrusion [8], spinning [9] and equal channel angular extrusion [10].

In particular, the mechanical behavior of NiTi SMA tube based on different loading modes is a critical factor in engineering applications. In general, NiTi SMA tube exhibits different mechanical behaviors as well as different transformation characteristics under different loading conditions. Sun and Li studied the mechanical behavior of B2 austenite NiTi SMA tube under simple tension and simple shear [11,12]. They found that when NiTi SMA tube with B2 austenite undergoes simple tension, SIM transformation occurs so that a martensitic spiral band can be observed. However, when NiTi

SMA tube with B2 austenite suffers from simple shear, a martensitic spiral band cannot be observed. SIM transformation of B2 austenite NiTi SMA tube under simple tension was further validated on the basis of experimental observation by Mao et al. [13]. Furthermore, Mao et al. also found that in the case of uniaxial compression, a homogeneous martensite phase transformation appears in B2 austenite NiTi SMA tube, where a martensitic spiral band is unable to be observed. Jiang et al. investigated buckling and recovery of B2 austenite NiTi SMA tube under compressive loading by combining digital image correlation (DIC) with finite element simulation, where SIM transformation plays a significant role [14,15]. Bechle and Kyriakides studied the mechanical behavior of B2 austenite NiTi SMA tube subjected to bending load. Consequently, they found that bending can also lead to localized nucleation of martensite phase, where deformation patterns exhibit the considerable difference between the compression and tension sides [16]. Wang et al. investigated the mechanical behavior of austenitic NiTi SMA tube subjected to biaxial load, which involves tension loading and torsion loading, and has been widely studied in the literature [17]. Bechle and Kyriakides also investigated localization evolution of B2 austenite NiTi SMA tube undergoing biaxial stress loading by means of DIC, which is able to track the evolution and degree of inhomogeneous deformation resulting from transformation from austenite into martensite under loading as well as from martensite into austenite under unloading [18,19]. Rivin et al. investigated the recoverable deformation of hollow thin-walled NiTi tube subjected to radial loading which acts on the outer wall of NiTi tube [20]. Zhang et al. studied the mechanical behavior of hollow thin-walled NiTi tube under radial quasi-static compression, where the compressive loading is also imposed on the outer wall of NiTi tube [21].

In the present study, a new radial loading mode is put forward for NiTi SMA tube by selecting mild steel cylinder as loading media. The radial loading is implemented by imposing inner pressure on the inner wall of NiTi tube, where inner pressure results from plastic deformation of the steel cylinder.

2. Materials and Methods

NiTi SMA bars—which have chemical composition of Ni-50 at % Ti and Ni-49.1 at % Ti, respectively—were used for manufacturing NiTi SMA tubes. The inner diameter, outer diameter, and height of NiTi SMA tubes were 8, 10, and 10 mm, respectively. In order to investigate the mechanical behavior and deformation mechanism, the mild steel cylinder (with diameter and height of 8 and 20 mm, respectively) was symmetrically inserted into NiTi SMA tube. The assembled sample experienced a compression test on the INSTRON-5500R equipment (Instron Corporation, Norwood, MA, USA), as shown in Figure 1. The pressure, p which is imposed on NiTi SMA tube can be solved according to plastic yield and plastic mechanics of the mild steel cylinder. The deformation mechanism of NiTi SMA tube can be analyzed based on the pressure p and according to the dimension variation during plastic deformation.

Figure 1. Photograph of NiTi shape memory alloy (SMA) tube undergoing radial loading test.

Microstructures of NiTi SMA samples were characterized using transmission electron microscope (TEM) Specimens for TEM experiment were mechanically ground to 70 μm. Subsequently, they were thinned via twin-jet polishing in an electrolyte composed of 6% $HClO_4$, 34% $C_4H_{10}O$, and 60% CH_3OH, according to volume percentage. TEM observation was performed on a FEI TECNAI G2 F30 microscope (FEI Corporation, Hillsboro, OR, USA). To further acquire the microstructure of NiTi SMA samples, electron back-scattered diffraction (EBSD) experiments were carried out on NiTi SMA specimens using a Zeiss Supra 55 scanning electron microscope (University of South Carolina, Columbia, SC, USA) coupled with an OXFORD EBSD instrument (Oxford Instruments, Oxford, UK). The samples for EBSD were mechanically polished and subsequently electropolished in a solution of 30% HNO_3 and 70% CH_3OH by volume fraction at $-30\,°C$. A step size of 1.5 μm was used to capture area scans for NiTi SMA specimens

3. Mechanical Analysis of NiTi SMA Tube under Radial Loading

The mechanical model of NiTi SMA tube under radial loading can be shown in Figure 2. NiTi SMA tube is viewed as the thick-walled tube under the action of the pressure p. The deformation mechanics of NiTi SMA tube belongs to the axisymmetric problem as well as the plane stress problem, where the axial stress σ_z is equal to zero without the axial loading, and the radial stress σ_p and the tangential stress σ_θ are irrespective of the axial coordinate axis. Simultaneously, it is assumed that all the total material particles within NiTi SMA tube experience deformation. σ_p and σ_θ are both the principal stresses due to the absence of shear stresses in NiTi SMA tube.

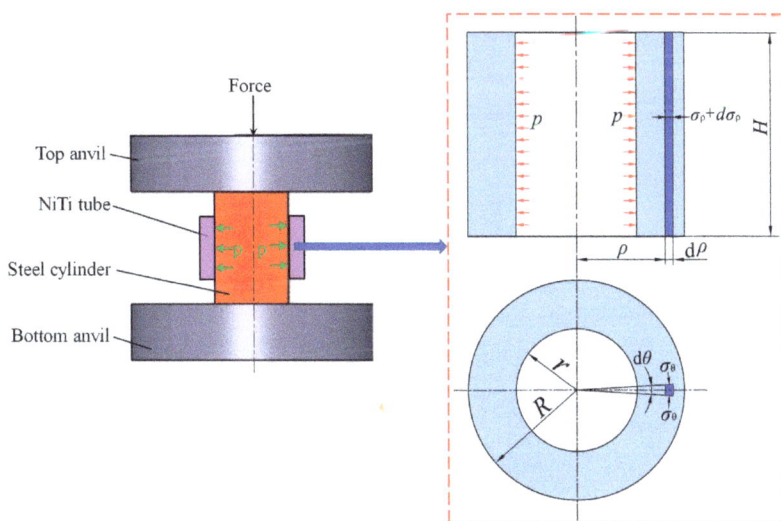

Figure 2. Mechanical model of NiTi SMA tube subjected to radial loading, where p is the pressure, R the outer radius, r the inner radius, and H the height.

3.1. Geometric Equation between Strain and Displacement

The tangential strain ε_θ and the radial strain ε_p of any point in NiTi SMA tube are expressed as follows, respectively

$$\begin{cases} \varepsilon_\rho = \dfrac{du}{d\rho} \\ \varepsilon_\theta = \dfrac{u}{\rho} \end{cases} \tag{1}$$

where u is the displacement of any point along the radius direction in NiTi SMA tube.

3.2. Equilibrium Equation under Static Inner Pressure

By intercepting an element slab in NiTi SMA tube in Figure 2, the equilibrium differential equation along the radial direction is obtained as follows.

$$(\sigma_\rho + d\sigma_\rho)(\rho + d\rho)hd\theta - \sigma_\rho \rho hd\theta - 2\sigma_\theta(sin\frac{d\theta}{2})hd\rho = 0 \tag{2}$$

By ignoring the high-order items of Equation (2), the following equation can be obtained.

$$\frac{d\sigma_\rho}{d\rho} + \frac{\sigma_\rho - \sigma_\theta}{\rho} = 0 \tag{3}$$

3.3. Constitutive Equation between Stress and Strain

According to generalized Hooke's law, the relationship between the stresses and the strains can be expressed by

$$\begin{cases} \varepsilon_\rho = \dfrac{1}{E}(\sigma_\rho - \mu\sigma_\theta) \\ \varepsilon_\theta = \dfrac{1}{E}(\sigma_\theta - \mu\sigma_\rho) \end{cases} \tag{4}$$

Substitution of Equation (1) into Equation (4) results in

$$\begin{cases} \dfrac{du}{d\rho} = \dfrac{1}{E}(\sigma_\rho - \mu\sigma_\theta) \\ \dfrac{u}{\rho} = \dfrac{1}{E}(\sigma_\theta - \mu\sigma_\rho) \end{cases} \tag{5}$$

According to Equation (5), the expressions of σ_p and σ_θ can be expressed by

$$\begin{cases} \sigma_\rho = \dfrac{E}{1-\mu^2}(\dfrac{du}{d\rho} + \mu\dfrac{u}{\rho}) \\ \sigma_\theta = \dfrac{E}{1-\mu^2}(\dfrac{u}{\rho} + \mu\dfrac{du}{d\rho}) \end{cases} \tag{6}$$

Substitution of Equation (6) into Equation (3) leads to

$$\frac{d^2u}{d\rho^2} + \frac{1}{\rho}\frac{du}{d\rho} - \frac{u}{\rho^2} = 1 \tag{7}$$

The displacement u is obtained as follows by solving Equation (7).

$$u = \frac{1-\mu}{E}\frac{r^2 p\rho}{R^2 - r^2} + \frac{1+\mu}{E}\frac{r^2 R^2 p}{R^2 - r^2}\frac{1}{\rho} \tag{8}$$

Consequently, σ_p and σ_θ can be further expressed by

$$\begin{cases} \sigma_\rho = \dfrac{r^2 p}{R^2 - r^2} - \dfrac{r^2 R^2 p}{R^2 - r^2}\dfrac{1}{\rho^2} \\ \sigma_\theta = \dfrac{r^2 p}{R^2 - r^2} + \dfrac{r^2 R^2 p}{R^2 - r^2}\dfrac{1}{\rho^2} \end{cases} \tag{9}$$

The expressions for the radial strain, the tangential strain, and the axial strain ε_z are determined as follows.

$$\begin{cases} \varepsilon_\rho = \dfrac{1-\mu}{E}\dfrac{r^2 p}{R^2 - r^2} - \dfrac{1+\mu}{E}\dfrac{r^2 R^2 p}{R^2 - r^2}\dfrac{1}{\rho^2} \\ \varepsilon_\theta = \dfrac{1-\mu}{E}\dfrac{r^2 p}{R^2 - r^2} + \dfrac{1+\mu}{E}\dfrac{r^2 R^2 p}{R^2 - r^2}\dfrac{1}{\rho^2} \\ \varepsilon_z = -\dfrac{\mu}{E}\dfrac{2r^2 p}{R^2 - r^2} \end{cases} \tag{10}$$

The effective strain $\bar{\varepsilon}$ is expressed as follows.

$$\bar{\varepsilon} = \frac{\sqrt{2}}{3}\sqrt{(\varepsilon_\rho - \varepsilon_\theta)^2 + (\varepsilon_\theta - \varepsilon_z)^2 + (\varepsilon_z - \varepsilon_\rho)^2} \tag{11}$$

The axial stress σ_z is equal to zero due to the absence of loading along the axial direction. The effective stress is expressed as follows:

$$\bar{\sigma} = \frac{\sqrt{2}}{2}\sqrt{(\sigma_\rho - \sigma_\theta)^2 + (\sigma_\theta - 0)^2 + (0 - \sigma_\rho)^2} = \frac{\sqrt{2}}{2}\sqrt{(\sigma_\rho - \sigma_\theta)^2 + \sigma_\theta^2 + \sigma_\rho^2} \tag{12}$$

4. Results and Discussion

Figure 3 shows the microstructure of Ni-49.1 at % Ti SMA, which belongs to B2 austenite. The microstructure of Ni-50 at % Ti SMA is illustrated in Figure 4. It is found from Figure 4 that NiTi SMA belongs to B19′ martensite, where martensite twins can be captured. Furthermore, the martensite twins belong to type I twins. It is generally accepted that type I twin plays a crucial role in SME of NiTi SMA.

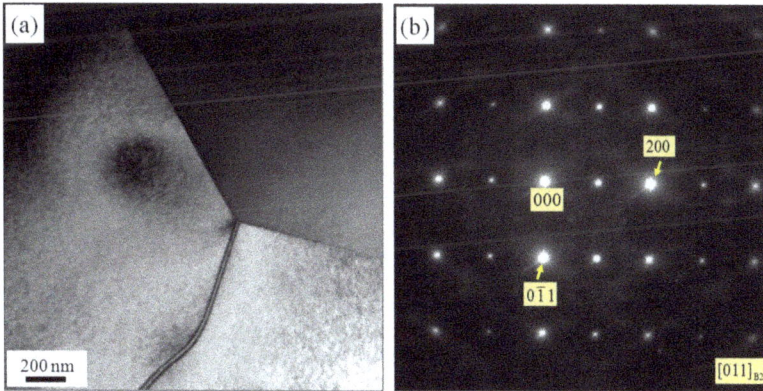

Figure 3. Transmission electron microscope (TEM) micrographs of Ni-49.1 at % Ti SMA: (**a**) bright field image; (**b**) diffraction pattern of (**a**).

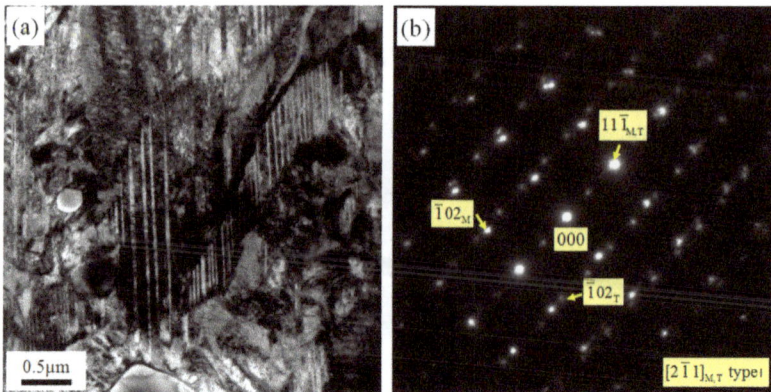

Figure 4. TEM micrographs of Ni-50 at % Ti SMA: (**a**) bright field image; (**b**) diffraction pattern of (**a**).

For the purpose of better understanding microstructure and phase composition of NiTi SMA EBSD maps of Ni-49.1 at % Ti and Ni-50 at % Ti SMAs are shown in Figures 5 and 6, respectively. It can be found from Figure 5 that Ni-49.1 at % Ti SMA consists of complete B2 austenite structure. However, regarding Ni-50 at % Ti SMA, a small amount of residual B2 austenite appears in the matrix of B19′ martensite. It is generally accepted that the residual B2 austenite phase has little influence on the mechanical behavior of Ni-50 at % Ti SMA. In other words, the mechanical behavior of Ni-50 at % Ti SMA is dominated by B19′ martensite rather than B2 austenite. In addition, the distribution of twin boundaries can be observed in Figure 6a, and the frequency distributions of misorientation angles between B19′ martensite and B19′ martensite, as well as between B19′ martensite and B2 austenite, are given in Figure 6b.

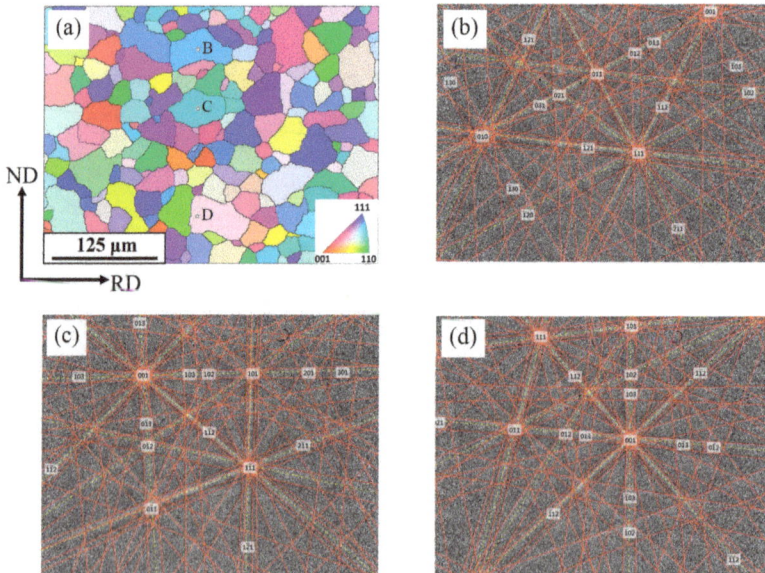

Figure 5. Electron back-scattered diffraction (EBSD) maps of Ni-49.1 at % Ti SMA: (**a**) microstructure; (**b**) kikuchi pattern of point B in (**a**); (**c**) kikuchi pattern of point C in (**a**); (**d**) kikuchi pattern of point D in (**a**).

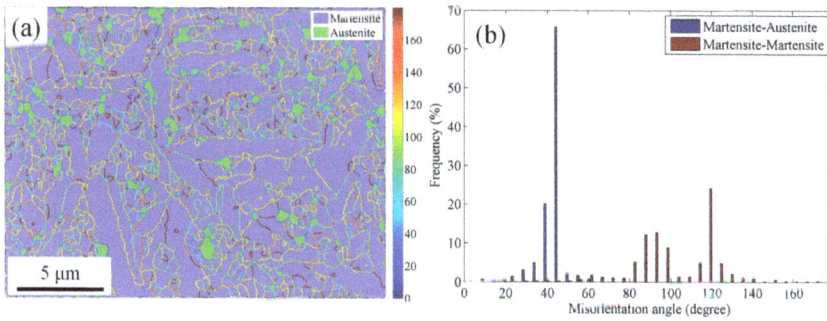

Figure 6. EBSD maps of Ni-50 at % Ti SMA: (**a**) microstructure; (**b**) frequency distribution of misorientation angle.

Figure 7 illustrates compression stress–strain curves of Ni-49.1 at % Ti and Ni-50 at % Ti SMAs, respectively. In general, in terms of compression loading, NiTi SMA with B2 austenite structure experiences elastic deformation of austenite, SIM transformation, elastic deformation of SIM, and plastic deformation of SIM based on dislocation slip [22]. However, in the case of compression loading, NiTi SMA with B19′ martensite structure undergoes elastic deformation of twinned martensite, reorientation and detwinning of twinned martensite, elastic deformation of reoriented and detwinned martensite, and plastic deformation of reoriented and detwinned martensite based on dislocation slip [23,24].

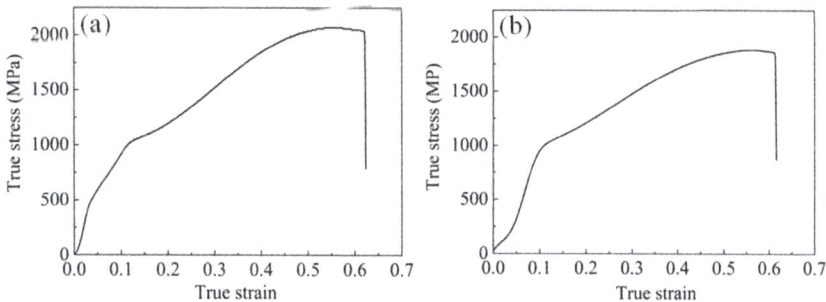

Figure 7. Compression stress–strain curves of NiTi SMAs: (**a**) Ni-49.1 at % Ti; (**b**) Ni-50 at % Ti.

According to the aforementioned mechanical analysis for NiTi SMA tube under radial loading, it can be accepted that the effective stress $\bar{\sigma}$ and the effective strain $\bar{\varepsilon}$ become the two key parameters that analyze the mechanical behavior of NiTi SMA tube during radial loading. As for NiTi SMA tube with B2 austenite structure, when the effective stress $\bar{\sigma}$ is more than a critical value σ_{sim} above which SIM transformation can be induced, the crystal structure of NiTi SMA is converted from B2 structure to B19′ structure, as shown in Figure 8. SIM transformation is of great significance in the superelastic behavior of NiTi SMA. Plastic deformation for dislocation slip—which has an adverse influence on superelasticity of NiTi SMA—occurs with the further progression of radial loading. As for NiTi SMA tube with B19′ martensite structure, when the effective stress σ is greater than a critical value σ_{rd} above which reorientation and detwinning of twinned martensite is capable of arising, the crystal structure of NiTi SMA remains as B19′ martensite, but the orientation of crystal exhibits a significant change, as shown in Figure 9. Reorientation and detwinning of twinned martensite lays a profound foundation for SME of NiTi SMA. With the further progression of radial loading, the plastic deformation of reoriented and detwinned martensite for dislocation slip occurs, which has a negative impact on the SME of NiTi SMA. Consequently, dislocation slip should be avoided for NiTi SMA tube undergoing radial loading in order to guarantee the perfect SME of NiTi SMA tube.

So far, because of the limitation of experimental techniques, SIM transformation of NiTi SMA tube under radial loading has been unable to be validated by means of experimental observation. Martensitic reorientation and detwinning of NiTi SMA tube subjected to radial loading should be observed experimentally in the future.

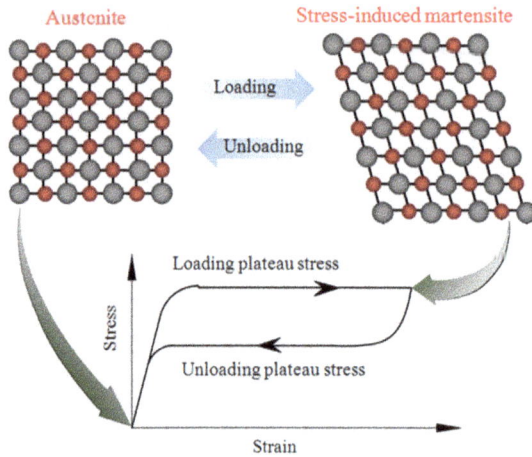

Figure 8. Schematic diagram of crystal structure evolution of NiTi SMA tube subjected to stress-induced martensite (SIM) transformation.

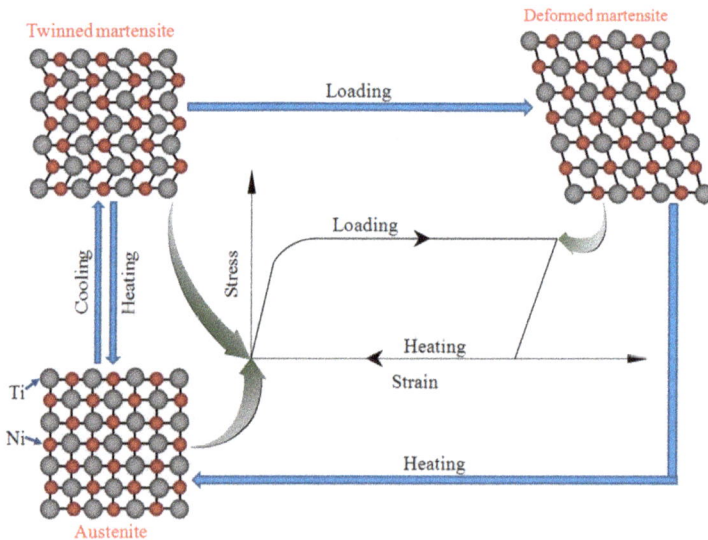

Figure 9. Schematic diagram of crystal structure evolution of NiTi SMA tube subjected to martensite reorientation and detwinning.

5. Conclusions

(1) A new radial loading mode is put forward for NiTi SMA tube, where NiTi SMA tube is subjected to the inner pressure which results from plastic deformation of the steel cylinder. The effective stress $\bar{\sigma}$ and the effective strain $\bar{\varepsilon}$ are regarded as the two key parameters that analyze the mechanical behavior of NiTi SMA tube during radial loading. As for NiTi SMA tube with B2 austenite structure, when the effective stress $\bar{\sigma}$ is more than a critical value σ_{sim} above which SIM transformation can be induced, the crystal structure of NiTi SMA is converted from B2 structure to B19' structure. As for NiTi SMA tube with B19' martensite structure, when the effective stress $\bar{\sigma}$ is more than a critical value σ_{rd} above which reorientation and detwinning of twinned martensite is capable of arising, the crystal structure of NiTi SMA remains as B19' martensite, but the crystal orientation exhibits a significant change.

(2) When NiTi SMA tube is subjected to radial loading, SIM transformation is of great significance in the superelasticity of NiTi SMA, and reorientation and detwinning of twinned martensite lays a profound foundation for the SME of NiTi SMA. Plastic deformation for dislocation slip has an adverse influence on the superelasticity and SME of NiTi SMA tube. Because of the limitation of experimental techniques, SIM transformation of NiTi SMA tube under radial loading has been unable to be validated by means of TEM observation since shape recovery of NiTi SMA tube can occur after unloading. Martensitic reorientation and detwinning of NiTi SMA tube subjected to radial loading should be observed by means of TEM techniques, since shape recovery of NiTi SMA tube cannot occur after unloading. We will perform the corresponding investigation in the future.

Acknowledgments: The work was financially supported by National Natural Science Foundation of China (Nos. 51475101, 51305091 and 51305092).

Author Contributions: Shuyong Jiang wrote the manuscript; Junbo Yu performed EBSD analysis; Li Hu performed TEM analysis; Yanqiu Zhang performed mechanical analysis.

Conflicts of Interest: The authors declare no conflict of interest.

References

1. Otsuka, K.; Ren, X. Physical metallurgy of Ti–Ni-based shape memory alloys. *Prog. Mater. Sci.* **2005**, *50*, 511–678. [CrossRef]
2. Robertson, S.W.; Ritchie, R.O. In vitro fatigue–crack growth and fracture toughness behavior of thin-walled superelastic Nitinol tube for endovascular stents: A basis for defining the effect of crack-like defects. *Biomaterials* **2007**, *28*, 700–709. [CrossRef] [PubMed]
3. Jee, K.; Han, J.; Jang, W. A method of pipe joining using shape memory alloys. *Mater. Sci. Eng. A* **2006**, *438*, 1110–1112. [CrossRef]
4. Wang, L.; Rong, L.; Yan, D.; Jiang, Z.; Li, Y. DSC study of the reverse martensitic transformation behavior in a shape memory alloy pipe-joint. *Intermetallics* **2005**, *13*, 403–407. [CrossRef]
5. Wu, M.H. Fabrication of nitinol materials and components. *Mater. Sci. Forum* **2002**, 285–292. [CrossRef]
6. Yoshida, K.; Watanabe, M.; Ishikawa, H. Drawing of Ni–Ti shape-memory-alloy fine tubes used in medical tests. *J. Mater. Process. Technol.* **2001**, *118*, 251–255. [CrossRef]
7. Yoshida, K.; Furuya, H. Mandrel drawing and plug drawing of shape-memory-alloy fine tubes used in catheters and stents. *J. Mater. Process. Technol.* **2004**, *153*, 145–150. [CrossRef]
8. Müller, K. Extrusion of nickel–titanium alloys Nitinol to hollow shapes. *J. Mater. Process. Technol.* **2001**, *111*, 122–126. [CrossRef]
9. Jiang, S.Y.; Zhang, Y.Q.; Zhao, Y.N.; Tang, M.; Li, C.F. Finite element simulation of ball spinning of NiTi shape memory alloy tube based on variable temperature field. *Trans. Nonferrous Met. Soc. China* **2013**, *23*, 781–787. [CrossRef]
10. Jiang, S.Y.; Zhao, Y.N.; Zhang, Y.Q.; Tang, M.; Li, C.F. Equal channel angular extrusion of NiTi shape memory alloy tube. *Trans. Nonferrous Met. Soc. China* **2013**, *23*, 2021–2028. [CrossRef]
11. Sun, Q.-P.; Li, Z.-Q. Phase transformation in superelastic NiTi polycrystalline micro-tubes under tension and torsion–from localization to homogeneous deformation. *Int. J. Solids Struct.* **2002**, *39*, 3797–3809. [CrossRef]

12. Li, Z.; Sun, Q. The initiation and growth of macroscopic martensite band in nano-grained niti microtube under tension. *Int. J. Plast.* **2002**, *18*, 1481–1498. [CrossRef]

13. Mao, S.; Luo, J.; Zhang, Z.; Wu, M.; Liu, Y.; Han, X. Ebsd studies of the stress-induced B2–B19′ martensitic transformation in NiTi tubes under uniaxial tension and compression. *Acta Mater.* **2010**, *58*, 3357–3366. [CrossRef]

14. Jiang, D.; Bechle, N.J.; Landis, C.M.; Kyriakides, S. Buckling and recovery of NiTi tubes under axial compression. *Int. J. Solids Struct.* **2016**, *80*, 52–63. [CrossRef]

15. Jiang, D.; Landis, C.M.; Kyriakides, S. Effects of tension/compression asymmetry on the buckling and recovery of NiTi tubes under axial compression. *Int. J. Solids Struct.* **2016**, *100*, 41–53. [CrossRef]

16. Bechle, N.J.; Kyriakides, S. Localization in NiTi tubes under bending. *Int. J. Solids Struct.* **2014**, *51*, 967–980. [CrossRef]

17. Wang, Y.; Yue, Z.; Wang, J. Experimental and numerical study of the superelastic behavior on NiTi thin-walled tube under biaxial loading. *Comput. Mater. Sci.* **2007**, *40*, 246–254. [CrossRef]

18. Bechle, N.J.; Kyriakides, S. Evolution of localization in pseudoelastic NiTi tubes under biaxial stress states. *Int. J. Plast.* **2016**, *82*, 1–31. [CrossRef]

19. Bechle, N.J.; Kyriakides, S. Evolution of phase transformation fronts and associated thermal effects in a NiTi tube under a biaxial stress state. *Extreme Mech. Lett.* **2016**, *8*, 55–63. [CrossRef]

20. Rivin, E.I.; Sayal, G.; Singh Johal, P.R. "Giant superelasticity effect" in NiTi superelastic materials and its Applications. *J. Mater. Civ. Eng.* **2006**, *18*, 851–857. [CrossRef]

21. Zhang, K.; Zhang, H.J.; Tang, Z.P. Experimental study of thin-walled TiNi tubes under radial quasi-static compression. *J. Intell. Mater. Syst. Struct.* **2011**, *22*, 2113–2126.

22. Elibol, G.; Wagner, M.F.X. Investigation of the stress-induced martensitic transformation in pseudoelastic NiTi under uniaxial tension, compression and compression–shear. *Mater. Sci. Eng. A* **2015**, *621*, 76–81. [CrossRef]

23. Liu, Y.; Xie, Z.; Van Humbeeck, J.; Delaey, L. Asymmetry of stress-strain curves under tension and compression for NiTi shape memory alloys. *Acta Mater.* **1998**, *46*, 4325–4338. [CrossRef]

24. Liu, Y.; Xie, Z.; Van Humbeeck, J.; Delaey, L. Some results on the detwinning process in NiTi shape memory alloys. *Scr. Mater.* **1999**, *41*, 1273–1281. [CrossRef]

metals

MDPI

Article

Infrared Dissimilar Joining of $Ti_{50}Ni_{50}$ and 316L Stainless Steel with Copper Barrier Layer in between Two Silver-Based Fillers

Ren-Kae Shiue [1], Shyi-Kaan Wu [1,2,*], Sheng-Hao Yang [2] and Chun-Kai Liu [2]

[1] Department of Materials Science and Engineering, National Taiwan University, Taipei 106, Taiwan;
    rkshiue@ntu.edu.tw
[2] Department of Mechanical Engineering, National Taiwan University, Taipei 106, Taiwan;
    r02522707@ntu.edu.tw (S.-H.Y.); r03522731@ntu.edu.tw (C.-K.L.)
*   Correspondence: skw@ntu.edu.tw; Tel.: +886-23-366-2732

Received: 1 July 2017; Accepted: 17 July 2017; Published: 19 July 2017

Abstract: Infrared dissimilar joining $Ti_{50}Ni_{50}$ and 316L stainless steel using Cu foil in between Cusil-ABA and BAg-8 filler metals has been studied. The Cu foil serves as a barrier layer with thicknesses of 70 µm and 50 µm, and it successfully isolates the interfacial reaction between Ti and Fe at the 316L SS (stainless steel) substrate side. In contrast, the Cu foil with 25 µm in thickness is completely dissolved into the braze melt during brazing and fails to be a barrier layer. A layer of $(Cu_xNi_{1-x})_2Ti$ intermetallic is formed at the $Ti_{50}Ni_{50}$ substrate side, and the Cu interlayer is dissolved into the Cusil-ABA melt to from a few proeutectic Cu particles for all specimens. For the 316L SS substrate side, no interfacial layer is observed and (Ag, Cu) eutectic dominates the brazed joint for 70 µm/50 µm Cu foil. The average shear strength of the bond with Cu barrier layer is greatly increased compared with that without Cu. The brazed joints with a 50 µm Cu layer demonstrate the highest average shear strengths of 354 MPa and 349 MPa for samples joined at 820 °C and 850 °C, respectively. Cracks are initiated/propagated in (Ag, Cu) eutectic next to the 316L substrate side featured with ductile dimple fracture. It shows great potential for industrial application.

Keywords: shape memory alloy; stainless steel; barrier layer; infrared brazing; microstructure

1. Introduction

Equal atomic TiNi shape memory alloy (SMA) is one of the best and intensively studied SMAs with excellent shape memory properties [1,2]. At present, equiatomic TiNi SMA has many industrial applications, such as mechanical actuators and micro sensors in intelligent material systems [3,4]. AISI (American Iron and Steel Institute) 316L stainless steel (316L SS) has the chemical composition in wt % of Fe-0.03C-17Cr-12Ni-2.5Mo and exhibits high oxidation and corrosion resistance [5]. Joining of $Ti_{50}Ni_{50}$ SMA and 316L SS has the opportunity for industrial applications. For instance, shape-memory actuators are used to replace exploding bolts utilized in petrochemical equipment and nuclear power plants [6]. The shape-memory actuator is joined with an infrastructure made of stainless steel or nickel-based alloy. In our reported study, dissimilar joining of $Ti_{50}Ni_{50}$ SMA and 316L SS by infrared brazing using two Ag-based filler metals, Ticusil (Ag-26.7Cu-4.5Ti, wt %) and Cusil-ABA (Ag-35.25Cu-1.75Ti, wt %), was evaluated [6]. Test results demonstrated that the maximum joining strength of 237 MPa shear stress can be obtained for Ticusil filler brazed at 950 °C for 60 s, and 66 MPa for Cusil-ABA one joined at 870 °C for 300 s. The formation of Ti-Fe-(Cr) interfacial layer was found to be detrimental to all joint strengths [6].

Dissimilar joining of Ti and many engineering alloys has been extensively studied in recent years [7–9]. Titanium (Ti) alloys directly bonded to SS will form $TiFe/TiFe_2$ intermetallics and result in inherent brittleness of the joint [10–12]. For avoiding the occurrence of brittle intermetallics, an interlayer of diffusion barrier was proposed to add in between the joining substrates [13–15]. A nickel interlayer with 30 μm in thickness had been used as a diffusion barrier in diffusion bonding of 304 SS and Ti-6Al-4V, so a robust joint was obtained [16]. A nickel barrier layer was also used to braze 17-4 PH (precipitation hardened) SS and Ti-6Al-4V successfully [17]. However, although the nickel interlayer can avoid the formation of brittle $TiFe/TiFe_2$ intermetallics in brazing the Ti alloy and SS, but another brittle one, such as $TiNi_3$, can occur in the joint. Therefore, the nickel interlayer may be not the best choice for the diffusion barrier in brazing $Ti_{50}Ni_{50}$ SMA and 316L SS, because $Ti_{50}Ni_{50}$ substrate can provide Ti and Ni atoms to enhance the brittle Ti-Ni intermetallic(s) to form.

It is well known that the solubility of Fe in Cu is very low [18]. In addition, the interfacial reaction of $Ti_{50}Ni_{50}$ SMA and Cusil-ABA/Ticusil filler metals forms a TiNiCu intermetallic in the $Ti_{50}Ni_{50}$ substrate side, and this intermetallic is regarded as not so brittle [6,19]. Thus, an interlayer of Cu foil can be selected as a diffusion barrier in between Cusil-ABA and BAg-8 filler metals to prevent the interaction between Ti and Fe atoms in brazing $Ti_{50}Ni_{50}$ SMA and 316L SS. The Cusil-ABA filler metal has good wettability with $Ti_{50}Ni_{50}$ SMA and thus is chosen to contact with $Ti_{50}Ni_{50}$ substrate [6]. BAg-8 filler metal is selected to neighbor the 316L SS substrate due having no Ti atom to form brittle Ti-Fe intermetallic(s) at the 316L SS substrate side, although the wettability of BAg-8 one with 316L SS is not so good [20].

The alloys' joining brazed by an infrared power source can quickly heat up the specimen with a maximum heating rate of 50 °C/s, and it is much faster than that of ordinary electric furnace [21,22]. Under the condition of temperature being precisely controlled, the joining process by infrared brazing is quite suitable to investigate the early-stage evolution of the reaction kinetics exhibited in the brazed joint. Therefore, microstructural evolution, metallurgical reaction, and bonding strength of brazed joints under different brazing variables have been extensively evaluated.

2. Materials Used and Experimental Procedure

Equal atomic TiNi SMA used in the experiment was prepared from titanium and nickel pellets (both with 99.9 wt % purity) with six times remelting in the vacuum arc remelter (Series 5 Bell Jar, Centorr Vacuum Ind., Nashua, NH, USA) under pure argon gas. The solution-treated $Ti_{50}Ni_{50}$ sample was cut by a diamond saw into small templates at a size of 15 mm × 15 mm × 3 mm for metallographical observations, and 15 mm × 7 mm × 4 mm for shear tests. The surface used for brazing was ground and polished by SiC papers and Al_2O_3 suspension solution, and then cleaned in an ultrasonic acetone container. Foils of Cusil-ABA, BAg-8, and Cu fillers were all purchased from Wesgo Metals (Hayward, CA, USA), and the former two foils' thickness was 50 μm and those of Cu foils were 25 μm, 50 μm, and 75 μm for comparison. Table 1 lists chemical compositions and melting temperatures of three filler foils.

In the experiment, infrared heating was used to braze $Ti_{50}Ni_{50}$ SMA and 316L SS using a Cu barrier layer in between Cusil-ABA (active braze alloy) and BAg-8 filler metals, as shown in Figure 1. An infrared heating furnace made from ULVAC Company (Tokyo, Japan) with the model of RHL-P816C was utilized as the main component. The furnace was kept at 5×10^{-5} mbar vacuum and set at 15 °C/s heating rate in the test. Figure 2 shows the schematic diagram of infrared brazing furnace. Graphite holders on both substrates were used to increase the infrared ray absorption on the specimen. The dissimilar brazed sample was preheated at 700 °C for 300 s before infrared brazing in order to equilibrate thermal gradient of the brazed sample. 820 °C and 850 °C for 300 s were selected as the brazing condition.

Table 1. Compositions and melting behaviors of three filler foils.

Braze Foil	Cusil-ABA	BAg-8	Copper
Chemical composition	63Ag-35.25Cu-1.75Ti (wt %)	72Ag-28Cu (wt %)	99.95% purity (wt %)
Solidus temperature	780 °C	780 °C	1083 °C
Liquidus temperature	815 °C	780 °C	1083 °C

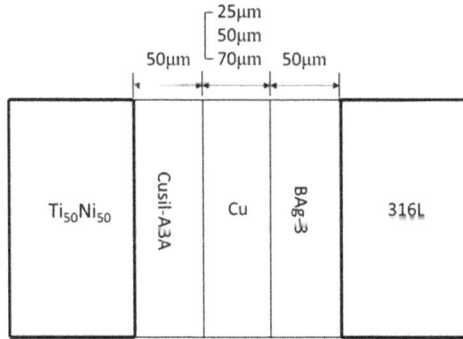

Figure 1. The schematic diagram shows the infrared brazed $Ti_{50}Ni_{50}$ SMA/Cusil-ABA/Cu foil/BAg-8/316L SS joint. Thicknesses of the copper foils are 25, 50, 70 μm and those of Cusil-ABA and BAg-8 filler metals are 50 μm; SMA: shape memory alloy; ABA: active braze alloy.

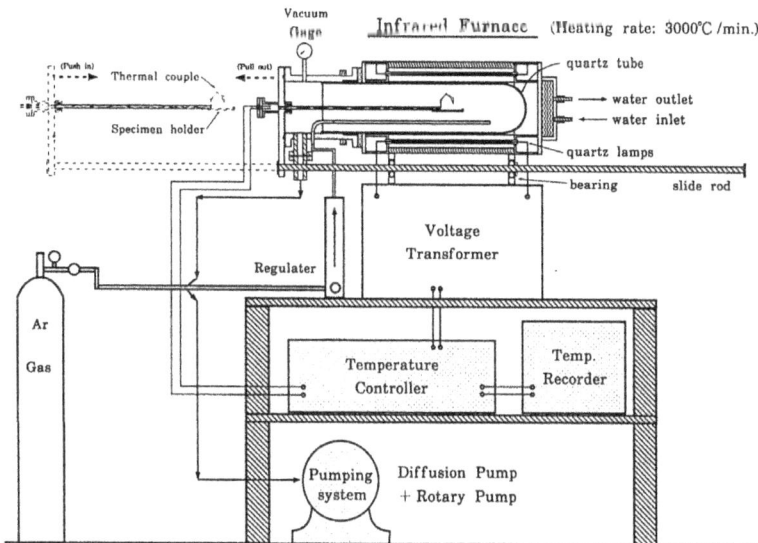

Figure 2. The schematic diagram of infrared brazing furnace.

To evaluate joint strength of the brazed sample in this study, shear tests were conducted with three specimens being performed in each brazing condition. According to our past experience, a double lap joint of 316L SS/$Ti_{50}Ni_{50}$/316L SS was used in shear test [21,22]. A Shimadzu universal testing machine with AG-10 model was used to conduct shear tests with a strain rate of 0.0167 mm/s. For the microstructural observations, the cross-section of infrared brazed joints were cut and then prepared by standard metallographic procedures, thereafter examined using a scanning electron microscope (SEM) with NOVA NANO 450 model.

3. Experimental Results and Discussion

3.1. Microstructures of Brazed Joints

Figure 3 and Table 2 show the cross-sectional results of SEM backscattered election images (BEIs) and energy dispersive spectroscope (EDS) chemical analysis results of $Ti_{50}Ni_{50}$/Cusil-ABA/70 μm Cu foil/BAg-8/316L SS specimens brazed by infrared rays at 820 °C for 300 s. In EDS analyses, the electron beam in a point with 1 μm in diameter was used for all single phase analysis, and a larger beam diameter of 5 μm was applied for eutectic analysis as marked by D in Figure 3. Figure 3a illustrates the cross-sectional SEM BEIs results, in which the area on the $Ti_{50}Ni_{50}$ side and that on the 316L SS side are displayed in Figure 3b,c, respectively. From Figure 3a, the thickness of the Cu interlayer is about 70 μm (labelled by C and F). The layer marked by C in the left side of the Cu interlayer contains 90.5 at % Cu and has some Cu atoms dissolved into Cusil-ABA melt to form the proeutectic Cu (labelled by G in Figure 3b). Figure 3b indicates that a layer of $(Cu_xNi_{1-x})_2Ti$ intermetallic (labelled by B) is formed at the interface of Cusil-ABA filler and $Ti_{50}Ni_{50}$ base metal. It is similar to the $Ti_{50}Ni_{50}$ side of 316L SS/Cusil-ABA/$Ti_{50}Ni_{50}$ joint brazed by infrared rays at 870 °C and 900 °C for 300 s [6]. In contrast, there is no interaction layer formed in between the BAg-8 braze and the 316L SS substrate, as displayed in Figure 3c. The Cu interlayer is dissolved into both molten brazes resulting in irregular interfaces among Cu and two brazes.

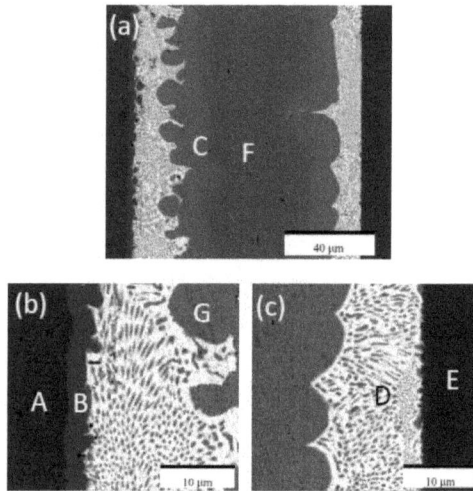

Figure 3. SEM (scanning electron microscope) backscattered electron images (BEIs) of $Ti_{50}Ni_{50}$/Cusil-ABA/70 μm Cu foil/BAg-8/316L SS specimen infrared brazed at 820 °C for 300 s: (**a**) the cross-section; (**b**) the $Ti_{50}Ni_{50}$ substrate side; (**c**) the 316L SS substrate side.

Table 2. EDS chemical analysis results in Figure 3.

at %	A	B	C	D	E	F	G
Ti	49.8	30.7	3.9	-	-	-	2.5
Cu	-	51.6	90.5	47.2	-	100	89.3
Ni	50.2	17.7	-	-	10.8	-	-
Fe	-	-	-	-	68.1	-	-
Ag	-	-	5.6	52.8	-	-	8.2
Cr	-	-	-	-	21.1	-	-
Phase	$Ti_{50}Ni_{50}$	$(Cu_xNi_{1-x})_2Ti$	Cu-rich	Ag-Cu eutectic	316L	Cu	Cu-rich

The thickness effect of the Cu barrier layer on the infrared brazed joint was studied. Figure 4 displays SEM BEIs cross sections of $Ti_{50}Ni_{50}$/Cusil-ABA/Cu/BAg-8/316L SS specimens brazed by infrared rays at 850 °C for 300 s with various thicknesses of the Cu interlayer. From Figure 4a, one can find that the continuous Cu barrier layer disappears and replaced by many huge proeutectic Cu-rich blocks. In contrast, the thickness of Cu barrier layer above 50 μm is enough to isolate the interaction between $Ti_{50}Ni_{50}$ and 316L SS, as illustrated in Figure 4b,c. The dissolution of Cu barrier layer into Cusil-ABA melt becomes more prominent in Figure 4c than that in Figure 3a due to the higher brazing temperature. Additionally, one can find that the width of BAg-8 melt (in the right side of the Cu interlayer) in Figure 4c is less than that in Figure 3a, since BAg-8 melt flows out of the joint during brazing at a higher temperature. According to Table 1, the BAg-8 eutectic temperature is less than the liquidus temperature of Cusil-ABA braze. The width of BAg-8 filler is significantly decreased with increasing the brazing temperature.

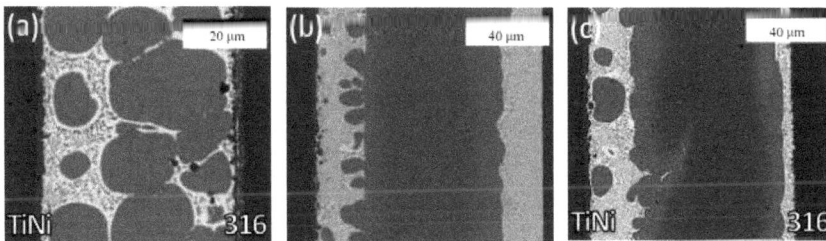

Figure 4. SEM BEI cross sections of $Ti_{50}Ni_{50}$/Cusil-ABA/Cu/BAg-8/316L SS specimens infrared brazed at 850 °C for 300 s with the Cu interlayer of (**a**) 25 μm; (**b**) 50 μm; (**c**) 70 μm. (TiNi: $Ti_{50}Ni_{50}$ substrate, 316: 316L SS substrate).

Figure 5 and Table 3 illustrate the cross-sectional results of SEM BEIs and EDS chemical analyses of $Ti_{50}Ni_{50}$/Cusil-ABA/25 μm Cu foil/BAg-8/316L SS specimen brazed by infrared rays at 850 °C for 300 s. Figure 5a,b indicate the area on the $Ti_{50}Ni_{50}$ side and that on 316L SS side, respectively. In Figure 5a, a $(Cu_xNi_{1-x})_2Ti$ intermetallic layer (labelled by B) is formed at the $Ti_{50}Ni_{50}$ side which is similar to that in Figure 3b [6,20,22]. The silver penetrated into the $(Cu_xNi_{1-x})_2Ti$ layer is also observed. According to Figure 5b, two interfacial layers are formed at the 316L SS side, one is Ti(Fe, Ni) intermetallic layer (labelled by D), and the other is regarded as a layer of (Fe, Cr)-rich phase (labelled by E) [6]. The existence of continuous Ti(Fe, Ni) reaction layer will be discussed later. Both D and E layers have silver agglomerations (white spots in Figure 5b) with the layer E having more. In addition, there is a TiCu precipitate in the braze, as labelled by G in Figure 5b [19]. The existence of the Ti-Fe interfacial layer in Figure 5b is quite dissimilar to that in Figure 3c. Because the 25 μm Cu barrier layer is completely dissolved into the braze melt, Ti atoms transport into the 316L SS side forming the interfacial Ti(Fe, Ni) intermetallic layer [6].

Figure 6 and Table 4 display the cross-sectional inspection of SEM BEIs and EDS analyses of $Ti_{50}Ni_{50}$/Cusil-ABA/50 μm Cu foil/BAg-8/316L SS specimen brazed by infrared rays at 850 °C for 300 s. Interfacial reactions shown in Figure 6 are similar to those in Figure 3. Additionally, the experimental results of the infrared brazed $Ti_{50}Ni_{50}$/Cusil-ABA/Cu foil/BAg-8/316L SS joint with 70 μm thickness of the Cu foil at 850 °C are similar to those of the 50 μm one illustrated in Figure 6, and therefore are not shown here.

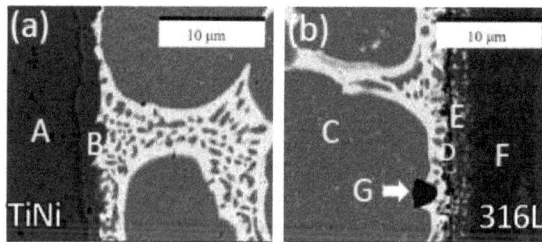

Figure 5. SEM BEI cross sections of Ti$_{50}$Ni$_{50}$/Cusil-ABA/25 μm Cu foil/BAg-8/316L SS specimen infrared brazed at 850 °C for 300 s: (**a**) the Ti$_{50}$Ni$_{50}$ substrate side; (**b**) the 316L SS substrate side. (TiNi: Ti$_{50}$Ni$_{50}$ substrate, 316L: 316L SS substrate).

Table 3. EDS chemical analysis results in Figure 5.

at %	A	B	C	D	E	F	Ğ
Ti	50.2	31.6	2.2	40.0	9.9	-	50.1
Cu	-	49.8	92.6	4.8	-	-	46.4
Ni	49.8	18.6	-	12.7	3.3	10.5	3.5
Fe	-	-	-	37.0	56.1	69.1	-
Ag	-	-	5.2	2.9	9.1	-	-
Cr	-	-	-	2.6	21.6	20.4	-
Phase	Ti$_{50}$Ni$_{50}$	(Cu$_x$Ni$_{1-x}$)$_2$Ti	Cu-rich	Ti(Fe, Ni)	(Fe, Cr)-rich	316L	TiCu

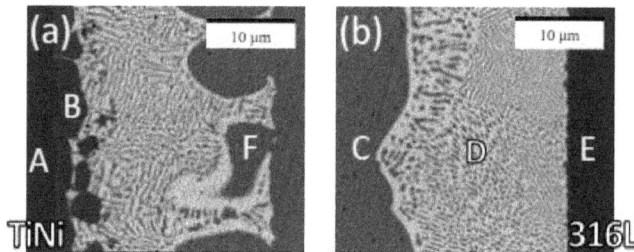

Figure 6. SEM BEI cross sections of Ti$_{50}$Ni$_{50}$/Cusil-ABA/50 μm Cu foil/BAg-8/316L SS specimen infrared brazed at 850 °C for 300 s: (**a**) the Ti$_{50}$Ni$_{50}$ substrate side; (**b**) the 316L SS substrate side. (TiNi: Ti$_{50}$Ni$_{50}$ substrate, 316L: 316L SS substrate).

Table 4. EDS chemical analysis results in Figure 6.

at %	A	B	C	D	E	F
Ti	49.7	30.6	-	-	-	4.5
Cu	-	48.6	100	27.7	-	90.4
Ni	50.3	20.8	-	-	13.1	-
Fe	-	-	-	-	68.9	-
Ag	-	-	-	72.3	-	5.1
Cr	-	-	-	-	18.0	-
Phase	Ti$_{50}$Ni$_{50}$	(Cu$_x$Ni$_{1-x}$)$_2$Ti	Cu	Ag-Cu eutectic	316L	Cu-rich

3.2. Microstructures of the Brazed Joints

Phase identification using the EDS chemical analysis results was assisted by related phase diagrams and citing references. From Figure 3, Figure 5, Figure 6 and Tables 2–4, a continuous layer of the (Cu$_x$Ni$_{1-x}$)$_2$Ti intermetallic, as labelled by B, is formed at the interface between Ti$_{50}$Ni$_{50}$ substrate and Cusil-ABA filler. Based on Ti-Cu-Ni ternary phase diagram isothermally sectionalized at

870 °C, the stoichiometric composition of this intermetallic layer is near the TiCuNi compound [19]. It is a nonstoichiometric compound, and can be indicated by $(Cu_xNi_{1-x})_2Ti$ with x ranging from 0.23 to 0.75 [19]. The EDS chemical analyses of layer B indicated in Tables 2–4 are in accordance with Ti-Cu-Ni diagram [18].

From Figure 5b, there are two continuous layers observed at the interface between 316L SS and the braze. One is the Ti(Fe, Ni) intermetallic layer labelled by D, and the other is the (Fe, Cr)-rich layer indicated by E. Figure 7 illustrates the Fe-Ni-Ti phase diagram isothermally sectionalized at 900 °C [19,23]. The $Ti_{50}Ni_{50}$ substrate dissolved into the braze melt causes high Ni and Ti contents in it. The complete dissolution of 25 μm Cu barrier layer makes the Ti and Ni ingredients readily react with the 316L SS substrate in brazing, and yields a continuous reaction layer of Ti(Fe, Ni) intermetallic. According to Figure 7, the composition of layer D is regarded as the TiFe intermetallic compound alloyed with Ni. The layer E shown in Figure 5b, which is next to the 316L SS substrate, contains about 56% Fe, 22% Cr, 3% Ni, 9% Ag, and 10% Ti (in at %). The transport of Ti from the $Ti_{50}Ni_{50}$ substrate and Cusil-ABA melt now acts as an active brazing element to react with the surface of 316L SS substrate. Thereafter, Ti, Ag, and Ni atoms interact with 316L SS substrate to form a (Fe, Cr)-rich layer solid-solution alloyed with Ag, Ti, and Ni. In addition, some Ag atoms are not alloyed into (Fe, Cr)-rich phase, and form white Ag-rich precipitates in layer E (Figure 5b).

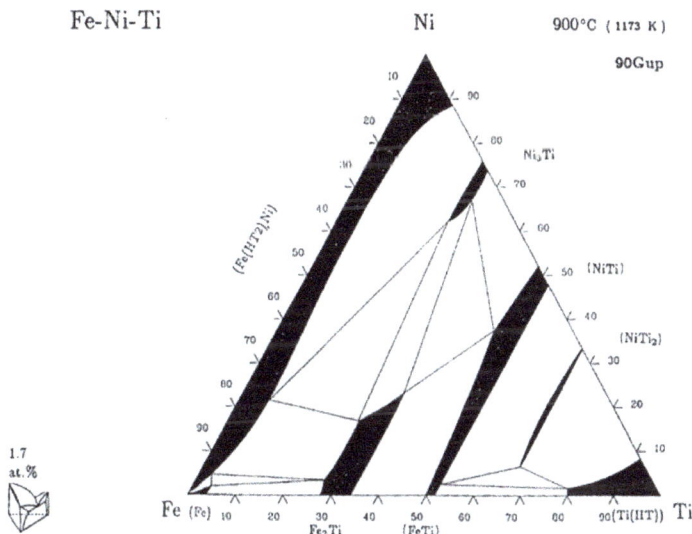

Figure 7. Isothermal section of Fe-Ni-Ti ternary phase diagram at 900 °C [19,23].

3.3. The Shear Strengths of Brazed Joints

Table 5 indicates average shear strengths of the $Ti_{50}Ni_{50}$/Cusil-ABA/Cu foil/BAg-8/316L SS joints with different Cu barrier layers brazed by infrared rays under different brazing conditions. For comparison, the shear strengths of $Ti_{50}Ni_{50}$/Cusil-ABA or Ticusil/316L SS joint without the Cu barrier layer infrared brazed at 870 °C for 300 s or at 950 °C for 60 s, respectively, are also tabulated [6]. From Table 5, average shear strengths of the specimens with 50 μm Cu foil can reach 354 MPa for brazing at 820 °C for 300 s and 349 MPa for brazing at 850 °C for 300 s. These values are significantly increased as compared with those without the Cu barrier layer, as indicated in Table 5.

Table 5. Average shear strengths of infrared brazed Ti$_{50}$Ni$_{50}$ SMA and 316L SS joints.

Filler Metal	Brazing Temperature	Brazing Time	Copper Foil Thickness	Average Shear Strength
Cusil-ABA/Cu foil/BAg-8	820 °C	300 s	25 μm	140 ± 3 MPa
			50 μm	354 ± 35 MPa
			70 μm	292 ± 37 MPa
	850 °C	300 s	25 μm	236 ± 38 MPa
			50 μm	349 ± 21 MPa
			70 μm	211 ± 39 MPa
Cusil-ABA *	870 °C	300 s	-	66 ± 15 MPa
Ticusil *	950 °C	60 s	-	237 ± 16 MPa

* From reference [6] in which the thickness of the filler metal is 50 μm.

All brazed joints failed along the interface between BAg-8 braze and 316L substrate. Figure 8a illustrates the cross-sectional results of SEM BEIs and the SEI fractograph of the fractured Ti$_{50}$Ni$_{50}$/Cusil-ABA/25 μm Cu foil/BAg-8/316L SS joined at 850 °C for 300 s. The brazed specimen failed along the interfacial reaction layer of Ti(Fe, Ni) (labelled by D in Figure 5b) next to the 316L SS substrate side. The SEM SEI fractograph shown in Figure 8a is features brittle fracture, and the EDS analysis result is consistent with that of layer D in Figure 5b. Figure 8b illustrates the cross-sectional results of SEM BEIs and the SEI fractograph of the fractured Ti$_{50}$Ni$_{50}$/Cusil-ABA/50 μm Cu foil/BAg-8/316L SS joined at 850 °C for 300 s. Different from Figure 8a, the fracture appearing in the brazed specimen is along the (Ag, Cu) eutectic (labelled by D in Figure 6b) near the 316L SS substrate side. The fracture of fine (Ag, Cu) eutectic, as shown in Figure 6b, results in many miniature dimples in Figure 8b. For the specimen with 70 μm Cu foil, the brazed joint is fractured along the (Ag, Cu) eutectic next to the 316L SS substrate side as shown in Figure 8c. A few solidification voids are observed from the fractured surface.

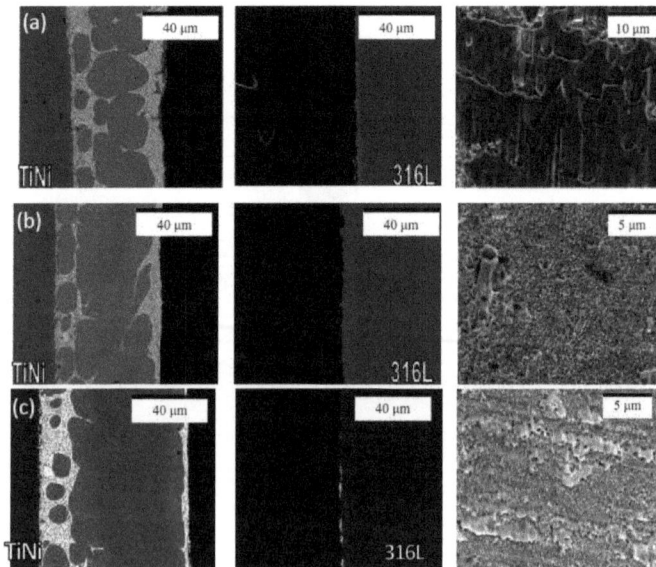

Figure 8. SEM BEI cross-sections and SEI fractographs of Ti$_{50}$Ni$_{50}$/Cusil-ABA/Cu foil/BAg-8/316L SS joint infrared brazed at 850 °C for 300 s after shear test with the Cu interlayer of (**a**) 25; (**b**) 50; and (**c**) 70 μm in thickness. (TiNi: Ti$_{50}$Ni$_{50}$ substrate, 316L: 316L SS substrate).

Shear strengths revealed in Table 5 indicate that the optimal thickness of the Cu barrier layer is 50 μm. The Cu barrier foil 25 μm in thickness is dissolved into the melt during brazing, and a continuous Ti(Fe, Ni) interfacial layer next to the 316L SS substrate side is formed which is the fracture location in shear test. In addition, the BAg-8 filler metal brazed at 820 °C is not high enough to sufficiently wet the 316L SS substrate. These reasons cause the brazed joint with 25 μm Cu foil to have a lower shear strength than that with 50 μm foil, especially the joint brazed at 820 °C. Shear strength of the joint with 50 μm Cu foil demonstrates the highest joint strength due to the presence of very fine (Ag, Cu) eutectic and free of interfacial Ti(Fe, Ni) reaction layer. The shear strength of the joint with 70 μm Cu foil is lower than that of 50 μm one. The BAg-8 melt is easier to flow out of the joint during brazing than the Cusil-ABA one. The width of BAg-8 braze melt in 70 μm Cu foil is much less than that in 50 μm one, as compared between Figure 8b,c. Because solidification shrinkage voids are prone to be formed in the joint with the 70 μm Cu layer, the joint's shear strength is deteriorated. Shear test is the first stage in evaluating bonding strength of the brazed joint. Dynamic tests, such as fatigue testing, of such a brazed joint will be performed in the future in order to evaluate the effect of interfacial intermetallics on the reliability of the brazed joint.

4. Conclusions

Shear strength and microstructural evolution of $Ti_{50}Ni_{50}$ SMA/Cusil-ABA/Cu foil/BAg-8/316L SS joints brazed by infrared rays at 820 °C and 850 °C for 300 s with the Cu foil as the diffusion barrier have been investigated. The introduction of the Cu barrier layer shows great potential in joining $Ti_{50}Ni_{50}$ SMA and 316L SS for industrial application. Important conclusions are as follows:

1. The Cu foil serves as a barrier layer at thicknesses of 70 μm and 50 μm, and it successfully isolates the interfacial reaction between Ti and Fe at the 316L SS substrate side. In contrast, the Cu foil with 25 μm in thickness is completely dissolved into the braze melt during brazing and fails to be a barrier layer.

2. A layer of $(Cu_xNi_{1-x})_2Ti$ intermetallic is formed at the $Ti_{50}Ni_{50}$ substrate side, and the Cu interlayer is dissolved into the Cusil-ABA melt to from a few proeutectic Cu particles for all specimens. For the 316L SS substrate side, no interfacial layer is observed and (Ag, Cu) eutectic dominates the brazed joint for 70 μm/50 μm Cu foil. However, an interfacial Ti(Fe, Ni) intermetallic layer and (Fe, Cr)-rich layer are formed in the brazed joint with the 25 μm Cu layer.

3. The joint with the 50 μm Cu barrier layer demonstrates the best average shear strengths of 354 MPa and 349 MPa for samples brazed at 820 °C and 850 °C, respectively. All specimens are fractured along the interface between the BAg-8 braze and 316L SS side. Cracks are initiated/propagated in (Ag, Cu) eutectic for the 70 μm and 50 μm thickness Cu foils and at the interfacial Ti(Fe, Ni) reaction layer for the 25 μm foil. The formation of a brittle Ti(Fe, Ni) intermetallic layer deteriorates the bonding strength of the joint.

Acknowledgments: This research financial-supported from the Ministry of Science and Technology, Taiwan under grand number MOST 105-2221-E-002-044-MY2 was kindly acknowledged.

Author Contributions: Shyi-Kaan Wu and Ren-Kae Shiue designed and planned the experiment. Sheng-Hao Yang and Chun-Kai Liu made the test.

Conflicts of Interest: The authors announce no conflict of interest.

References

1. Funakubo, H. *Shape Memory Alloys*; Gordon & Breach Science: New York, NY, USA, 1987.
2. Otsuka, K.; Kakeshita, T. Science and technology of shape-memory alloys: New developments. *MRS Bull.* **2002**, *27*, 91–100. [CrossRef]
3. Braun, S.; Sandstrom, N.; Stemme, G.; Wijngaart, W. Wafer-scale manufacturing of bulk shape-memory-alloy microactuators based on adhesive bonding of titanium-nickel sheets to structured silicon wafers. *J. Microelectromech. Syst.* **2009**, *18*, 1309–1317. [CrossRef]

4. Tomozawa, M.; Kim, H.Y.; Miyazaki, S. Shape memory behavior and internal structure of Ti-Ni-Cu shape memory alloy thin films and their application for microactuators. *Acta Mater.* **2009**, *57*, 441–452.
5. Smith, W.F. *Structure and Properties of Engineering Alloys*; McGraw-Hill: New York, NY, USA, 1993; pp. 312–316.
6. Shiue, R.K.; Chen, C.P.; Wu, S.K. Infrared brazing $Ti_{50}Ni_{50}$ shape memory alloy and 316L stainless steel with two sliver-based fillers. *Metall. Mater. Trans. A* **2015**, *46*, 2364–2371. [CrossRef]
7. Kumar, A.; Ganesh, P.; Kaul, R.; Sindal, B.K. New brazing recipe for ductile niobium-316L stainless steel joints. *Weld. J.* **2015**, *94*, 241–249.
8. Laik, A.; Shirzadi, A.A.; Sharma, G.; Tewari, R.; Jayakumar, T.; Dey, G.K. Microstructure and interfacial reactions during vacuum brazing of stainless steel to titanium using Ag-28 pct Cu alloy. *Metall. Mater. Trans. A* **2015**, *46*, 771–782.
9. Lee, M.K.; Park, J.J.; Lee, J.G.; Rhee, C.K. Phase-dependent corrosion of titanium-to-stainless steel joints brazed by Ag–Cu eutectic alloy filler and Ag interlayer. *J. Nucl. Mater.* **2013**, *439*, 168–173. [CrossRef]
10. Elrefaey, A.; Tillmann, W. Brazing of titanium to steel with different filler metals: Analysis and comparison. *J. Mater. Sci.* **2010**, *45*, 4332–4338. [CrossRef]
11. Lee, M.K.; Lee, J.G.; Lee, J.K.; Hong, S.M.; Lee, O.H.; Park, J.J.; Kim, I.W.; Rhee, C.K. Formation of interfacial brittle phases sigma phase and IMC in hybrid titanium-to-stainless steel joint. *Trans. Nonferrous Met. Soc. China* **2011**, *21*, 7–11. [CrossRef]
12. Liu, C.C.; Ou, C.L.; Shiue, R.K. The microstructural observation and wettability study of brazing Ti-6Al-4V and 304 stainless steel using three braze alloys. *J. Mater. Sci.* **2002**, *37*, 2225–2235. [CrossRef]
13. Morizono, Y.; Mizobata, A. Explosive coating of Ag-Cu filler alloy on metal substrates and its effect on subsequent brazing process. *ISIJ Int.* **2010**, *50*, 1200–1204. [CrossRef]
14. Lee, J.G.; Hong, S.J.; Lee, M.K.; Rhee, C.K. High strength bonding of titanium to stainless steel using an Ag interlayer. *J. Nucl. Mater.* **2009**, *395*, 145–149. [CrossRef]
15. Kamat, G.R. Solid-state diffusion welding of nickel to stainless steel. *Weld. J.* **1988**, *67*, 44–46.
16. He, P.; Zhang, J.H.; Zhou, R.L.; Li, X.Q. Diffusion bonding technology of a titanium alloy to a stainless steel web with a Ni interlayer. *Mater. Charact.* **1999**, *43*, 287–292. [CrossRef]
17. Shiue, R.K.; Wu, S.K.; Chan, C.H.; Huang, C.S. Infrared brazing Ti-6Al-4V and 17-4 PH stainless steel with a nickel barrier layer. *Metall. Mater. Trans. A* **2006**, *37*, 2207–2217. [CrossRef]
18. Massalski, T.B.; Okamoto, H.; Subramanian, P.R.; Kacprzak, L. Binary Alloy Phase Diagrams. In *ASM Metal Handbook*; ASM International: Geauga County, OH, USA, 1992; Volume 3, pp. 2–168.
19. Villars, P.; Prince, A.; Okamoto, H. *Handbook of Ternary Alloy Phase Diagrams*; ASM International: Geauga County, OH, USA, 1995.
20. Yang, S.H. *The Study of Infrared Brazing $Ti_{50}Ni_{50}$ SMA and 316L Stainless Steel/Inconel 600*; Department of Mechanical Engineering, National Taiwan University: Taipei, Taiwan, 2015; pp. 5–35.
21. Lee, S.J.; Wu, S.K.; Lin, R.Y. Infrared joining of TiAl intermetallics using Ti-15Cu-15Ni foil—Part I the microstructure morphologies of joint interfaces. *Acta Mater.* **1998**, *46*, 1283–1295. [CrossRef]
22. Shiue, R.K.; Chen, Y.H.; Wu, S.K. Infrared brazing $Ti_{50}Ni_{50}$ and invar using Ag-based filler foils. *Metall. Mater. Trans. A* **2013**, *44*, 4454–4460. [CrossRef]
23. Loo, F.; Vrolijk, J.; Bastin, G. Phase relations and diffusion paths in the Ti-Ni-Fe system at 900 °C. *J. Less-Common Met.* **1981**, *77*, 121–130.

metals

MDPI

Article

Deformation Behavior and Microstructure Evolution of NiTiCu Shape Memory Alloy Subjected to Plastic Deformation at High Temperatures

Shuyong Jiang [1,*], Dong Sun [2], Yanqiu Zhang [1] and Li Hu [2]

[1]   College of Mechanical and Electrical Engineering, Harbin Engineering University, Harbin 150001, China; zhangyq@hrbeu.edu.cn
[2]   College of Materials Science and Chemical Engineering, Harbin Engineering University, Harbin 150001, China; 18845143035@163.com (D.S.); heu_huli@126.com (L.H.)
*   Correspondence: jiangshuyong@hrbeu.edu.cn; Tel.: +86-451-8251-9710

Received: 15 July 2017; Accepted: 27 July 2017; Published: 3 August 2017

Abstract: Deformation behavior and microstructure evolution of NiTiCu shape memory alloy (SMA), which possesses martensite phase at room temperature, were investigated based on a uniaxial compression test at the temperatures of 700~1000 °C and at the strain rates of 0.0005~0.5 s^{-1}. The constitutive equation of NiTiCu SMA was established in order to describe the flow characteristic of NiTiCu SMA, which is dominated by dynamic recovery and dynamic recrystallization. Dislocations become the dominant substructures of martensite phase in NiTiCu SMA compressed at 700 °C. Martensite twins are dominant in NiTiCu SMA compressed at 800 and 900 °C. Martensite twins are not observed in NiTiCu SMA compressed at 1000 °C. The microstructures resulting from dynamic recovery or dynamic recrystallization significantly influences the substructures in the martensite phase of NiTiCu SMA at room temperature. Dislocation substructures formed during dynamic recovery, such as dislocation cells and subgrain boundaries, can suppress the formation of twins in the martensite laths of NiTiCu SMA. The size of dynamic recrystallized grains affects the formation of martensite twins. Martensite twins are not easily formed in the larger recrystallized grain, since the constraint of the grain boundaries plays a weak role. However, in the smaller recrystallized grain, martensite twins are induced to accommodate the transformation from austenite to martensite.

Keywords: shape memory alloy; NiTiCu alloy; constitutive behavior; phase transformation; microstructure

---

1. Introduction

NiTi shape memory alloys (SMAs) have deserved increasing attention in the engineering field because they have shape memory effect [1]. With a view to widening the application of NiTi SMA in the domain of engineering, third elements are added to the binary NiTi SMA in order to change the transformation temperature or hysteresis [2–6]. For example, Cu element is added to binary NiTi SMA so as to significantly lower the phase transformation hysteresis, which lays the foundation for the application of an actuator or micro-electro-mechanical system (MEMS) [7,8]. As a consequence, over the last few decades, many researchers have paid more attention to NiTiCu SMAs [9–12]. It is well known that plastic deformation substantially influences the microstructures and transformation behavior of NiTi-based SMAs [13,14]. In particular, thermomechanical processing, especially plastic working, is an important step in manufacturing NiTi-based SMA products [15]. Consequently, it is very important to understand deformation behavior as well as microstructural evolution of NiTi-based SMAs at high temperatures [16–20]. Up to date, many scholars have devoted themselves to studying the flow behavior of NiTi-based SMAs at high temperatures based on the Arrhenius-type constitutive equation [21–25], which lays the foundation for clarifying deformation mechanisms of NiTi-based SMAs.

In the present study, the deformation behavior and microstructure evolution of $Ni_{45}Ti_{50}Cu_5$ (at %) SMA were investigated based on a uniaxial compression test, where the temperatures range from 700 to 1000 °C and the strain rates range from 0.0005 to 0.5 s^{-1}. In particular, the $Ni_{45}Ti_{50}Cu_5$ SMA of interest possesses martensite phase at room temperature. Therefore, it is of great significance to investigate the deformation behavior and microstructure evolution of $Ni_{45}Ti_{50}Cu_5$ SMA.

2. Materials and Methods

As-rolled $Ni_{45}Ti_{50}Cu_5$ (at %) SMA bar with a diameter of 30 mm was commercially received from Xi'an Saite Metal Materials Development Co., Ltd. (Xi'an, China). The phase transformation of the as-rolled NiTiCu SMA was measured using Pyris Diamond type differential scanning calorimetry (DSC, Perkin Elmer Inc., Waltham, MA, USA). The DSC test was carried out in the range of −150~150 °C, where the heating and cooling steps were 10 °C/min. The DSC curve of the as-rolled NiTiCu SMA can be found in Reference [26]. The phase transformation temperatures of the as-rolled NiTiCu SMA were as follows: M_s = 53.8 °C, M_f = 8.3 °C, A_s = 73.1 °C and A_f = 113.5 °C.

NiTiCu SMA samples, which possess a height of 9 mm and a diameter of 6 mm, were electro-discharge machined from the as-rolled NiTiCu SMA bar. Subsequently, they were used for the compression test. An INSTRON-5500R universal material testing machine (Instron Corporation, Norwood, MA, USA) was used for implementing the compression tests. The NiTiCu SMA samples were compressed by the deformation extent of 60%, where the temperatures range from 700 to 1000 °C and the strain rates range from 0.0005 to 0.5 s^{-1}. Subsequently, all of the compressed NiTiCu SMA specimens were put into ice water for the purpose of guaranteeing complete martensite phase transformation.

As for the as-rolled and compressed NiTiCu SMA samples, the microstructures were captured by transmission electron microscopy (TEM). The NiTiCu SMA samples used for TEM observation were made into foils with the thickness of 70 μm by means of mechanical grinding. Subsequently, the foils were thinned by twin-jet polishing in an electrolyte which is composed of 6% $HClO_4$, 34% $CH_3(CH_2)_3OH$, and 60% CH_3OH by volume fraction. Finally, the NiTiCu SMA samples for TEM observation were characterized by virtue of an FEI TECNAI G2 F30 microscope (FEI Corporation, Hillsboro, OR, USA). TEM observation results indicate that the substructure of martensite phase contains martensite laths and martensite twins, as shown in Figure 1.

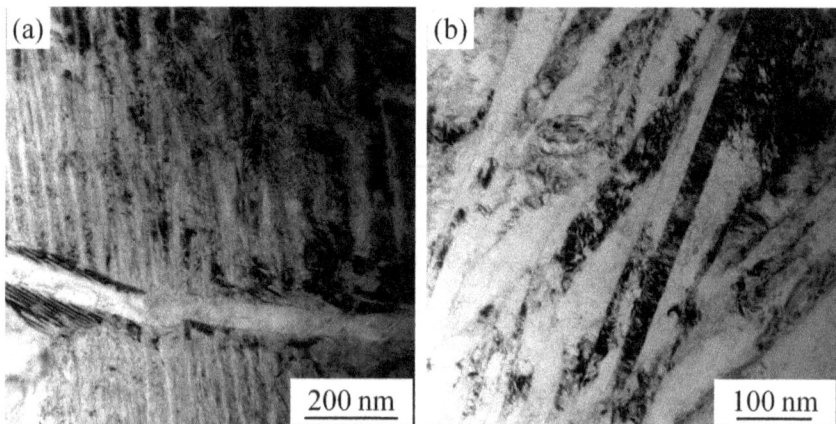

Figure 1. Transmission electron microscope (TEM) micrographs of as-rolled NiTiCu shape memory alloy (SMA): (**a**) Bright field image indicating martensite twins; (**b**) Bright field image indicating martensite laths.

3. Results and Discussion

3.1. Deformation Behavior of NiTiCu SMA

Figure 2 indicates the true stress-strain curves of NiTiCu SMA undergoing uniaxial compression at the temperatures ranging from 700 to 1000 °C and the strain rates ranging from 0.0005 to 0.5 s^{-1}. As can be seen in Figure 3, the flow stress of NiTiCu SMA increases with increasing strain rate. Furthermore, elevating the temperature contributes to lowering the flow stress of NiTiCu SMA. It can be noted from the stress-strain curves of NiTiCu SMA that dynamic recovery (DRV) or dynamic recrystallization (DRX) can occur during compression deformation.

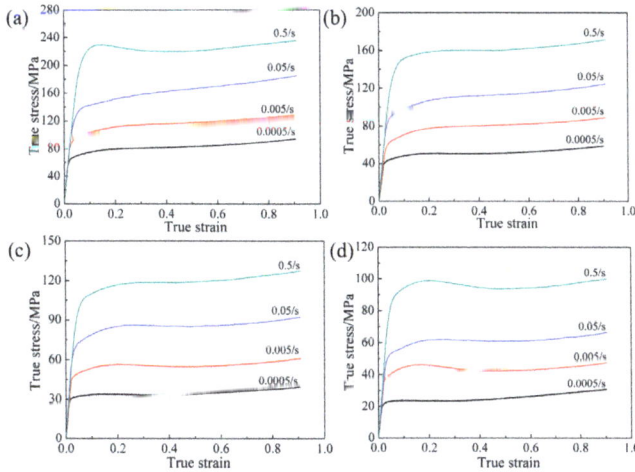

Figure 2. True stress-strain curves of NiTiCu SMA undergoing uniaxial compression at various strain rates and temperatures: (**a**) 700 °C; (**b**) 800 °C; (**c**) 900 °C; (**d**) 1000 °C.

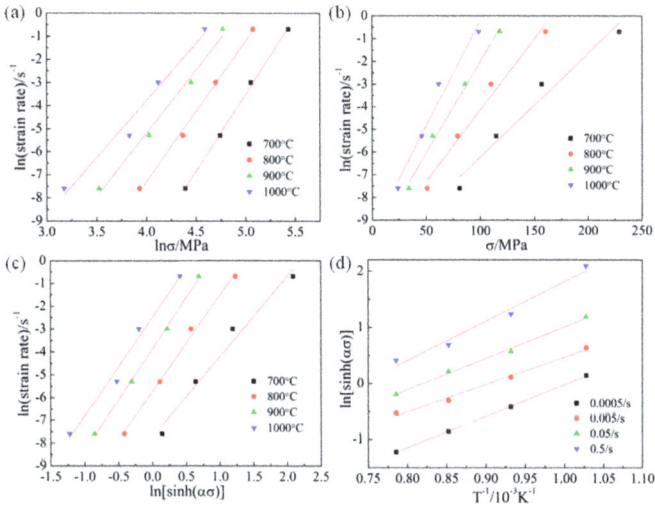

Figure 3. Determination of material constants by means of fitting method based on experimental data: (**a**) Solving the value of *n*; (**b**) Solving the value of *β*; (**c**) Modifying the value of *n*; (**d**) Solving the value of *Q*.

Based on the stress-strain curves of NiTiCu SMA at high temperatures, it can be found that the plastic flow of NiTiCu SMA at high temperatures is dependent on the strain rates as well as the deformation temperatures. As a consequence, the constitutive equation of NiTiCu SMA at high temperatures is established according to the Arrhenius type equation [22–25], namely:

$$\dot{\varepsilon} = A[\sinh(\alpha\sigma)]^n \exp(-\frac{Q}{RT}) \tag{1}$$

where $\dot{\varepsilon}$ is the strain rate, υ the flow stress, T the absolute temperature, Q the activation energy, R the universal gas constant (8.314 J·mol^{-1}·K^{-1}), and A, α and n the material constants.

At a low stress level, Equation (1) is approximately expressed by:

$$\dot{\varepsilon} = A_1 \sigma^n \exp\left(-\frac{Q}{RT}\right), \alpha\sigma \leq 0.83373 \tag{2}$$

where A_1 is still a material constant and $A_1 = A\alpha^n$.

At a high stress level, Equation (1) is approximately simplified as:

$$\dot{\varepsilon} = A_2 \exp(\beta\sigma) \exp\left(-\frac{Q}{RT}\right), \alpha\sigma \leq 1.60944 \tag{3}$$

where A_2 and β remain the material constants and $A_2 = \frac{A}{2^n}$, $\beta = n\alpha$.

It is generally accepted that the Zener-Hollomon parameter Z, which is viewed as the function of strain rate and temperature, can be expressed as follows:

$$Z = \dot{\varepsilon}\exp\left(\frac{Q}{RT}\right) \tag{4}$$

Substituting the parameter Z into Equation (1) leads to:

$$Z = A[\sinh(\alpha\sigma)]^n \tag{5}$$

The following equation can be obtained by transforming Equation (5):

$$\sigma = \frac{1}{\alpha}\ln\left\{\left(\frac{Z}{A}\right)^{\frac{1}{n}} + \sqrt{\left(\frac{Z}{A}\right)^{\frac{2}{n}} + 1}\right\} \tag{6}$$

As a consequence, Equation (6) is the constitutive equation which is expressed by the parameter Z.

The constitutive equation of NiTiCu SMA can be obtained after the values of A, α, n, and Q can be determined on the basis of the experimental data. In the present work, the peak stress is used to determine the corresponding parameters, as shown in Table 1.

Table 1. Peak stress of NiTiCu SMA at the various strain rates and deformation temperatures (MPa).

$\dot{\varepsilon}$/s^{-1}	ln$\dot{\varepsilon}$	T/K			
		973	1073	1173	1273
0.0005	−7.6009	80.878	50.8828	33.95	23.8579
0.005	−5.2983	114.8027	79.0071	56.1214	46.1041
0.05	−2.9957	156.7221	109.9611	85.8336	61.5655
0.5	−0.6931	229.1717	160.2669	118.3127	98.6517

To solve the value of n, the following equation can be obtained by employing the natural logarithm for Equation (2):

$$\ln \dot{\varepsilon} = \ln A_1 + n \ln \sigma - \frac{Q}{RT} \tag{7}$$

According to Equation (7), the value of n can be obtained by means of the linear fitting method on the basis of experimental data, as shown in Figure 3a. The value of n is determined as 5.7988.

Prior to obtaining the value of α, the value of β needs to be determined by employing the natural logarithm of Equation (3), and consequently the following equation is obtained:

$$\ln \dot{\varepsilon} = \ln A_2 + \beta \sigma - \frac{Q}{RT} \tag{8}$$

In the same manner, the value of n can be obtained by virtue of the linear fitting method on the basis of experimental data, as shown in Figure 3b. As a consequence, the value of β is calculated as 0.07059. Finally, according to the values of n and β, the value of α can be determined as $\alpha = \beta/n = 1.217 \times 10^{-2}$ MPa.

In order to further modify the value of n, the following equation can be obtained by employing the natural logarithm of Equation (1):

$$\ln \dot{\varepsilon} = \ln A + n \ln[\sinh(\alpha\sigma)] - \frac{Q}{RT} \tag{9}$$

Similarly, the value of n can be modified as 4.1318 according to the linear fitting method on the basis of experimental data, as shown in Figure 3c.

For the purpose of obtaining the value of Q, in the case of the given strain rates, the following equation can be obtained by differentiating T^{-1} in Equation (9):

$$Q = nR \left(\frac{\partial \ln[\sinh(\alpha\sigma)]}{\partial T^{-1}} \right)_{\dot{\varepsilon}} \tag{10}$$

The value of Q can be determined as 198.84×10^3 J·mol^{-1} by combining the modified value of n and the value of R with the fitting value derived from Figure 3d.

For the purpose of obtaining the value of A, the following equation can be obtained by employing the natural logarithm of Equation (5):

$$\ln Z = \ln A + n \ln[\sinh(\alpha\sigma)] \tag{11}$$

According to Equation (11), the value of $\ln A$ can be determined as 16.3994 based on the experimental data, as shown in Figure 4, and consequently the value of A is further calculated to be 1.3249×10^7.

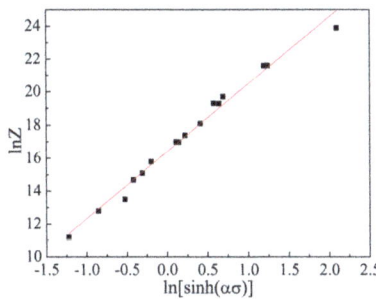

Figure 4. Determination of material constant A based on the relationship between ln Z and ln[sinh($\alpha\sigma$)].

As a consequence, the constitutive equation of NiTiCu SMA is represented by the following two equations:

$$\dot{\varepsilon} = 1.3249 \times 10^7 [\sinh(1.217 \times 10^{-2}\sigma)]^{4.1318} \exp\left(\frac{-1.9884 \times 10^5}{RT}\right) \tag{12}$$

$$\sigma = 82.17 \ln\left[\left(\frac{Z}{A}\right)^{\frac{1}{4.1318}} + \sqrt{\left(\frac{Z}{A}\right)^{\frac{2}{4.1318}} + 1}\right] \tag{13}$$

3.2. Microstructural Evolution of NiTiCu SMA

The microstructural evolution of several representative NiTiCu SMA samples, which undergo compression deformation, is captured in order to reveal the corresponding plastic deformation mechanisms at high temperatures. It can be generally accepted that DRV or DRX can take place when metallic alloy experiences plastic deformation at elevated temperatures. In the case of DRX, in particular, the size of the recrystallized grains increases with increasing the deformation temperature or decreasing the strain rate. According to the phase transformation temperatures of NiTiCu SMA, it can be noted that when NiTiCu SMA samples are subjected to compression deformation at high temperatures and are quenched into the ice water, they are transformed from B2 austenite into B19′ martensite because of complete martensite phase transformation.

For the purpose of better revealing the martensite structure of NiTiCu SMA, TEM micrographs of the NiTiCu SMA samples, which are subjected to compression at the temperatures of 700~1000 °C at the strain rate of 0.005 s^{-1}, were captured, as shown in Figures 5–8.

It can be found that dislocation substructures, such as dislocation cells and subgrain boundaries, appear in the martensite phase of NiTiCu SMA subjected to compression deformation at 700 °C, as shown in Figure 5a. It can thus be concluded that dislocations become the dominant substructures of martensite phase in the NiTiCu SMA specimen undergoing compression at 700 °C. Martensite twins are dominant in the NiTiCu SMA specimen undergoing compression at 800 and 900 °C, as shown in Figures 6 and 7. However, martensite twins are not observed in the NiTiCu SMA specimen undergoing compression at 1000 °C, as shown in Figure 8. According to the aforementioned analysis, it is noted that DRV or DRX can occur in NiTiCu SMA samples subjected to compression deformation at high temperatures, which depends on the deformation temperatures. It can be deduced that DRV or DRX microstructures have an influence on the substructures of martensite phase of NiTiCu SMA. In general, DRV is the dominant mechanism when NiTi-based SMAs are subjected to compression deformation at 700 °C [21]. As a consequence, DRV leads to the deformation of grains and thus dislocation substructures, such as dislocation cells and subgrain boundaries, are formed in the grain interior. These dislocation substructures are kept in the martensite phase of the NiTiCu SMA sample during subsequent martensite phase transformations. These retained dislocation substructures suppress the formation of martensite twins. It can be generally accepted that complete DRX can take place in NiTi-based SMAs suffering from compression at temperatures above 800 °C [18,21]. It is proposed that the size of the dynamic recrystallized grains influences the formation of martensite twins in NiTiCu SMAs. In general, the size of the dynamic recrystallized grains increases with increasing the deformation temperature [18]. Furthermore, it is generally accepted that the occurrence of martensite twins aims to accommodate the formation of martensite phase during the transformation from B2 austenite to B19′ martensite. It can be inferred that when the dynamic recrystallized grains possess a smaller size, the grain boundaries are able to suppress the formation of martensite phase. Consequently, the occurrence of martensite twins contributes to the formation of martensite phase. When the dynamic recrystallized grains possess a larger size, there is sufficient space to guarantee the formation of martensite phase in the grain interior. Therefore, martensite twins are not easily formed in NiTiCu SMA samples subjected to compression deformation at 1000 °C.

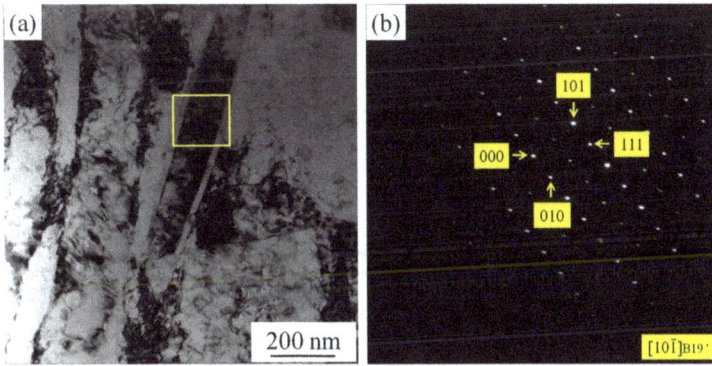

Figure 5. TEM micrographs of NiTiCu SMA subjected to compression deformation at 700 °C: (**a**) Bright field image showing the existence of dislocation substructures in martensite laths; (**b**) Diffraction pattern of (**a**).

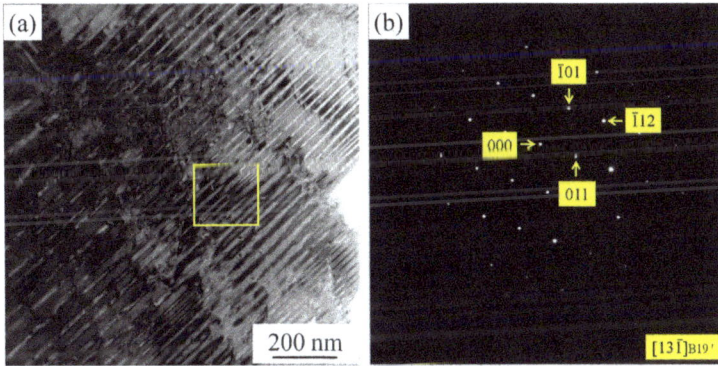

Figure 6. TEM micrographs of NiTiCu SMA subjected to compression deformation at 800 °C: (**a**) Bright field image showing the existence of martensite twins; (**b**) Diffraction pattern of (**a**).

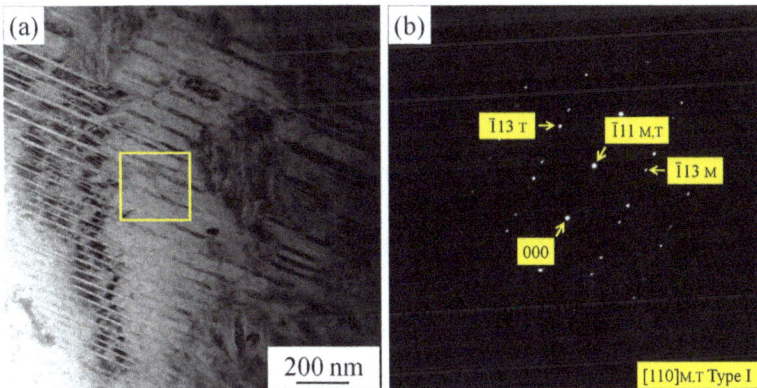

Figure 7. TEM micrographs of NiTiCu SMA subjected to compression deformation at 900 °C: (**a**) Bright field image showing the existence of martensite twins; (**b**) Diffraction pattern of (**a**).

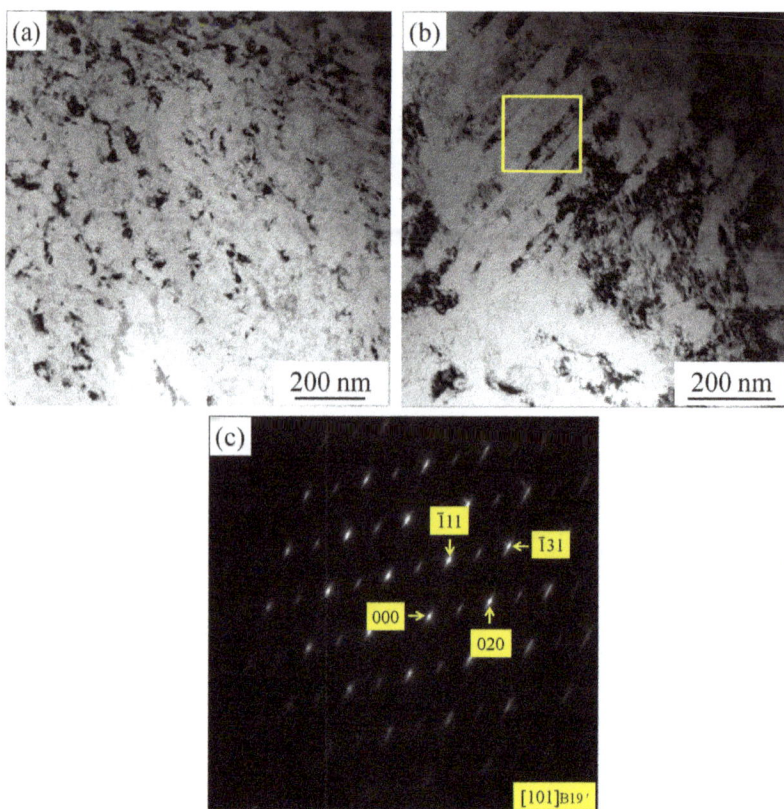

Figure 8. TEM micrographs of NiTiCu SMA subjected to compression deformation at 1000 °C: (**a**) Bright field image showing no martensite laths; (**b**) Bright field image showing the existence of martensite laths; (**c**) Diffraction pattern of (**b**).

4. Conclusions

The deformation behavior and microstructure evolution of NiTiCu SMA, which possesses martensite phase at room temperature, were investigated based on a uniaxial compression test at the temperatures of 700~1000 °C and at the strain rates of 0.0005~0.5 s^{-1}. The following conclusions can be drawn:

(1) The constitutive equation of NiTiCu SMA based on the Zener-Hollomon parameter was established in order to describe the flow characteristic of NiTiCu SMA. The results show that the flow stress of NiTiCu SMA depends on the strain rates. Depending on temperatures, DRV or DRX are the main mechanisms for the plastic deformation of NiTiCu SMA at elevated temperatures.

(2) The microstructures resulting from DRV or DRX have a significant influence on the substructures in the martensite phase of the NiTiCu SMA sample at room temperature. Dislocations become the dominant substructures of martensite in the NiTiCu SMA specimen undergoing compression at 700 °C. Martensite twins are dominant in the NiTiCu SMA specimen undergoing compression at 800 and 900 °C. Martensite twins are not observed in the NiTiCu SMA specimen undergoing compression at 1000 °C.

(3) Dislocation substructures resulting from DRV, such as dislocation cells and subgrain boundaries, are able to suppress the formation of martensite twins in the martensite laths of NiTiCu SMA.

The size of dynamic recrystallized grains has an effect on the formation of martensite twins. Martensite twins are not easily formed in larger dynamic recrystallized grain, since the constraint of the grain boundaries plays a weak role. However, in smaller dynamic recrystallized grain, martensite twins are induced in order to accommodate the occurrence of the transformation from austenite phase to martensite phase.

Acknowledgments: The work was financially supported by National Natural Science Foundation of China (Nos. 51475101, 51305091 and 51305092).

Author Contributions: Shuyong Jiang wrote the manuscript; Dong Sun performed TEM analysis; Yanqiu Zhang established constitutive equation; Li Hu processed experimental data.

Conflicts of Interest: The authors declare no conflict of interest.

References

1. Jani, J.M.; Leary, M.; Subic, A.; Gibson, M.A. A review of shape memory alloy research, applications and opportunities. *Mater. Des.* **2014**, *56*, 1078–1113. [CrossRef]

2. Basu, R.; Eskandari, M.; Upadhayay, L.; Mohtadi-Bonab, M.A.; Szpunar, J.A. A systematic investigation on the role of microstructure on phase transformation behavior in Ni–Ti–Fe shape memory alloys. *J. Alloys Compd.* **2015**, *645*, 213–222. [CrossRef]

3. Choi, E.; Hong, H.K.; Kim, H.S.; Chung, Y.S. Hysteretic behavior of NiTi and NiTiNb SMA wires under recovery or pre-stressing stress. *J. Alloys Compd.* **2013**, *577S*, S444–S447. [CrossRef]

4. Wang, M.J.; Jiang, M.Y.; Liao, G.Y.; Guo, S.; Zhao, X.Q. Martensitic transformation involved mechanical behaviors and wide hysteresis of NiTiNb shape memory alloys. *Prog. Natl. Sci. Mater. Int.* **2012**, *22*, 130–138. [CrossRef]

5. Nespoli, A.; Passaretti, F., Villa, E. Phase transition and mechanical damping properties: A DMTA study of NiTiCu shape memory alloys. *Intermetallics* **2013**, *32*, 394–400. [CrossRef]

6. Basu, R.; Mohtadi-Bonab, M.A.; Wang, X.; Eskandari, M.; Szpunar, J.A. Role of microstructure on phase transformation behavior in Ni-Ti-Fe shape memory alloys during thermal cycling. *J. Alloys Compd.* **2015**, *652*, 459–469. [CrossRef]

7. Choudhary, N.; Kaur, D. Shape memory alloy thin films and heterostructures for MEMS applications: A review. *Sens. Actuators A* **2016**, *242*, 162–181. [CrossRef]

8. Kaur, N.; Kaur, D. NiTiCu/AlN/NiTiCu shape memory thin film heterostructures for vibration damping in MEMS. *J. Alloys Compd.* **2014**, *590*, 116–124. [CrossRef]

9. Kotil, T.; Sehitoglu, H.; Maier, H.J.; Chumlyakov, Y.I. Transformation and detwinning induced electrical resistance variations in NiTiCu. *Mater. Sci. Eng. A* **2003**, *359*, 280–289. [CrossRef]

10. Saikrishna, C.N.; Ramaiah, K.V.; Bhaumik, S.K. Effects of thermo-mechanical cycling on the strain response of Ni–Ti–Cu shape memory alloy wire actuator. *Mater. Sci. Eng. A* **2006**, *428*, 217–224. [CrossRef]

11. Nespoli, A.; Villa, E.; Besseghini, S. Characterization of the martensitic transformation in Ni50-xTi50Cux alloys through pure thermal measurements. *J. Alloys Compd.* **2011**, *509*, 644–647. [CrossRef]

12. Colombo, S.; Cannizzo, C.; Gariboldi, F.; Airoldi, G. Electrical resistance and deformation during the stress-assisted two-way memory effect in Ni45Ti50Cu5 alloy. *J. Alloys Compd.* **2006**, *422*, 313–320. [CrossRef]

13. Sharifi, E.M.; Karimzadeh, F.; Kermanpur, A. The effect of cold rolling and annealing on microstructure and tensile properties of the nanostructured Ni50Ti50 shape memory alloy. *Mater. Sci. Eng. A* **2014**, *607*, 33–37. [CrossRef]

14. Tadayyon, G.; Guo, Y.; Mazinani, M.; Zebarjad, S.M.; Tiernan, P.; Tofail, S.A.M.; Biggs, M.J.P. Effect of different stages of deformation on the microstructure evolution of Ti-rich NiTi shape memory alloy. *Mater. Charact.* **2017**, *125*, 51–66. [CrossRef]

15. Yeom, J.T.; Kim, J.H.; Hong, J.K.; Kim, S.W.; Park, C.H.; Nam, T.H.; Lee, K.Y. Hot forging design of as-cast NiTi shape memory alloy. *Mater. Res. Bull.* **2014**, *58*, 234–238. [CrossRef]

16. Sehitoglu, H.; Karaman, I.; Zhang, X.; Hong, K.; Chumlyakov, Y.; Kireeva, I. Deformation of NiTiCu shape memory single crystals in compression. *Metall. Mater. Trans. A* **2001**, *32A*, 477–489. [CrossRef]

17. Morakabati, M.; Kheirandish, S.; Aboutalebi, M.; Taheri, A.K.; Abbasi, S.M. The effect of Cu addition on the hot deformation behavior of NiTi shape memory alloys. *J. Alloys Compd.* **2010**, *499*, 57–62. [CrossRef]

18. Morakabati, M.; Aboutalebi, M.; Kheirandish, S.; Taheri, A.K.; Abbasi, S.M. Hot tensile properties and microstructural evolution of as cast NiTi and NiTiCu shape memory alloys. *Mater. Des.* **2011**, *32*, 406–413. [CrossRef]

19. Mirzadeh, H.; Parsa, M.H. Hot deformation and dynamic recrystallization of NiTi intermetallic compound. *J. Alloys Compd.* **2014**, *614*, 56–59. [CrossRef]

20. Yin, X.Q.; Park, C.H.; Li, Y.F.; Ye, W.J.; Zuo, Y.T.; Lee, S.W.; Yeom, J.T.; Mi, X.J. Mechanism of continuous dynamic recrystallization in a 50Ti-47Ni-3Fe shape memory alloy during hot compressive deformation. *J. Alloys Compd.* **2017**, *693*, 426–431. [CrossRef]

21. Jiang, S.Y.; Zhang, Y.Q.; Zhao, Y.N. Dynamic recovery and dynamic recrystallization of NiTi shape memory alloy under hot compression deformation. *Trans. Nonferrous Met. Soc. China* **2013**, *23*, 140–147. [CrossRef]

22. Etaati, A.; Dehghani, K. A study on hot deformation behavior of Ni-42.5Ti-7.5Cu alloy. *Mater. Chem. Phys.* **2013**, *140*, 208–215. [CrossRef]

23. Shamsolhodaei, A.; Zarei-hanzaki, A.; Ghambari, M.; Moemeni, S. The high temperature flow behavior modeling of NiTi shape memory alloy employing phenomenological and physical based constitutive models: A comparative study. *Intermetallics* **2014**, *53*, 140–149. [CrossRef]

24. Jiang, S.Y.; Zhang, Y.Q.; Zhao, Y.N.; Tang, M.; Yi, W.L. Constitutive behavior of Ni-Ti shape memory alloy under hot compression. *J. Cent. South Univ.* **2013**, *20*, 24–29. [CrossRef]

25. Zhang, Y.Q.; Jiang, S.Y.; Zhao, Y.N.; Liu, S.W. Constitutive equation and processing map of equiatomic NiTi shape memory alloy under hot plastic deformation. *Trans. Nonferrous Met. Soc. China* **2016**, *26*, 2152–2161. [CrossRef]

26. Zhang, Y.; Jiang, S.; Chen, C.; Hu, L.; Zhu, X. Hot workability of a NiTiCu shape memory alloy with acicular martensite phase based on processing maps. *Intermetallics* **2017**, *86*, 94–103. [CrossRef]

![metals logo] *metals*

MDPI

Article

Microstructure, Mechanical Property, and Phase Transformation of Quaternary NiTiFeNb and NiTiFeTa Shape Memory Alloys

Yulong Liang [1,2], Shuyong Jiang [1,*], Yanqiu Zhang [1] and Junbo Yu [2]

1   College of Mechanical and Electrical Engineering, Harbin Engineering University, Harbin 150001, China;
    yulongliang380@126.com (Y.L.); zhangyq@hrbeu.edu.cn (Y.Z.)
2   College of Materials Science and Chemical Engineering, Harbin Engineering University,
    Harbin 150001, China; yujunbo@hrbeu.edu.cn
*   Correspondence: jiangshuyong@hrbeu.edu.cn; Tel.: +86-451-8251-9710

Received: 3 July 2017; Accepted: 9 August 2017; Published: 12 August 2017

Abstract: Based on ternary $Ni_{45}Ti_{51.8}Fe_{3.2}$ (at %) shape memory alloy (SMA), Nb and Ta elements are added to an NiTiFe SMA by replacing Ni element, and consequently quaternary $Ni_{44}Ti_{51.8}Fe_{3.2}Nb_1$ and $Ni_{44}Ti_{51.8}Fe_{3.2}Ta_1$ (at %) SMAs are fabricated. The microstructure, mechanical property, and phase transformation of NiTiFeNb and NiTiFeTa SMAs are further investigated. Ti_2Ni and β-Nb phases can be observed in NiTiFeNb SMA, whereas Ti_2Ni and Ni_3Ti phases can be captured in NiTiFeTa SMA. As compared to NiTiFe SMA, quaternary NiTiFeNb and NiTiFeTa SMAs possess the higher strength, since solution strengthening plays a considerable role. NiTiFeNb and NiTiFeTa SMAs exhibit a one-step transformation from B2 austenite to B19′ martensite during cooling, but they experience a two-step transformation of B19′-R-B2 during heating.

Keywords: shape memory alloy; mechanical property; microstructures; phase transformation

1. Introduction

NiTi shape memory alloys (SMAs) have deserved increasing attention in the domain of engineering since they possess shape memory effect and superelasticity. In general, phase transformation temperature and mechanical property are the two critical factors influencing the application of NiTi SMA in engineering [1–3]. In particular, a third element can be added to the binary NiTi SMA so as to change its phase transformation temperature and mechanical property [4,5]. As a consequence, typical ternary NiTi-based SMAs, such as NiTiCu, NiTiNb, and NiTiFe, have gone toward the engineering application. For instance, NiTiCu SMA possesses a narrow transformation temperature hysteresis, so it can be used in actuators [6–8]. However, NiTiNb SMA possesses a broad transformation temperature hysteresis, so it is suitable for pipe coupling [9–11]. NiTiFe SMA is typically used for coupling pipe since it possesses a lower martensite transformation start temperature [12–14]. NiTiPt, NiTiPd, NiTiZr, and NiTiHf SMAs become candidates for high temperature SMAs since they possess a higher reverse transformation temperature [15–18]. So far, quaternary NiTi-based SMAs have deserved more attention. Some quaternary SMAs have also become potential candidates for high temperature SMAs, such as TiNiPdCu [19], NiTiHfCu [20], NiTiHfZr [21], and NiTiHfTa [22] SMAs. In addition, some new quaternary NiTi-based SMAs, which possess special properties, have been put forward as well. These new quaternary NiTi-based SMAs include NiTiHfPd SMA with high strength [23,24], TiNiCuPd SMA with near-zero hysteresis [25], and NiTiCuV SMA with high thermal stabilization [26]. However, only a few studies have reported the investigations of quaternary NiTi-based SMAs.

In the present study, Nb and Ta elements were added to NiTiFe SMA in order to prepare NiTiFeNb and NiTiFeTa SMAs. Furthermore, the microstructure, mechanical property, and phase transformation of NiTiFeNb and NiTiFeTa SMAs were investigated.

2. Materials and Methods

On the basis of ternary $Ni_{45}Ti_{51.8}Fe_{3.2}$ (at %) SMA, Nb and Ta elements were added by replacing the Ni element, and consequently quaternary $Ni_{44}Ti_{51.8}Fe_{3.2}Nb_1$ and $Ni_{44}Ti_{51.8}Fe_{3.2}Ta_1$ (at %) SMAs were fabricated via the vacuum arc melting method. Then, three as-cast NiTi-based SMAs were heated to 1000 °C and maintained for 12 h. Subsequently, the three heat-treated NiTi-based SMA samples were quenched with ice water. NiTi-based SMA samples, whose height and diameter were 6 mm and 4 mm, respectively, were removed from the heat-treated NiTi-based SMA ingots using electro-discharge machining (EDM) for the purpose of compression tests. The compression tests were performed with INSTRON-5500R equipment (Instron Corporation, Norwood, MA, USA) at room temperature, where the compression strain rate was determined as $0.001\ s^{-1}$.

Differential scanning calorimetry (DSC) was used to determine the phase transformation temperatures of the three heat-treated NiTi-based SMA specimens by using a Pyris Diamond type differential scanning calorimeter (Perkin Elmer Inc., Waltham, MA, USA). Therein, DSC measurement temperature ranged from −150 °C to 150 °C, and the heating and cooling rates were determined as 10 °C/min.

Optical microscopy (OM) as well as transmission electron microscopy (TEM) was employed to characterize the microstructures of the three heat-treated NiTi-based SMA samples. The samples for the OM experiment were etched in a solution whose composition was determined as $HF:HNO_3:H_2O = 1:2:10$. Then, the OM experiment was performed using an OLYMPUS311 type optical microscope (Olympus Corporation Tokyo, Japan). The TEM experiment was carried out on an FEI TECNAI G2 F30 microscope (FEI Corporation, Hillsboro, OR, USA) with a side-entry and double-tilt specimen stage with an angular range of ± 40° at an accelerating voltage of 300 kV. Foils for TEM observation were mechanically ground to 70 μm and then thinned by twin-jet polishing in an electrolyte containing 90% C_2H_5OH and 10% $HClO_4$ by volume fraction.

Employing a Philips X'Pert Pro diffractometer (Royal Dutch Philips Electronics Ltd. Amsterdam, Netherlands) with CuKα radiation at ambient temperature, X-ray diffraction (XRD) analysis was used to identify the phase composition of the three heat-treated NiTi-based SMA samples. The samples were scanned over 2θ, which ranges from 20° to 90° by continuous scanning based on a tube voltage of 40 kV and a tube current of 40 mA.

3. Results and Discussion

3.1. Microstructure Analysis of NiTi-based SMAs

Figure 1 shows the microstructures of $Ni_{45}Ti_{51.8}Fe_{3.2}$, $Ni_{44}Ti_{51.8}Fe_{3.2}Nb_1$, and $Ni_{44}Ti_{51.8}Fe_{3.2}Ta_1$ SMAs determined by OM. The microstructures of the three NiTi-based SMAs are dominated by equiaxed grains rather than dendrites. In addition, there are some precipitates in the grain interior as well as at the grain boundary.

To further determine the phase composition of the three NiTi-based SMAs, XRD maps of $Ni_{45}Ti_{51.8}Fe_{3.2}$, $Ni_{44}Ti_{51.8}Fe_{3.2}Nb_1$, and $Ni_{44}Ti_{51.8}Fe_{3.2}Ta_1$ SMAs are illustrated in Figure 2. It is evident that the three NiTi-based SMAs are composed of B2 austenite and Ti_2Ni precipitate.

In order to gain an in-depth insight into the microstructures of the three NiTi-based SMAs, TEM micrographs of $Ni_{45}Ti_{51.8}Fe_{3.2}$, $Ni_{44}Ti_{51.8}Fe_{3.2}Nb_1$, and $Ni_{44}Ti_{51.8}Fe_{3.2}Ta_1$ SMAs are shown in Figures 3–5, respectively. It can be observed from Figure 3 that $Ni_{45}Ti_{51.8}Fe_{3.2}$ SMA consists of B2 austenite matrix and Ti_2Ni precipitate. It can be seen in Figure 4 that in terms of $Ni_{44}Ti_{51.8}Fe_{3.2}Nb_1$ SMA, Ti_2Ni and β-Nb precipitates occur in the matrix of B2 austenite. It can be seen in Figure 5 that the matrix of $Ni_{44}Ti_{51.8}Fe_{3.2}Ta_1$ SMA belongs to B2 austenite, where both Ti_2Ni and Ni_3Ti precipitates can be observed.

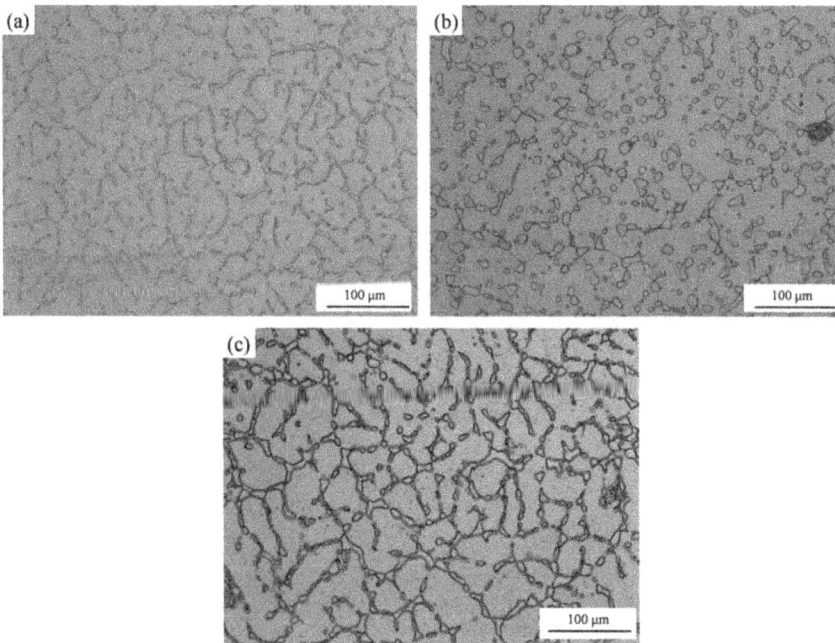

Figure 1. OM (Optical microscopy) micrographs of the three NiTi-based SMAs: (**a**) $Ni_{45}Ti_{51.8}Fe_{3.2}$; (**b**) $Ni_{44}Ti_{51.8}Fe_{3.2}Nb_1$; (**c**) $Ni_{44}Ti_{51.8}Fe_{3.2}Ta_1$.

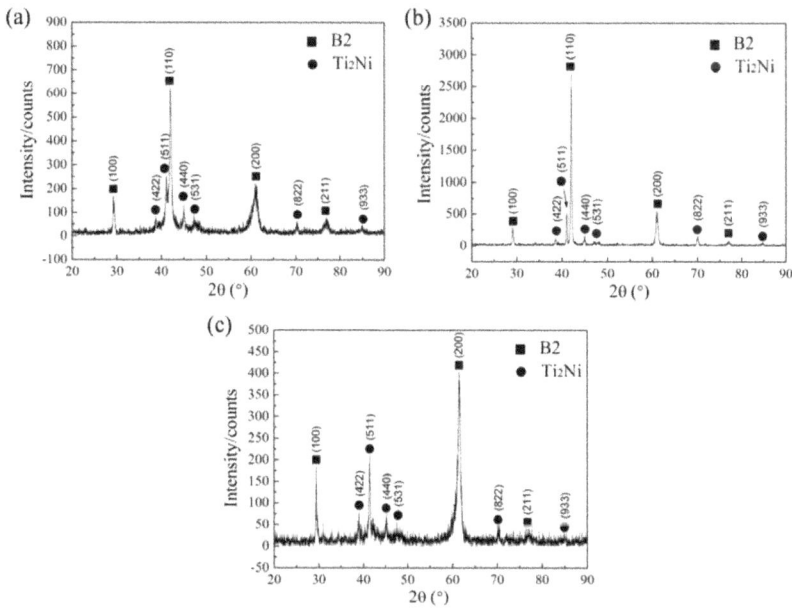

Figure 2. XRD (X-ray diffraction) maps of the three NiTi-based SMAs: (**a**) $Ni_{45}Ti_{51.8}Fe_{3.2}$; (**b**) $Ni_{44}Ti_{51.8}Fe_{3.2}Nb_1$; (**c**) $Ni_{44}Ti_{51.8}Fe_{3.2}Ta_1$.

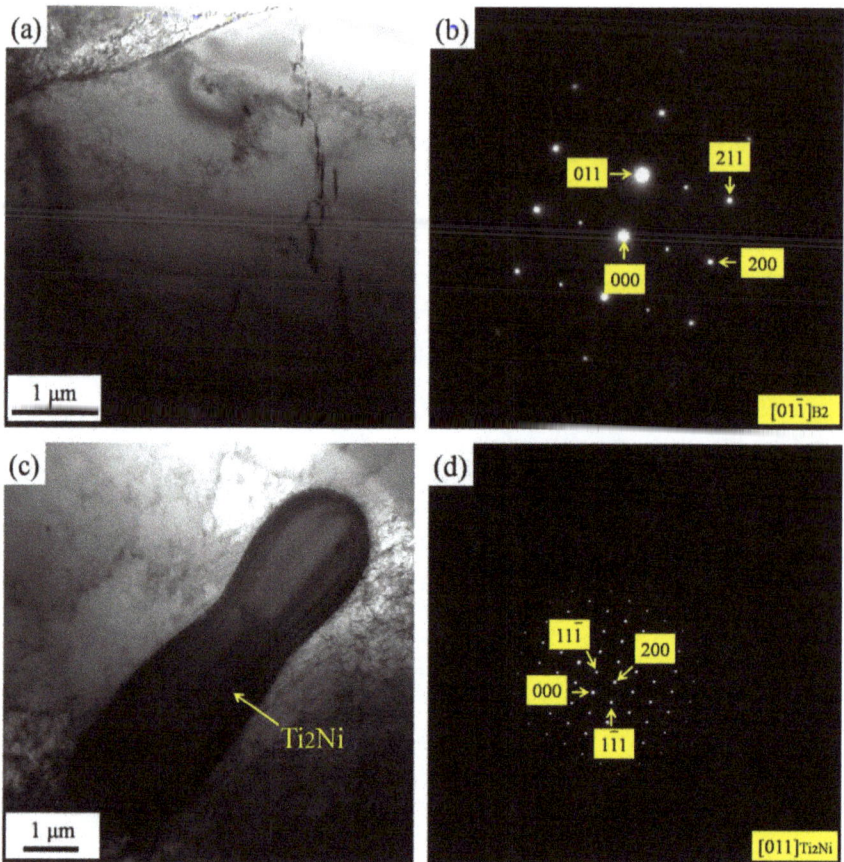

Figure 3. TEM (Transmission electron microscopy) micrographs of $Ni_{45}Ti_{51.8}Fe_{3.2}$ SMA (Shape memory alloy): (**a**) Bright field image showing B2 austenite matrix; (**b**) Diffraction pattern of B2 austenite matrix in (**a**); (**c**) Bright field image showing Ti_2Ni precipitate; (**d**) Diffraction pattern of Ti_2Ni precipitate in (**c**).

3.2. Mechanical Property of NiTi-based SMAs

Figure 6 illustrates the stress-strain curves of $Ni_{45}Ti_{51.8}Fe_{3.2}$, $Ni_{44}Ti_{51.8}Fe_{3.2}Nb_1$, and $Ni_{44}Ti_{51.8}Fe_{3.2}Ta_1$ SMAs under uniaxial compression. It is observed that the addition of Nb and Ta elements results in the increasing yield strength of NiTi-based SMAs, but also leads to the decreasing plasticity of NiTi-based SMAs. Furthermore, $Ni_{44}Ti_{51.8}Fe_{3.2}Nb_1$ SMA possesses a higher yield strength than $Ni_{44}Ti_{51.8}Fe_{3.2}Ta_1$ SMA. In other words, the Nb element plays a more substantial role in strengthening NiTi-based SMAs compared with the Ta element. According to the aforementioned microstructural analysis, no metallic compounds of Nb and Ta elements are observed in the case of $Ni_{44}Ti_{51.8}Fe_{3.2}Nb_1$ and $Ni_{44}Ti_{51.8}Fe_{3.2}Ta_1$ SMAs. Consequently, it can be concluded that Nb and Ta exist in the solid solution of B2 austenite as the solute atoms. In particular, compared to $Ni_{45}Ti_{51.8}Fe_{3.2}$ and $Ni_{44}Ti_{51.8}Fe_{3.2}Nb_1$ SMAs, $Ni_{44}Ti_{51.8}Fe_{3.2}Ta_1$ SMA exhibits a steady strain hardening ability during plastic deformation. It is well known that the strain hardening ability of metal materials is related closely to the dislocation density. In general, the strain hardening ability increases with increasing dislocation denstiy. Therefore, as for $Ni_{44}Ti_{51.8}Fe_{3.2}Ta_1$ SMA, plenty of dislocations need to be enhanced so as to guarantee the compatiblity of plastic deformation.

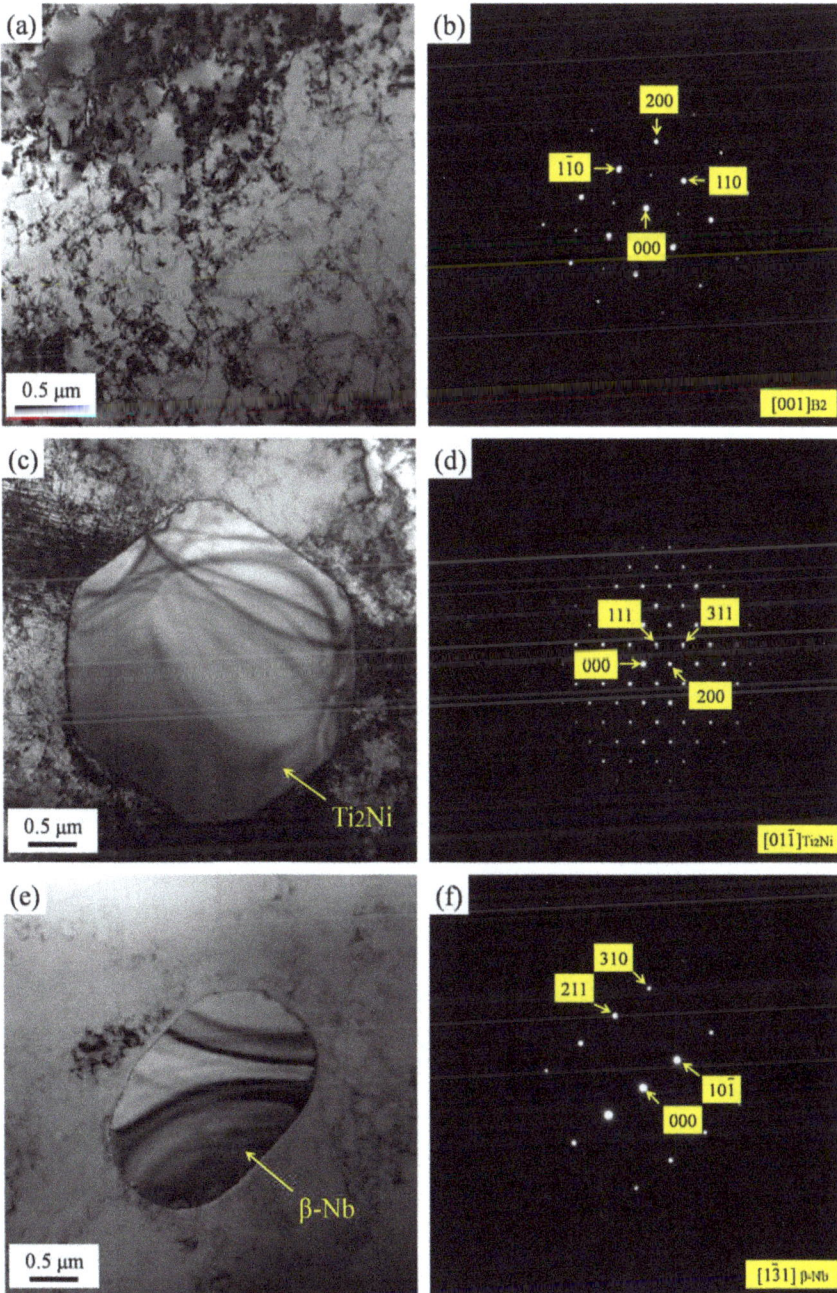

Figure 4. TEM micrographs of $Ni_{44}Ti_{51.8}Fe_{3.2}Nb_1$ SMA: (**a**) Bright field image showing B2 austenite matrix; (**b**) Diffraction pattern of B2 austenite matrix in (**a**); (**c**) Bright field image showing Ti_2Ni precipitate; (**d**) Diffraction pattern of Ti_2Ni precipitate in (**c**); (**e**) Bright field image showing β-Nb precipitate; (**f**) Diffraction pattern of β-Nb precipitate in (**e**).

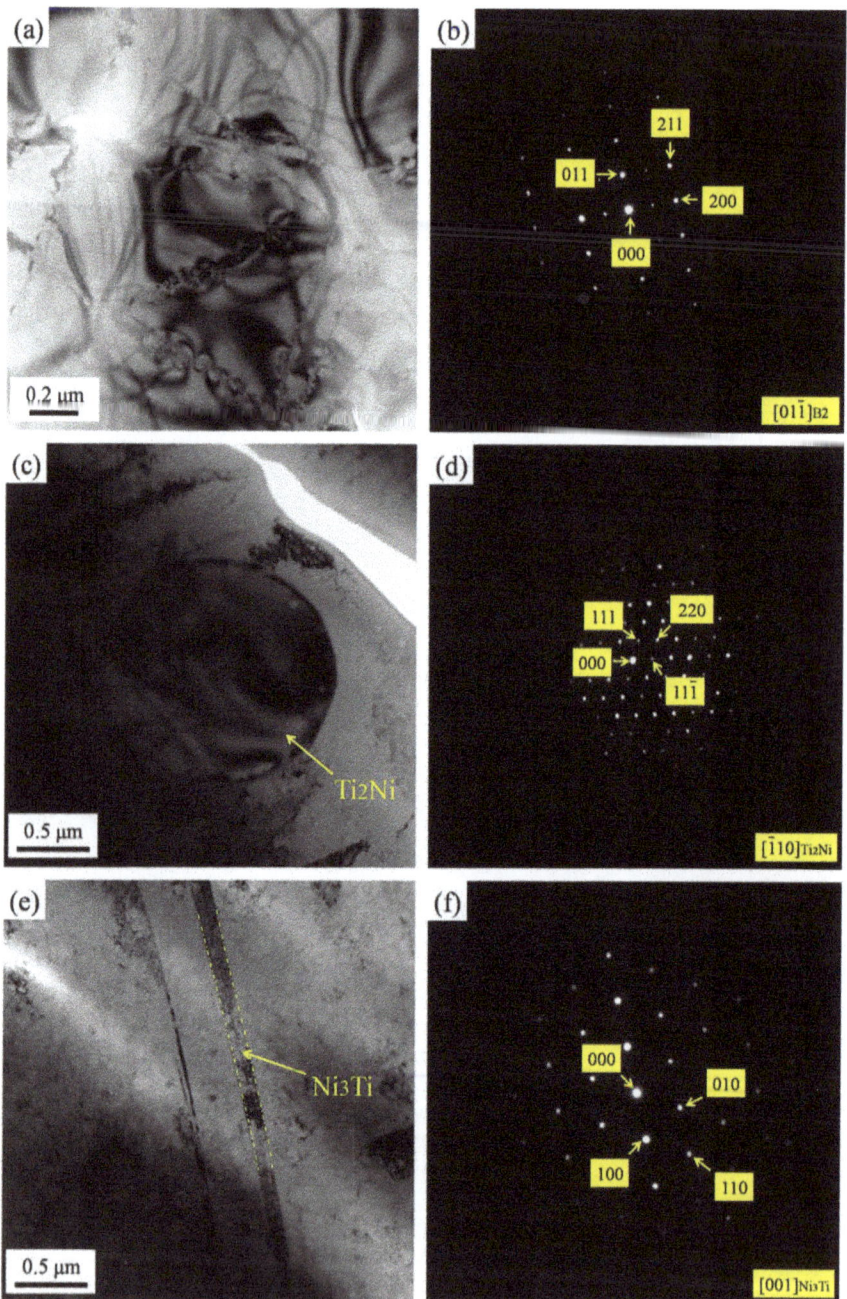

Figure 5. TEM micrographs of $Ni_{44}Ti_{51.8}Fe_{3.2}Ta_1$ SMA: (**a**) Bright field image showing B2 austenite matrix; (**b**) Diffraction pattern of B2 austenite matrix in (**a**); (**c**) Bright field image showing Ti_2Ni precipitate; (**d**) Diffraction pattern of Ti_2Ni precipitate in (**c**); (**e**) Bright field image showing Ni_3Ti precipitate; (**f**) Diffraction pattern of Ni_3Ti precipitate in (**e**).

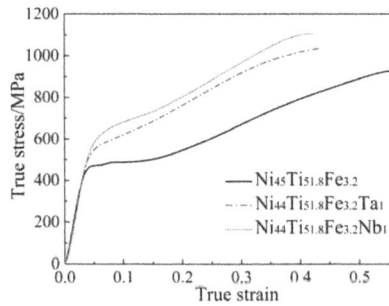

Figure 6. Compressive stress-strain curves of the three NiTi-based SMAs.

3.3. Phase Transformation of NiTi based SMAs

Figure 7 shows the DSC curves of $Ni_{45}Ti_{51.8}Fe_{3.2}$, $Ni_{44}Ti_{51.8}Fe_{3.2}Nb_1$, and $Ni_{44}Ti_{51.8}Fe_{3.2}Ta_1$ SMAs. All three NiTi-based SMAs exhibit a one-step phase transformation during cooling. The one-step phase transformation deals with the transformation from B2 austenite (A) to B19′ martensite (M). However, all three NiTi-based SMAs exhibit a two-step phase transformation during heating. First, they are converted from B19′ martensite to an R-phase. Subsequently, they are transformed from the R-phase into B2 austenite. It can be noted that the addition of Nb and Ta elements does not change the phase transformation path of NiTiFe SMA, but it does have a certain effect on the transformation temperatures of NiTiFe SMA. As a consequence, all of the transformation temperatures are diminished. In particular, as for $Ni_{44}Ti_{51.8}Fe_{3.2}Nb_1$ SMA, the addition of the Nb element results in the severe diminishment of the martensite and austenite transformation temperatures.

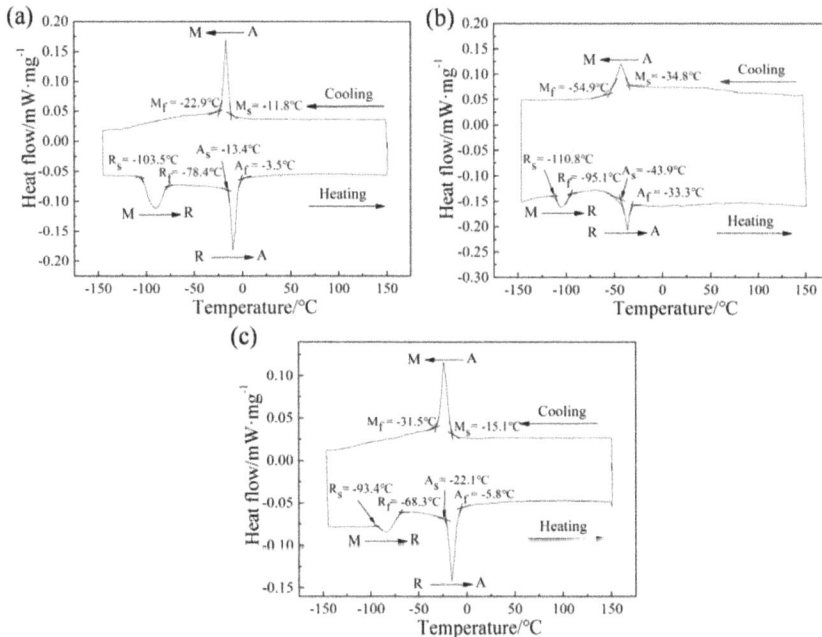

Figure 7. DSC (Differential scanning calorimetry) curves of the three NiTi-based SMAs: (**a**) $Ni_{45}Ti_{51.8}Fe_{3.2}$; (**b**) $Ni_{44}Ti_{51.8}Fe_{3.2}Nb_1$; (**c**) $Ni_{44}Ti_{51.8}Fe_{3.2}Ta_1$.

Compared to $Ni_{45}Ti_{51.8}Fe_{3.2}$ SMA, $Ni_{44}Ti_{51.8}Fe_{3.2}Nb_1$ and $Ni_{44}Ti_{51.8}Fe_{3.2}Ta_1$ SMAs possess a lower phase transformation temperature. In particular, in the case of $Ni_{44}Ti_{51.8}Fe_{3.2}Nb_1$ SMA, the addition of the Nb element influences the phase transformation temperature considerably, and consequently all the phase transformation temperatures are lowered substantially compared with $Ni_{45}Ti_{51.8}Fe_{3.2}$ SMA. In general, for the purpose of lowering the martensite transformation start temperature M_s, the transformation resistance should be enhanced when B2 austenite is transformed into B19′ martensite. Simultaneously, in order to enhance the austenite transformation start temperature A_s, the mechanical driving force should be diminished when B19′ martensite is transformed into B2 austenite.

It can be deduced that the addition of Nb and Ta elements contributes to the increase of the elastic strain energy of B2 austenite. Consequently, the elastic strain energy becomes the resistance, which prevents B2 austenite from being transformed into B19′ martensite, so that the martensite phase transformation temperature is lowered. Conversely, the elastic strain energy kept in the martensite interface becomes the mechanical driving force, which facilitates the transformation from B19′ martensite to B2 austenite, so that the austenite phase transformation temperature is enhanced.

4. Conclusions

Based on ternary $Ni_{45}Ti_{51.8}Fe_{3.2}$ (at %) SMA, two new quaternary $Ni_{44}Ti_{51.8}Fe_{3.2}Nb_1$ and $Ni_{44}Ti_{51.8}Fe_{3.2}Ta_1$ (at %) SMAs were designed and fabricated by adding Nb and Ta elements to replace the Ni element. Furthermore, the microstructure, mechanical property, and phase transformation of NiTiFeNb and NiTiFeTa SMAs were investigated. The following conclusions are drawn:

(1) The microstructures of three NiTi-based SMAs are dominated by equiaxed grains rather than dendrites. $Ni_{45}Ti_{51.8}Fe_{3.2}$ SMA comprises a B2 austenite matrix and a Ti_2Ni precipitate. In the case of $Ni_{44}Ti_{51.8}Fe_{3.2}Nb_1$ SMA, Ti_2Ni and β-Nb precipitates occur in the matrix of B2 austenite. The matrix of $Ni_{44}Ti_{51.8}Fe_{3.2}Ta_1$ SMA belongs to B2 austenite, where Ti_2Ni and Ni_3Ti precipitates can be observed.

(2) The addition of Nb and Ta elements results in the increasing yield strength of NiTi-based SMA, but leads to the decreasing plasticity of NiTi-based SMA. Furthermore, $Ni_{44}Ti_{51.8}Fe_{3.2}Nb_1$ SMA possesses a higher yield strength than $Ni_{44}Ti_{51.8}Fe_{3.2}Ta_1$ SMA, where the Nb element plays a more substantial role in strengthening NiTi-based SMAs compared with the Ta element. Nb and Ta exist in the solid solution of B2 austenite as the solute atoms, since no metallic compounds of Nb and Ta elements are observed in the case of $Ni_{44}Ti_{51.8}Fe_{3.2}Nb_1$. In particular, $Ni_{44}Ti_{51.8}Fe_{3.2}Ta_1$ SMA exhibits a steady strain hardening ability during plastic deformation.

(3) All three NiTi-based SMAs exhibit a one-step phase transformation during cooling, which is involved in the transformation from B2 austenite to B19′ martensite. However, all three NiTi-based SMAs exhibit a two-step phase transformation during heating, where they are converted from B19′ martensite to the R-phase and subsequently they are transformed from the R-phase into the B2 phase. The addition of Nb and Ta elements does not change the phase transformation path of NiTiFe SMA, but it does have a certain effect on the transformation temperatures of NiTiFe SMA. As a consequence, all the transformation temperatures are diminished.

Acknowledgments: The work was financially supported by National Natural Science Foundation of China (Nos. 51475101, 51305091 and 51305092).

Author Contributions: Yulong Liang wrote the manuscript; Shuyong Jiang supervised the manuscript; Yanqiu Zhang performed the OM and TEM analysis; Junbo Yu performed the DSC analysis and compression test.

Conflicts of Interest: The authors declare no conflict of interest.

References

1. Elahinia, M.H.; Hashemi, M.; Tabesh, M.; Bhaduri, S.B. Manufacturing and processing of NiTi implants: A review. *Prog. Mater. Sci.* **2012**, *57*, 911–946. [CrossRef]
2. Sun, L.; Huang, W.M.; Ding, Z.; Zhao, Y.; Wang, C.C.; Purnawali, H.; Tang, C. Stimulus-responsive shape memory materials: A review. *Mater. Des.* **2012**, *33*, 577–640. [CrossRef]
3. Meng, Q.; Yang, H.; Liu, Y.; Nam, T.H. Transformation intervals and elastic strain energies of B2-B19′ martensitic transformation of NiTi. *Intermetallics* **2010**, *18*, 2431–2434. [CrossRef]
4. Otsuka, K.; Ren, X. Physical metallurgy of Ti–Ni-based shape memory alloys. *Prog. Mater. Sci.* **2005**, *50*, 511 678. [CrossRef]
5. Mohd Jani, J.; Leary, M.; Subic, A.; Gibson, M.A. A review of shape memory alloy research, applications and opportunities. *Mater. Des.* **2014**, *56*, 1078–1113. [CrossRef]
6. Etaati, A.; Dehghani, K. A study on hot deformation behavior of Ni–42.5Ti–7.5Cu alloy. *Mater. Chem. Phys.* **2013**, *140*, 208–215. [CrossRef]
7. Nespoli, A.; Villa, E.; Besseghini, C. Characterization of the martensitic transformation in $Ni_{50-x}Ti_{50}Cu_x$ alloys through pure thermal measurements. *J. Alloys Compd.* **2011**, *509*, 644–647. [CrossRef]
8. Goryczka, T.; Ochin, P. Microstructure, texture and shape memory effect in $Ni_{25}Ti_{50}Cu_5$ ribbons and strips. *Mater. Sci. Eng. A* **2006**, *438–440*, 714–718. [CrossRef]
9. He, X.M.; Rong, L.J.; Yan, D.S.; Li, Y.Y. TiNiNb wide hysteresis shape memory alloy with low niobium content. *Mater. Sci. Eng. A* **2004**, *371*, 193–197. [CrossRef]
10. Zhao, X.; Yan, X.; Yang, Y.; Xu, H. Wide hysteresis NiTi (Nb) shape memory alloys with low Nb content (4.5 at %). *Mater. Sci. Eng. A* **2006**, *438*, 575–578. [CrossRef]
11. Shu, X.Y.; Lu, S.Q.; Li, G.F.; Liu, J.W.; Peng, P. Nb solution influencing on phase transformation temperature of Ni47Ti44Nb9 alloy. *J. Alloys Compd.* **2014**, *609*, 156–161. [CrossRef]
12. Xue, G.; Wang, W.; Wu, D.; Zhai, Q.; Zheng, H. On the explanation for the time-dependence of B2 to R martensitic transformation in $Ti_{50}Ni_{47}Fe_3$ shape memory alloy. *Mater. Lett.* **2012**, *72*, 119–121. [CrossRef]
13. Basu, R.; Eskandari, M.; Upadhayay, L.; Mohtadi-Bonab, M.A.; Szpunar, J.A. A systematic investigation on the role of microstructure on phase transformation behavior in Ni–Ti–Fe shape memory alloys. *J. Alloys Compd.* **2015**, *645*, 213–222. [CrossRef]
14. Matsuda, M.; Yamashita, R.; Tsurekawa, S.; Takashima, K.; Mitsuhara, M.; Nishida, M. Antiphase boundary-like structure of B19′ martensite via R-phase transformation in Ti–Ni–Fe alloy. *J. Alloys Compd.* **2014**, *586*, 87–93. [CrossRef]
15. Kovarik, L.; Yang, F.; Garg, A.; Diercks, D.; Kaufman, M.; Noebe, R.D.; Mills, M.J. Structural analysis of a new precipitate phase in high-temperature TiNiPt shape memory alloys. *Acta Mater.* **2010**, *58*, 4660–4673. [CrossRef]
16. Atli, K.C.; Karaman, I.; Noebe, R.D.; Maier, H.J. Comparative analysis of the effects of severe plastic deformation and thermomechanical training on the functional stability of $Ti_{50.5}Ni_{24.5}Pd_{25}$ high-temperature shape memory alloy. *Scr. Mater.* **2011**, *64*, 315–318. [CrossRef]
17. Santamarta, R.; Arróyave, R.; Pons, J.; Evirgen, A.; Karaman, I.; Karaca, H.E.; Noebe, R.D. TEM study of structural and microstructural characteristics of a precipitate phase in Ni-rich Ni–Ti–Hf and Ni–Ti–Zr shape memory alloys. *Acta Mater.* **2013**, *61*, 6191–6206. [CrossRef]
18. Kockar, B.; Karaman, I.; Kim, J.I.; Chumlyakov, Y. A method to enhance cyclic reversibility of NiTiHf high temperature shape memory alloys. *Scr. Mater.* **2006**, *54*, 2203–2208. [CrossRef]
19. Rehman, S.U.; Khan, M.; Nusair Khan, A.; Ali, L.; Zaman, S.; Waseem, M.; Ali, L.; Jaffery, S.H.I. Transformation behavior and shape memory properties of $Ti_{50}Ni_{15}Pd_{25}Cu_{10}$ high temperature shape memory alloy at various aging temperatures. *Mater. Sci. Eng. A* **2014**, *619*, 171–179. [CrossRef]
20. Karaca, H.E.; Acar, E.; Ded, G.S.; Saghaian, S.M.; Basaran, B.; Tobe, H.; Kok, M.; Maier, H.J.; Noebe, R.D.; Chumlyakov, Y.I. Microstructure and transformation related behaviors of a $Ni_{45.3}Ti_{29.7}Hf_{20}Cu_5$ high temperature shape memory alloy. *Mater. Sci. Eng. A* **2015**, *627*, 82–94. [CrossRef]
21. Hong, S.H.; Kim, J.T.; Park, H.J.; Kim, Y.S.; Suh, J.Y.; Na, Y.S.; Lim, K.R.; Shim, C.H.; Park, J.M.; Kim, K.B. Influence of Zr content on phase formation, transition and mechanical behavior of Ni–Ti–Hf–Zr high temperature shape memory alloys. *J. Alloys Compd.* **2017**, *692*, 77–85. [CrossRef]

22. Prasad, R.V.S.; Park, C.H.; Kim, S.W.; Hong, J.K.; Yeom, J.T. Microstructure and phase transformation behavior of a new high temperature NiTiHf-Ta shape memory alloy with excellent formability. *J. Alloys Compd.* **2017**, *697*, 55–61. [CrossRef]
23. Acar, E.; Karaca, H.E.; Tobe, H.; Noebe, R.D.; Chumlyakov, Y.I. Characterization of the shape memory properties of a $Ni_{45.3}Ti_{39.7}Hf_{10}Pd_5$ alloy. *J. Alloys Compd.* **2013**, *578*, 297–302. [CrossRef]
24. Karaca, H.E.; Acar, E.; Ded, G.S.; Basaran, B.; Tobe, H.; Noebe, R.D.; Bigelow, G.; Chumlyakov, Y.I. Shape memory behavior of high strength NiTiHfPd polycrystalline alloys. *Acta Mater.* **2013**, *61*, 5036–5049. [CrossRef]
25. Meng, X.L.; Li, H.; Cai, W.; Hao, S.J.; Cui, L.S. Thermal cycling stability mechanism of $Ti_{50.5}Ni_{33.5}Cu_{11.5}Pd_{4.5}$ shape memory alloy with near-zero hysteresis. *Scr. Mater.* **2015**, *103*, 30–33. [CrossRef]
26. Schmidt, M.; Ullrich, J.; Wieczorek, A.; Frenzel, J.; Schütze, A.; Eggeler, G.; Seelecke, S. Thermal stabilization of NiTiCuV shape memory alloys: Observations during elastocaloric training. *Shape Mem. Superelast.* **2015**, *1*, 132–141. [CrossRef]

metals

MDPI

Article

Processing Map of NiTiNb Shape Memory Alloy Subjected to Plastic Deformation at High Temperatures

Yu Wang [1,2], Shuyong Jiang [1,*] and Yanqiu Zhang [1]

1   College of Mechanical and Electrical Engineering, Harbin Engineering University, Harbin 150001, China; wangyuhrbeu@126.com (Y.W.); zhangyq@hrbeu.edu.cn (Y.Z.)
2   College of Materials Science and Chemical Engineering, Harbin Engineering University, Harbin 150001, China
*   Correspondence: jiangshuyong@hrbeu.edu.cn; Tel.: +86-451-8251-9710

Received: 29 July 2017; Accepted: 23 August 2017; Published: 25 August 2017

Abstract: The processing map of $Ni_{47}Ti_{44}Nb_9$ (at %) shape memory alloy (SMA), which possesses B2 austenite phases and β-Nb phases at room temperature, is established in order to optimize the hot working parameters. Based on true stress-strain curves of NiTiNb SMA during uniaxial compression deformation at the temperatures ranging from 700 to 1000 °C and at the strain rates ranging from 0.0005 to 0.5 s^{-1}, according to dynamic material model (DMM) principle, the processing map of NiTiNb SMA is obtained on the basis of power dissipation map and instability map. The instability region of NiTiNb SMA increases with increasing the true strain and it mainly focuses on the region with high strain rate. The workability of NiTiNb SMA becomes worse and worse with increasing plastic strain, as well as decreasing deformation temperature. There exist two stability zones which are suitable for hot working of NiTiNb SMA. In one stability region, the deformation temperature ranges from 750 to 840 °C and the strain rate ranges from 0.0003 to 0.001 s^{-1}. In the other stability region, the deformation temperature ranges from 930 to 1000 °C and the strain rate ranges from 0.016 to 0.1 s^{-1}. The severe microstructure defects, such as coarsening grains, band microstructure, and intercrystalline overfiring appear in the microstructures of NiTiNb SMA which is subjected to plastic deformation in the instability zone.

Keywords: shape memory alloy; NiTiNb alloy; plastic deformation; processing map

1. Introduction

Binary NiTi shape memory alloy (SMA) has been extensively used in the engineering field due to its unique phenomena, which include a shape memory effect and superelasticity [1–3]. It is of great importance to add the third element to the binary NiTi SMA so as to broaden the engineering application [4–6]. As a typical example, the addition of Nb element to the binary NiTi SMA contributes to enhancing phase transformation temperature hysteresis [7,8]. In particular, when the soft β-Nb phase in the NiTiNb SMA suffers from plastic deformation, the relaxation of elastic strain in the martensite interface contributes to lowering the driving force of reverse martensite transformation and, hence, facilitating the stability of martensite [9]. Therefore, NiTiNb SMA has been a perfect candidate for pipe coupling because the large phase transformation temperature hysteresis plays a predominant role in guaranteeing the reliability of pipe coupling in engineering applications [10–12].

It is well known that hot working, especially high-temperature plastic deformation, is an indispensable means to manufacture the product of NiTi-based SMAs [13–15]. Furthermore, it is of great importance in improving the microstructures and the properties of NiTi-based SMAs, as well [16–18]. Therefore, it is very necessary to explore an effective tool for optimizing the process parameters, which are suitable for hot working of NiTi-based SMAs. As we know, the processing map has been a reliable and effective tool to help optimize the hot working parameters of metal materials [19–23]. So far, no literature has reported the involved information with respect to processing maps of NiTiNb SMA. Therefore, in the present study, uniaxial compression deformation of $Ni_{47}Ti_{44}Nb_9$ (at %) SMA is carried out at the temperatures ranging from 600 to 1000 °C and at strain rates ranging from 0.0005 to 0.5 s^{-1}. The processing map of NiTiNb SMA is established according to dynamic material model (DMM) theory [24].

2. Materials and Methods

The commercially as-rolled $Ni_{47}Ti_{44}Nb_9$ (at %) SMA bar, which possesses the diameter of 20 mm, was obtained from Xi'an Saite Metal Materials Development Co., Ltd. (Xi'an, China). The phase composition of as-rolled NiTiNb SMA was characterized by X-ray diffraction (XRD) testing using a Philips X'Pert Pro diffractometer (Royal Dutch Philips Electronics Ltd., Amsterdam, The Netherlands) with CuKα radiation at ambient temperature. The involved sample was scanned on the basis of 2θ ranging from 20° to 90° by means of continuous scanning based on a tube voltage of 40 kV and tube current of 40 mA. Figure 1 shows the XRD diagram of the as-rolled NiTiNb SMA, where NiTiNb SMA consists of B2 austenite and β-Nb phases.

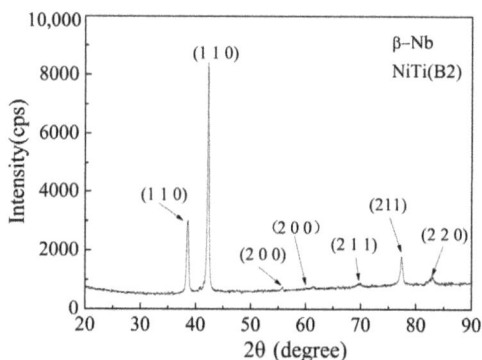

Figure 1. XRD map of as-rolled NiTiNb SMA.

Sixteen NiTiNb SMA samples, which possess diameters of 6 mm and heights of 9 mm, were removed from the as-rolled NiTiNb SMA bar using electro-discharge machining (EDM, DK7725, Jiangsu Dongqing CNC Machine Tool Co., Ltd., Taizhou, China). The NiTiNb SMA samples were placed between the top anvil and the bottom one of an INSTRON-5500R equipment (Instron Corporation, Norwood, MA, USA). Subsequently, they were compressed by the deformation degree of 60% at temperatures ranging from 700 to 1000 °C and at strain rates ranging from 0.0005 to 0.5 s^{-1}.

Optical microscopy (OM) observation was used to investigate the microstructures of as-rolled and compressed NiTiNb SMA samples by means of an OLYMPUS 311 (Olympus Corporation, Tokyo, Japan) optical microscope. The sample for OM observation was etched in a solution of HF:HNO$_3$:H$_2$O = 1:3:10. The microstructure of as-rolled NiTiNb SMA is shown in Figure 2. It can be observed that the as-rolled NiTiNb SMA exhibits a homogeneous worm-like microstructure.

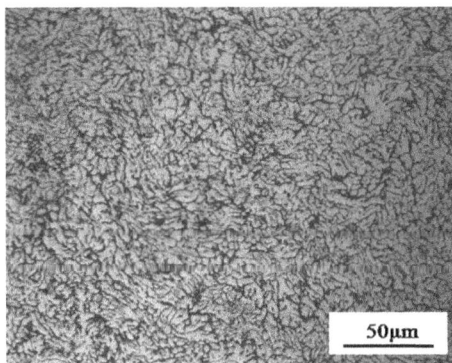

Figure 2. Microstructures of as-rolled NiTiNb SMA.

3. Principle for the Processing Map

A processing map of NiTiNb SMA is established on the basis of dynamic material model (DMM). According to DMM, when NiTiNb SMA is subjected to plastic deformation at high temperatures, the dissipation power P is composed of two parts. One part deals with the power (G) consumed due to plastic deformation and the other part refers to the energy (J) dissipated due to microstructural evolution. Therefore, the dissipation power P is expressed by [25]:

$$P = \sigma \cdot \dot{\varepsilon} = G + J = \int_0^{\dot{\varepsilon}} \sigma d\dot{\varepsilon} + \int_0^{\sigma} \dot{\varepsilon} d\sigma \qquad (1)$$

where G is defined as the dissipated content and J refers to the dissipated co-content.

When strain ε and temperature T are unchangeable, stress σ is regarded as a function of the strain rate $\dot{\varepsilon}$, which is described as a power law relationship [26–28], namely:

$$\sigma = K\dot{\varepsilon}^m \qquad (2)$$

where K refers to material coefficient and m stands for strain rate sensitivity. The value of m is expressed as:

$$m = \frac{dJ}{dG} = \frac{\dot{\varepsilon} d\sigma}{\sigma d\dot{\varepsilon}} = \frac{\dot{\varepsilon} \sigma d \ln \sigma}{\sigma \dot{\varepsilon} d \ln \dot{\varepsilon}} \approx \frac{\Delta \lg \sigma}{\Delta \lg \dot{\varepsilon}} \qquad (3)$$

When strain ε and temperature T are constant, the dissipated co-content J is represented by:

$$J = \int_0^{\sigma} \dot{\varepsilon} d\sigma = \frac{m\sigma \dot{\varepsilon}}{m+1} \qquad (4)$$

In general, the m value shows a linear dependence on the temperature T and the strain rate $\dot{\varepsilon}$. The metal material is considered to be an ideal linear dissipation state if the value of m is taken as 1. Then, the dissipated co-content J reaches the maximum value J_{max} [29], namely:

$$J_{max} = \frac{\sigma \dot{\varepsilon}}{2} \qquad (5)$$

Consequently, according to Equations (4) and (5), the power dissipation efficiency η is expressed as follows:

$$\eta = \frac{J}{J_{max}} = \frac{2m}{m+1} \qquad (6)$$

where η depends on temperature T, strain ε and strain rate $\dot{\varepsilon}$. In the case of a constant strain, the power dissipation map is established by drawing a contour map of η versus the strain rate $\dot{\varepsilon}$ and temperature T.

The power dissipation map is indicative of the microstructural evolution law resulting from the dissipated energy of the material. In general, the power dissipation map is of great importance in terms of determining the workability of metal material. However, the workability of metal material is not completely dependent on the power dissipation map since there is a larger η value in a region where the workability of metal material is very poor. Therefore, it is necessary to use a judging criterion for evaluating the workability of metal material. According to the maximum entropy principle, the unstable flow occurs during plastic deformation of metal material when the following equation is satisfied [30], namely:

$$\frac{dD}{d\dot{\varepsilon}} < \frac{D}{\dot{\varepsilon}} \tag{7}$$

where D is the power dissipation function, which depends on the specific temperature. If the total power is dissipated, D is identical to P. Based on DMM, if the partition in Equation (1) leads to the different instability parameters, D is identical to J. As a consequence, Equation (7) is expressed as:

$$\frac{dJ}{d\dot{\varepsilon}} < \frac{J}{\dot{\varepsilon}} \tag{8}$$

According to the mathematical transformation, Equation (8) is expressed by:

$$\frac{dJ}{d\dot{\varepsilon}} = \frac{J\dot{\varepsilon}}{\dot{\varepsilon}J} \cdot \frac{dJ}{d\dot{\varepsilon}} = \frac{J}{\dot{\varepsilon}} \cdot \frac{d(\int (1/J)dJ)}{d(\int (1/\dot{\varepsilon})d\dot{\varepsilon})} = \frac{J}{\dot{\varepsilon}} \cdot \frac{d\lg J}{d\lg\dot{\varepsilon}} \tag{9}$$

By combining Equation (9) with Equation (8), the following equation is acquired, namely:

$$\frac{d\lg J}{d\lg\dot{\varepsilon}} < 1 \tag{10}$$

Then, the substitution of Equation (4) into Equation (10) results in:

$$\frac{d\lg J}{d\lg\dot{\varepsilon}} = \frac{d\lg(m/m+1)}{d\lg\dot{\varepsilon}} + \frac{d\lg\sigma}{d\lg\dot{\varepsilon}} + \frac{d\lg\dot{\varepsilon}}{d\lg\dot{\varepsilon}} < 1 \tag{11}$$

As a result, the criterion judging the unstable flow of metal material during plastic deformation is expressed as follows [31]:

$$\xi(\dot{\varepsilon}) = \frac{\partial\lg\left(\frac{m}{m+1}\right)}{\partial\lg\dot{\varepsilon}} + m < 0 \tag{12}$$

where the instability parameter $\xi(\dot{\varepsilon})$ depends on m and $\dot{\varepsilon}$. In addition, m relies on T and $\dot{\varepsilon}$. Accordingly, $\xi(\dot{\varepsilon})$ is dependent upon T and $\dot{\varepsilon}$. As for a given strain, the instability map is constructed by plotting a contour map of $\xi(\dot{\varepsilon})$ versus $\dot{\varepsilon}$ and T. In general, $\xi(\dot{\varepsilon})$ is negative in the zone where metal material presents an unstable flow during plastic deformation. Therefore, the instability region is identified by means of the instability map. Finally, the processing map is established on the basis of the instability map and the power dissipation map.

4. Results and Discussion

Figure 3 indicates the true stress-strain curves of NiTiNb SMA undergoing uniaxial compression at the temperatures ranging from 700–1000 °C and at the strain rates ranging from 0.0005–0.5 s^{-1}. It is evident that flow stress is dependent upon the strain rate and temperature. In the case of a constant strain rate, the flow stress decreases with increasing temperature. As for a constant temperature, the flow stress increases with increasing strain rate. According to the various temperatures and strain rates, the values of flow stresses, which correspond to the true strains of 0.3, 0.6, and 0.9, respectively, are extracted from the true stress-strain data, as shown in Table 1.

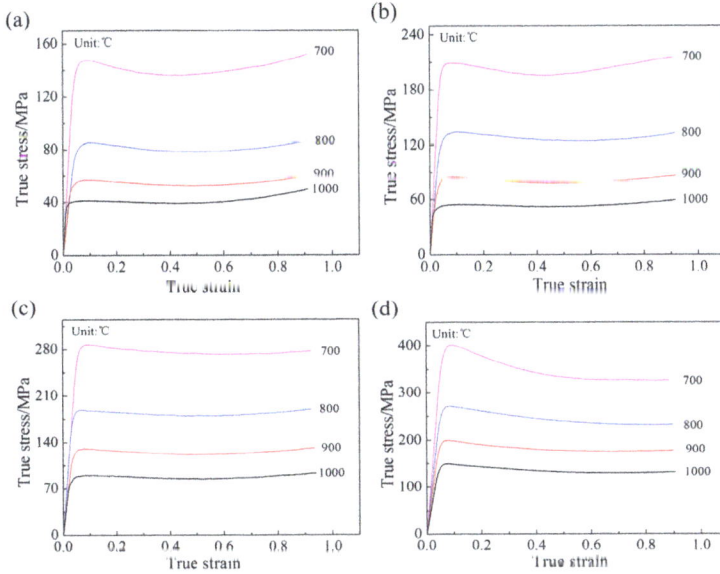

Figure 3. True stress-strain curves of NiTiNb SMA undergoing uniaxial compression based on the various temperatures and strain rates: (**a**) $\dot{\varepsilon} = 0.0005$ s^{-1}; (**b**) $\dot{\varepsilon} = 0.005$ s^{-1}; (**c**) $\dot{\varepsilon} = 0.05$ s^{-1}; and (**d**) $\dot{\varepsilon} = 0.5$ s^{-1}.

Table 1. Flow stresses of NiTiNb SMA (MPa).

ε	$\dot{\varepsilon}/s^{-1}$	$T/°C$			
		700	800	900	1000
0.3	0.0005	137.8010	80.4978	53.4650	39.7838
	0.005	199.2236	128.6067	79.6107	53.3478
	0.05	278.4276	182.6738	124.5443	87.1276
	0.5	362.9075	254.1071	185.9302	139.1716
0.6	0.0005	138.6029	78.6816	52.6534	40.1727
	0.005	199.7022	124.5523	78.5104	53.1897
	0.05	272.5415	179.6717	121.7723	85.0241
	0.5	328.3940	236.2028	176.2441	130.1123
0.9	0.0005	150.7980	85.8143	59.2176	48.7737
	0.005	214.5237	131.7976	85.6582	58.8096
	0.05	276.0619	187.4464	129.4072	91.1164
	0.5	325.8891	231.8062	177.1036	130.8063

According to the experimental data shown in Table 1, the curve of lgσ versus lg$\dot{\varepsilon}$ can be obtained by means of the linear fitting method, as shown in Figure 4. It is evident that there is an approximate linear relationship between lgσ and lg$\dot{\varepsilon}$. The approximate linear relationship indicates that NiTiNb SMA satisfies the conditions of DMM during plastic deformation.

(a)

(b)

(c)

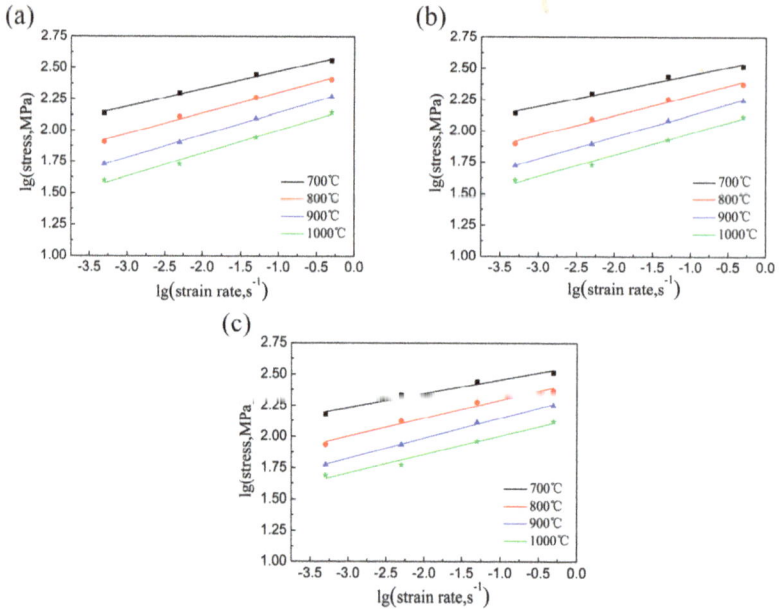

Figure 4. Linear relationship between $\log \sigma$ and $\log \dot\varepsilon$ based on various strains: (**a**) $\varepsilon = 0.3$; (**b**) $\varepsilon = 0.6$; and (**c**) $\varepsilon = 0.9$.

By performing cubic-spline fitting based on the aforementioned data, the fitted curves of $\lg\sigma$ versus $\lg\dot\varepsilon$ are acquired. Then, according to Equation (3), a series of m values are obtained by identifying the slopes of these fitted curves. Consequently, according to Equation (6), the power dissipation efficiency η is calculated at various plastic strains. Furthermore, the power dissipation maps of NiTiNb SMA based on various strains are obtained, as shown in Figure 5. On the one hand, the power dissipation maps are able to reflect relative variation rate of internal entropy in the metal material subjected to hot plastic deformation. On the other hand, the power dissipation maps can be used for roughly estimating the microstructure change of metal material undergoing plastic deformation at the various temperatures and strain rates. In general, the higher η values mean that the deformed microstructures probably possess better performance. It is noted that the value of η approximately increases with increasing deformation temperature, whereas it decreases with increasing strain rate. It can be found that, in the whole temperature range, there exist two regions where η possesses a peak value. One region is involved in the temperature range of 750–840 °C, as well as the strain rate range of 0.0003~0.001 s^{-1}. The other region deals with the temperature range of 930–1000 °C as well as the strain rate range of 0.016–0.1 s^{-1}. In addition, the maximum value of η decreases with increasing true strain. The phenomenon indicates that the hot workability of NiTiNb SMA becomes worse and worse along with the increase in plastic strain.

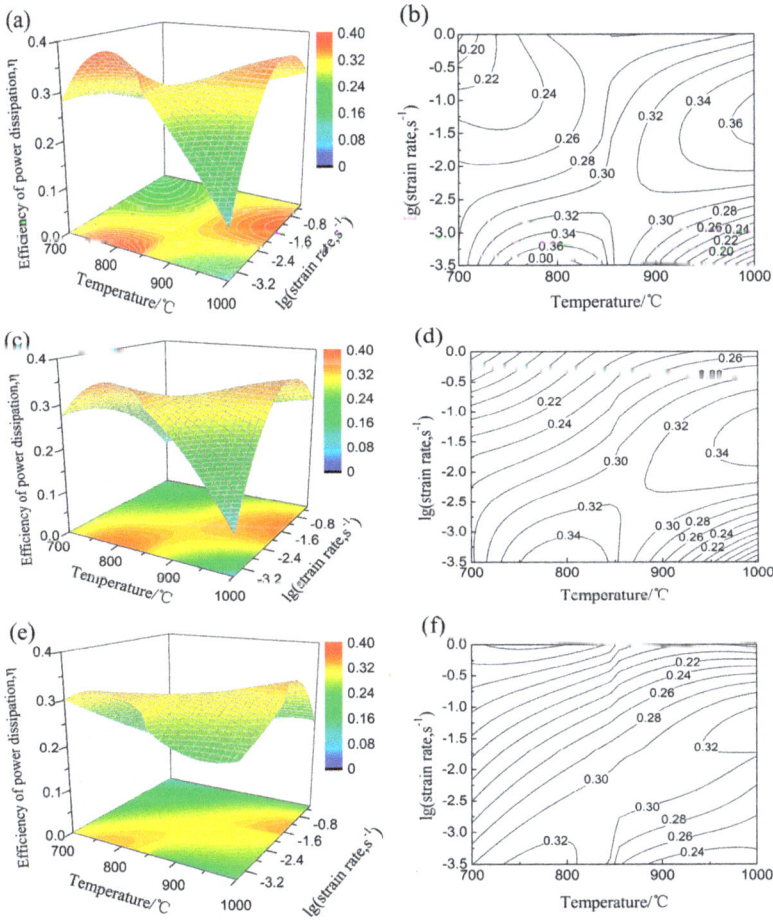

Figure 5. Power dissipation maps of NiTiNb SMA based on various strains: (**a**) $\varepsilon = 0.3$, 3D surface map; (**b**) $\varepsilon = 0.3$, 2D contour line map; (**c**) $\varepsilon = 0.6$, 3D surface map; (**d**) $\varepsilon = 0.6$, 2D contour line map; (**e**) $\varepsilon = 0.9$, 3D surface map; and (**f**) $\varepsilon = 0.9$, 2D contour line map.

The values of $\zeta(\dot{\varepsilon})$ under the various deformation conditions can be calculated by combining Equations (3) and (12). As a consequence, the instability maps are established, as shown in Figure 6. In general, the region where $\zeta(\dot{\varepsilon})$ possesses negative values in the instability maps is defined as the instability region where metal material exhibits an unstable flow during plastic deformation. In a similar manner, the region where $\zeta(\dot{\varepsilon})$ possesses positive values in the instability maps is defined as the stability region where metal material shows a stable flow during plastic deformation. It is observed from Figure 6 that the unstable flow mainly appears in the zone possessing high strain rate. Furthermore, the instability region increases with increasing true strain.

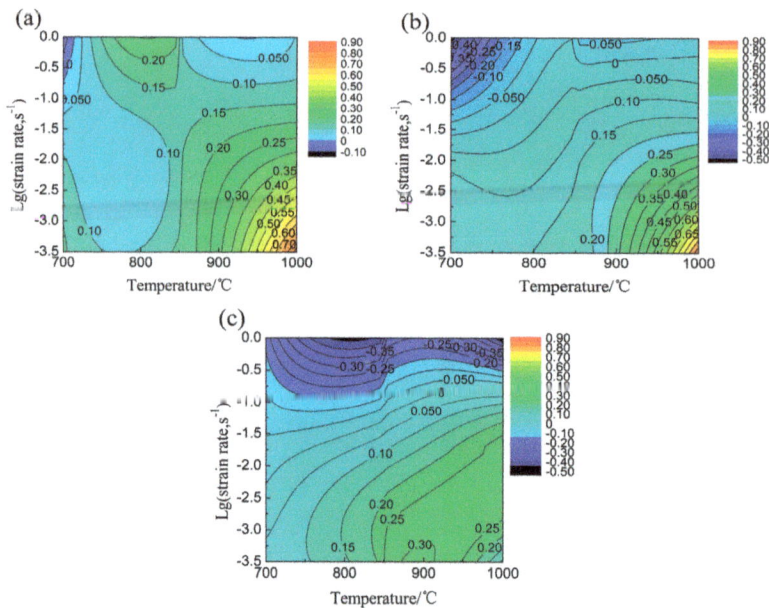

Figure 6. Instability maps of NiTiNb SMA based on various strains: (**a**) $\varepsilon = 0.3$; (**b**) $\varepsilon = 0.6$; and (**c**) $\varepsilon = 0.9$.

The processing map is established on the basis of the power dissipation map and the instability map, as illustrated in Figure 7. In Figure 7, the instability zone is designated in blue, but the stability zone is represented in white. It can be observed that the instability region of NiTiNb SMA increases with increasing true strain. The phenomenon further demonstrates that the workability of NiTiNb SMA becomes worse and worse along with increasing deformation extent. In addition, the instability zone of NiTiNb SMA mainly focuses on the region with high strain rate. In particular, as for the true strain of 0.9, the instability zone of NiTiNb SMA is mainly concentrated on the region with high strain rate. Furthermore, the strain rate range, which represents the stability zone, decreases with decreasing deformation temperature. This indicates that the lower deformation temperature leads to the poorer workability of NiTiNb SMA. However, the stability zone is not completely suitable for hot working of NiTiNb SMA, as well. In general, the high value of η in the stable working zone indicates that the larger fraction of energy is dissipated during microstructural evolution of NiTiNb SMA subjected to plastic deformation at high temperatures, such as dynamic recrystallization, dynamic recovery and phase transformation. Therefore, the higher η value is more suitable for hot working. In addition, it can be found from Figure 7c that the higher η value, which represents the stability zone, is located in two regions. One region means that NiTiNb SMA experiences hot working in the temperature range of 750–840 °C, as well as at the strain rate range of 0.0003–0.001 s^{-1}. The other region indicates that NiTiNb SMA is subjected to hot working in the temperature range of 930–1000 °C, as well as at the strain rate range of 0.016–0.1 s^{-1}. As a consequence, the aforementioned high η value in the stability zone is considered to represent the optimum hot working zone of NiTiNb SMA. In addition, there exist some zones, which possess very low η value in the stability region. The phenomenon indicates that when NiTiNb SMA is subjected to hot working in the regions with very low η values, although the severe working defects should not be formed, the inhomogeneous microstructure defects can be induced. Therefore, it is more appropriate for NiTiNb SMA not to be subjected to hot working in the stability regions with very low η value. In particular, when NiTiNb SMA is subjected to hot working in the instability regions, the severe microstructure defects are induced, as shown in Figure 8. It is obviously observed from Figure 8 that coarsening grains, band microstructure, and intercrystalline overfiring appear in the microstructures of the deformed NiTiNb SMA. These microstructure defects have an adverse impact on the properties of NiTiNb SMA.

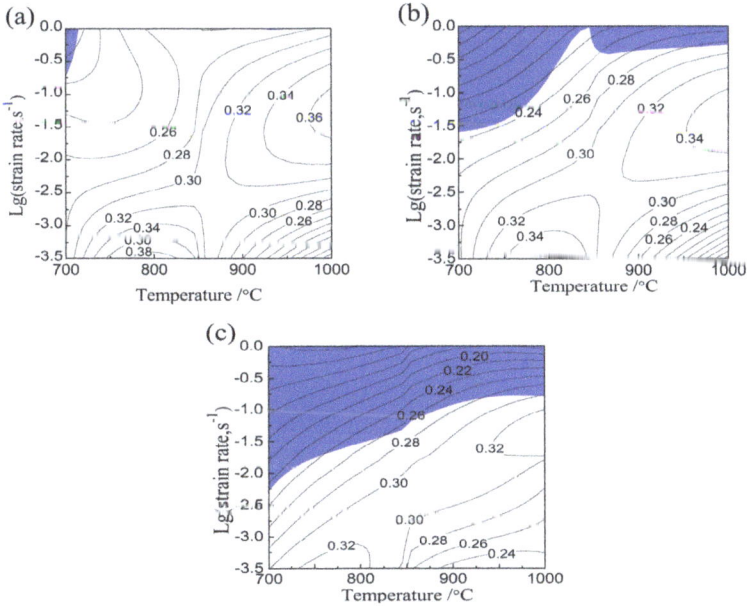

Figure 7. Processing maps of NiTiNb SMA based on various strains: (**a**) $\varepsilon = 0.3$; (**b**) $\varepsilon = 0.6$; and (**c**) $\varepsilon = 0.9$. The blue color in the figure represents the instability region.

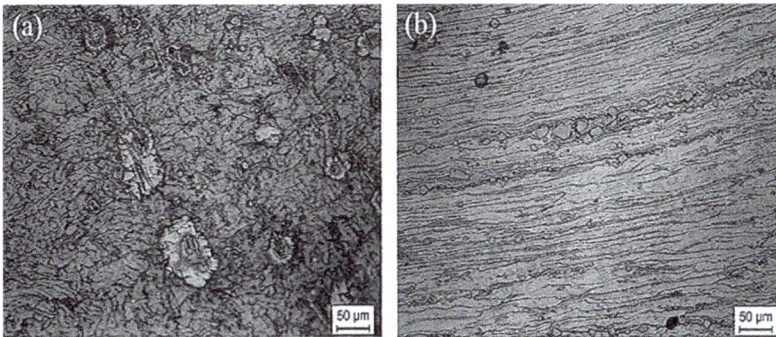

Figure 8. Microstructures of NiTiNb SMA undergoing compression at 1000 °C and at 0.05 s^{-1}: (**a**) coarsening grain; and (**b**) band microstructure and intercrystalline overfiring.

5. Conclusions

Based on true stress-strain curves of NiTiNb SMA during uniaxial compression deformation at temperatures ranging from 700 to 1000 °C, and at strain rates ranging from 0.0005 to 0.5 s^{-1}, according to the values of flow stresses corresponding to true strains of 0.3, 0.6, and 0.9, a processing map of NiTiNb SMA is established based on the dynamic material model (DMM) principle. As a consequence, the following conclusions are drawn:

(1) Flow stress of NiTiNb SMA is dependent upon the strain rate and temperature. In the case of a constant strain rate, flow stress decreases with increasing temperature. In the case of a constant temperature, flow stress increases with the increasing strain rate. The instability region of NiTiNb SMA increases with the increasing true strain and it mainly focuses on the region with high strain rate. The workability of NiTiNb SMA becomes worse and worse with increasing plastic strain, as well as decreasing the deformation temperature.

(2) There exist two stability zones which are suitable for hot working of NiTiNb SMA. One is the region where NiTiNb SMA experiences hot working in the temperature range of 750–840 °C, as well as at the strain rate range of 0.0003–0.001 s^{-1}. The other is the region where NiTiNb SMA is subjected to hot working in the temperature range of 930–1000 °C, as well as at the strain rate range of 0.016–0.1 s^{-1}. The processing map lays the foundation for optimizing the hot working parameters of NiTiNb SMA.

Acknowledgments: The work was financially supported by National Natural Science Foundation of China (nos. 51475101, 51305091 and 51305092).

Author Contributions: Yu Wang wrote the manuscript and performed the XRD analysis, OM observation, and compression test; Shuyong Jiang supervised the manuscript; and Yanqiu Zhang established the processing map.

Conflicts of Interest: The authors declare no conflict of interest.

References

1. Jani, J.M.; Leary, M.; Subic, A.; Gibson, M.A. A review of shape memory alloy research, applications and opportunities. *Mater. Des.* **2014**, *56*, 1078–1113. [CrossRef]
2. Elibol, C.; Wagner, M.F.X. Investigation of the stress-induced martensitic transformation in pseudoelastic NiTi under uniaxial tension, compression and compression-shear. *Mater. Sci. Eng. A* **2015**, *621*, 76–81. [CrossRef]
3. Kuang, C.H.; Chien, C.; Wu, S.K. Multistage martensitic transformation in high temperature aged Ti$_{48}$Ni$_{52}$ shape memory alloy. *Intermetallics* **2015**, *67*, 12–18. [CrossRef]
4. Jones, N.G.; Dye, D. Influence of applied stress on the transformation behaviour and martensite evolution of a Ti-Ni-Cu shape memory alloy. *Intermetallics* **2013**, *32*, 239–249. [CrossRef]
5. Basu, R.; Eskandari, M.; Upadhayay, L.; Mohtadi-Bonab, M.A.; Szpunar, J.A. A systematic investigation on the role of microstructure on phase transformation behavior in Ni-Ti-Fe shape memory alloys. *J. Alloys Compd.* **2015**, *645*, 213–222. [CrossRef]
6. Mohammad Sharifi, E.; Kermanpur, A.; Karimzadeh, F. The effect of thermomechanical processing on the microstructure and mechanical properties of the nanocrystalline TiNiCo shape memory alloy. *Mater. Sci. Eng. A* **2014**, *598*, 183–189. [CrossRef]
7. Jiang, P.C.; Zheng, Y.F.; Tong, Y.X.; Chen, F.; Tian, B.; Li, L.; Gunderov, D.V.; Valiev, R.Z. Transformation hysteresis and shape memory effect of an ultrafine-grained TiNiNb shape memory alloy. *Intermetallics* **2014**, *54*, 133–135. [CrossRef]
8. Choi, E.; Hong, H.K.; Kim, H.S.; Chung, Y.S. Hysteretic behavior of NiTi and NiTiNb SMA wires under recovery or pre-stressing stress. *J. Alloys Compd.* **2013**, *577*, 444–447. [CrossRef]
9. Zhao, L.C.; Duerig, T.W.; Justi, S. The study of niobium-rich precipitates in a Ni-Ti-Nb shape memory alloy. *Scr. Metall. Mater.* **1990**, *24*, 221–226. [CrossRef]
10. Dong, Z.Z.; Zhou, S.L.; Liu, W.X. A study of NiTiNb shape-memory alloy pipe-joint with improved properties. *Mater. Sci. Forum* **2002**, *394–395*, 107–110. [CrossRef]
11. Uchida, K.; Shigenaka, N.; Sakuma, T.; Sutou, Y.; Yamauchi, K. Effect of Nb content on martensitic transformation temperatures and mechanical properties of Ti-Ni-Nb shape memory alloys for pipe joint applications. *Mater. Trans.* **2007**, *48*, 445–450. [CrossRef]
12. Korostelev, A.B. Properties of a Ti-Ni-Nb Alloy for producing thermomechanical couplings. *Russ. Metall.* **2011**, *2011*, 576–578. [CrossRef]
13. Etaati, A.; Dehghani, K. A study on hot deformation behavior of Ni-42.5Ti-7.5Cu alloy. *Mater. Chem. Phys.* **2013**, *140*, 208–215. [CrossRef]

14. Mirzadeh, H.; Parsa, M.H. Hot deformation and dynamic recrystallization of NiTi intermetallic compound. *J. Alloys Compd.* **2014**, *614*, 56–59. [CrossRef]
15. Yeom, J.T.; Kim, J.H.; Hong, J.K.; Kim, S.W.; Park, C.H.; Nam, T.H.; Lee, K.Y. Hot forging design of as-cast NiTi shape memory alloy. *Mater. Res. Bull.* **2014**, *58*, 234–238. [CrossRef]
16. Morakabati, M.; Kheirandish, S.; Aboutalebi, M.; Taheri, A.K.; Abbasi, S.M. The effect of Cu addition on the hot deformation behavior of NiTi shape memory alloys. *J. Alloys Compd.* **2010**, *499*, 57–62. [CrossRef]
17. Morakabati, M.; Aboutalebi, M.; Kheirandish, S.; Taheri, A.K.; Abbasi, S.M. Hot tensile properties and microstructural evolution of as cast NiTi and NiTiCu shape memory alloys. *Mater. Des.* **2011**, *32*, 406–413. [CrossRef]
18. Shamsolhodaei, A.; Zarei-hanzaki, A.; Ghambari, M.; Moemeni, S. The high temperature flow behavior modeling of NiTi shape memory alloy employing phenomenological and physical based constitutive models: A comparative study. *Intermetallics* **2014**, *53*, 140–149. [CrossRef]
19. Zhang, Y.Q.; Jiang, S.Y.; Zhao, Y.N.; Liu, S.W. Constitutive equation and processing map of equiatomic NiTi shape memory alloy under hot plastic deformation. *Trans. Nonferr. Met. Soc.* **2016**, *26*, 2152–2161. [CrossRef]
20. Zhang, Y.; Jiang, S.; Chen, C.; Hu, L.; Zhu, X. Hot workability of a NiTiCu shape memory alloy with acicular martensite phase based on processing maps. *Intermetallics* **2017**, *86*, 94–103. [CrossRef]
21. Gangolu, S.; Gourav Rao, A.; Sabirov, I.; Kashyap, B.P.; Prabhu, N.; Deshmukh, V.P. Development of constitutive relationship and processing map for Al-6.65Si-0.44Mg alloy and its composite with B$_4$C particulates. *Mater. Sci. Eng. A* **2016**, *655*, 256–264. [CrossRef]
22. Rastegari, H.; Kermanpur, A.; Najafizadeh, A.; Somani, M.C.; Porter, D.A.; Ghassemali, E.; Jarfors, A.E.W. Determination of processing maps for the warm working of vanadium microalloyed eutectoid steels. *Mater. Sci. Eng. A* **2016**, *658*, 167–175. [CrossRef]
23. Rajput, S.K.; Chaudhari, G.P.; Nath, S.K. Characterization of hot deformation behavior of a low carbon steel using processing maps, constitutive equations and Zener-Hollomon parameter. *J. Mater. Process. Technol.* **2016**, *237*, 113–125. [CrossRef]
24. Momeni, A.; Dehghani, K. Hot working behavior of 2205 austenite-ferrite duplex stainless steel characterized by constitutive equations and processing maps. *Mater. Sci. Eng. A* **2011**, *528*, 1448–1454. [CrossRef]
25. Prasad, Y.V.R.K.; Sasidhara, S. *Hot Working Guide: A Compendium of Processing Maps*; ASM International: Materials Park, OH, USA, 1997.
26. Wu, H.; Wu, C.; Yang, J.; Lin, M. Hot workability analysis of AZ61 Mg alloys with processing maps. *Mater. Sci. Eng. A* **2014**, *607*, 261–268. [CrossRef]
27. Shang, X.; Zhou, J.; Wang, X.; Luo, Y. Optimizing and identifying the process parameters of AZ31 magnesium alloy in hot compression on the base of processing maps. *J. Alloys Compd.* **2015**, *629*, 155–161. [CrossRef]
28. Wu, H.; Wen, S.P.; Huang, H.; Gao, K.Y.; Wu, X.L.; Wang, W.; Nie, Z.R. Hot deformation behavior and processing map of a new type Al-Zn-Mg-Er-Zr alloy. *J. Alloys Compd.* **2016**, *685*, 869–880. [CrossRef]
29. Zeng, W.D.; Zhou, Y.G.; Zhou, J.; Yu, H.Q.; Zhang, X.M.; Xu, B. Recent development of processing map theory. *Rare Met. Mater. Eng.* **2006**, *35*, 673–677.
30. Ziegler, H.; Sneedon, I.N.; Hill, R. *Progress in Solid Mechanics*; Wiley: New York, NY, USA, 1963.
31. Łukaszek-Solek, A.; Krawczyk, J. The analysis of the hot deformation behaviour of the Ti-3Al-8V-6Cr-4Zr-4Mo alloy, using processing maps, a map of microstructure and of hardness. *Mater. Des.* **2015**, *65*, 165–173. [CrossRef]

![metals logo] *metals*

MDPI

Article

A Combined Experimental-Numerical Approach for Investigating Texture Evolution of NiTi Shape Memory Alloy under Uniaxial Compression

Li Hu, Shuyong Jiang * and Yanqiu Zhang

College of Mechanical and Electrical Engineering, Harbin Engineering University, Harbin 150001, China; heu_huli@126.com (L.H.); zhangyz@hrbeu.edu.cn (Y.Z.)
* Correspondence: jiangshuyong@hrbeu.edu.cn; Tel.: +86-451-8251-9710

Received: 26 July 2017; Accepted: 7 September 2017; Published: 9 September 2017

Abstract: Texture evolution of NiTi shape memory alloy was investigated during uniaxial compression deformation at 673 K (400 °C) by combining crystal plasticity finite element method with electron back-scattered diffraction experiment and transmission electron microscope experiment. Transmission electron microscope observation indicates that dislocation slip rather than deformation twinning plays a dominant role in plastic deformation of B2 austenite NiTi shape memory alloy at 673 K (400 °C). Electron back-scattered diffraction experiment illustrates heterogeneous microstructure evolution resulting from dislocation slip in NiTi shape memory alloy at 673 K (400 °C). {110}<100>, {010}<100> and {110}<111> slip systems are introduced into a crystal plasticity constitutive model. Based on the constructed representative volume element model and the extracted crystallographic orientations, particle swarm optimization algorithm is used to identify crystal plasticity parameters from experimental results of NiTi shape memory alloy. Using the fitted material parameters, a crystal plasticity finite element method is used to predict texture evolution of NiTi shape memory alloy during uniaxial compression deformation. The simulation results agree well with the experimental ones. With the progression of plastic deformation, a crystallographic plane of NiTi shape memory alloy gradually rotates to be vertical to the loading direction, which lays the foundation for forming the <111> fiber texture.

Keywords: shape memory alloy; NiTi alloy; plastic deformation; crystal plasticity; finite element method

1. Introduction

NiTi shape memory alloy (SMA) has attracted increasing attention in the field of materials science and engineering because of its excellent shape memory effect, outstanding superelasticity and perfect biological compatibility [1–3]. Plastic deformation plays a significant role in manufacturing the products of NiTi SMA. Over the past decades, a number of investigations have attempted to identify, quantify or at least clarify the nature and significance of various mechanisms that contribute to understanding plastic deformation of NiTi SMA under different deformation conditions. In general, a plastic deformation mechanism of NiTi SMA is temperature-dependent and thus exhibits a predominant distinction in the case of various temperatures, where multiple plastic deformation mechanisms can occur, including stress-induced martensite phase transformation, dislocation slip, deformation twinning, grain boundary slide, grain rotation, dislocation climb and grain boundary migration [4,5]. In particular, above the austenite finish temperature (A_f), stress-induced martensite transformation occurs up to the martensite desist temperature (M_d). Above the M_d temperature, stress-induced martensite transformation can no longer take place and NiTi SMA still exhibits high ductility (exceeding 30%), which is unusual for B2 intermetallics [6]. Therefore, it is of considerable importance to experimentally investigate plastic deformation of NiTi SMA above the M_d temperature. Moreover, it can be generally accepted that dislocation slip plays a dominant role in plastic deformation of NiTi SMA above the M_d temperature. The slip modes {110}<100> and {010}<100> are identified and commonly reported during plastic deformation of NiTi SMA [7–9]. However, these two slip modes provide only three independent slip systems, and thus they are unable to accommodate all the strains for generalized polycrystalline plasticity since at least five independent slip systems are required for dislocations to accommodate arbitrary plastic deformation [10]. Consequently, in addition to <100> slip processes, alternate deformation mechanisms must be activated to contribute to enhancing the plasticity. Benafan et al. [5] have proposed that there are two possibilities to help satisfy the requirements for generalized plasticity. One possibility is the activation of a secondary slip system, especially the {110}<111> slip system, which is also validated by Ezaz et al. [11]. The other possibility is deformation twinning, especially (114) compound twinning, which is also observed by Ezaz et al. [12]. Furthermore, it is very necessary to clarify and understand the influence of plastic deformation mechanisms on the overall mechanical response of NiTi SMA above the M_d temperature, which contributes to giving a deep insight into the microstructural evolution of NiTi SMA during plastic deformation and further broadening the engineering applications of NiTi SMA.

So far, many theoretical analyses and numerical simulations, which focus on coupling plasticity and phase transformation of NiTi SMA, have attracted a great deal of attention. Different macroscopic phenomenological models [13,14] and micromechanical models [8,15] have been constructed to describe corresponding thermo-mechanical properties of NiTi SMA and these models have played an important role in promoting the extensive engineering application of NiTi SMA. By contrast, in terms of simulating plastic deformation of NiTi SMA above the M_d temperature, our literature search has shown that both the microscale and macroscale simulations have not been fully addressed yet. In particular, the microscale simulations, which deal with various plastic deformation mechanisms of NiTi SMA, need to be considered.

The goal of the present study is to develop an experiment-based crystal plasticity finite element model to investigate the texture evolution of NiTi SMA subjected to uniaxial compression deformation at 673 K (400 °C). Particle swarm optimization (PSO) algorithm is used to calibrate crystal plasticity parameters from experimental data. Numerical results of polycrystalline model are then compared with correspondingly experimental results. It is worth noting that the procedure of calibrating crystal plasticity parameters on the basis of the PSO algorithm and the investigation of uniaxial compression deformation of NiTi SMA at 673 K (400 °C) by means of crystal plasticity finite element method (CPFEM) at the grain scale has never been reported in the literature.

2. Materials and Methods

The as-received NiTi SMA bar with a diameter of 12 mm, which possesses a nominal composition of Ni50.9Ti49.1 (at %), was prepared by virtue of vacuum induction melting method and subsequent rolling at 1073 K (800 °C). The NiTi SMA samples with the diameter of 4 mm and the height of 6 mm were all taken from the as-received NiTi SMA bar via electro-discharge machining (EDM, DK7725, Jiangsu Dongqing CNC Machine Tool Co. Ltd, Taizhou, China) The NiTi SMA samples were placed between the top anvil and the bottom one of INSTRON-5500R universal testing machine (Instron Corporation, Norwood, MA, USA) and then were compressed at the various deformation degrees by 20%, 30% and 40%, respectively, at the strain rate of 0.001 s^{-1} and at the temperature of 673 K (400 °C).

To acquire the plastic deformation mechanism of NiTi SMA subjected to uniaxial compression at the temperature of 673 K (400 °C), NiTi SMA sample subjected to compression deformation degree of 40% was characterized by transmission electron microscope (TEM). The foil for TEM observation was firstly ground to 70 μm thickness through mechanical polishing and then was thinned by twin-jet polishing in an electrolyte consisting of 6% $HClO_4$, 34% $C_4H_{10}O$ and 60% CH_3OH by volume fraction at 253 K (−20 °C) at a potential of 30 V. Finally, TEM observations were conducted on a FEI TECNAI $G^2$20 microscope (FEI Corporation, Hillsboro, OR, USA) with point resolution of 0.23 nm and linear resolution of 0.14 nm at an accelerating voltage of 200 kV. Furthermore, to investigate the evolution of microstructure and crystallographic orientation of NiTi SMA samples, electron back-scattered diffraction (EBSD) experiments were conducted on these samples using a Zeiss Supra 55 scanning electron microscope (SEM, Carle Carl Zeiss Company, Oberkochen, German) coupled with OXFORD EBSD (Oxford Instruments, Oxford, UK) instrument. To obtain a suitable surface for EBSD observation, electro-polishing was conducted in an electrolyte consisting of 10% HNO_3 and 90% CH_3OH by volume fraction at 253 K (-20 °C) at a potential of 18 V. Due to the relatively large scanning area, the scan step was chosen to be 2 μm in simultaneous consideration of scanning time and scanning resolution.

3. Results and Discussion

3.1. Investigation on Microstructure Evolution

Figure 1 demonstrates TEM bright field image, the corresponding dark field image and a selected area electron diffraction (SAED) pattern of an NiTi SMA sample subjected to uniaxial compression at the deformation degree of 40% at 673 K (400 °C). It can be observed from Figure 1 that the matrix of NiTi SMA belongs to B2 austenite according to the SAED pattern. In addition, plenty of dislocations appear in the matrix of NiTi SMA and deformation band is formed. This observation provides a direct experimental evidence that in the case of uniaxial compression at 673 K (400 °C), no deformation twins are found, and thus dislocation slip is responsible for plastic deformation mechanism of NiTi SMA.

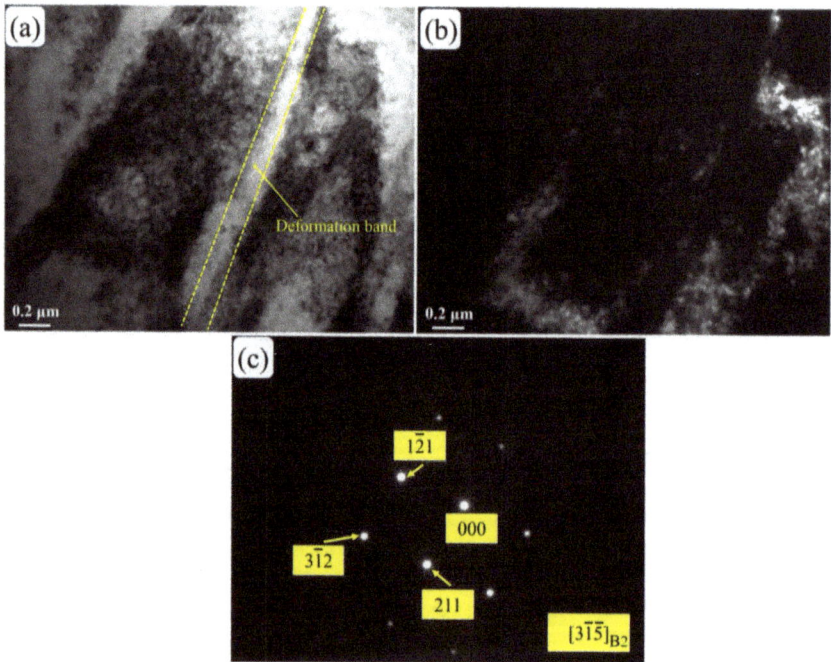

Figure 1. TEM micrographs of NiTi SMA sample subjected to uniaxial compression at the deformation degree of 40%: (**a**) bright field image; (**b**) dark field image; (**c**) SAED (selected area electron diffraction) pattern of (**b**).

Figure 2 demonstrates the initial microstructure of the cross-section of NiTi SMA specimen obtained from EBSD experiment, where RD and ND stand for rolling direction and normal direction, respectively. It can be seen from Figure 2 that the microstructure is characterized by the equiaxed ones.

In addition, the average grain diameter of individual grains is determined to be about 25 μm by means of the statistical analysis of individual grains in the scanning area. Moreover, these equiaxed grains indicate that recrystallization seems to occur during thermo-mechanical processing of the as-received NiTi SMA, which would result in a relatively weak texture compared with most alloys processed at room or moderate temperature [5].

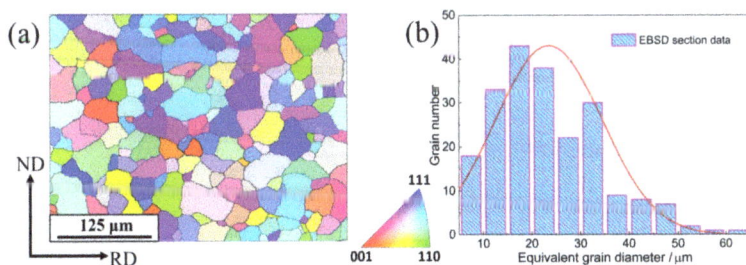

Figure 2. Initial microstructure of as-received NiTi SMA based on a EBSD (electron back-scattered diffraction) experiment: (**a**) EBSD scan area in Rolling-Normal (RD–ND) plane; (**b**) statistical analysis of equivalent grain diameter in the Rolling-Normal (RD–ND) plane.

Figure 3 shows the microstructure evolution of NiTi SMA sample subjected to uniaxial compression at the deformation degree of 20%, 30% and 40%, respectively. It can be noted from Figure 3 that, with the progression of plastic deformation, the morphologies of individual grains indicate the existence of some manner of similarity that the shape of individual grain is elongated in a way along the direction that is vertical to the loading direction. Furthermore, orientations in different parts of the same grain show some discrepancy and this observation is due to the fact that dislocation slip exhibits a certain difference in individual grains owing to the differences in terms of grain orientation and grain morphology within the polycrystalline NiTi SMA, and thus an inhomogeneous plastic response of individual grain emerges in order to accommodate arbitrary plastic deformation. Another interesting result in Figure 3 is that, as the deformation degree becomes larger, the quality of EBSD measurement gets worse, namely, more and more locations cannot be identified by EBSD scanning. A Kikuchi pattern is characteristic of the crystal structure and orientation of the scanning region from which it is generated, and then it is used in EBSD measurement to identify a grain orientation in the test area of the specimen. The indexed Kikuchi patterns in the center of the scanning area at various deformation degrees are illustrated in Figure 3, and they correspond to the regions with relatively low dislocation density. Moreover, regions having high dislocation density could lead to misindexing due to poor band contrast, hence a mistake in identifying the Kikuchi patterns [16]. Therefore, the poor identification of EBSD scanning is enhanced as the dislocation density increases with the progression of plastic deformation in uniaxial compression of NiTi SMA.

To more specifically investigate the microstructure evolution resulting from the dislocation slip during plastic deformation, maps of grain boundaries at various deformation degrees are illustrated in Figure 4, where the blue lines indicate that the misorientation between neighboring grains boundaries are greater than 15°, and the red lines indicate that the misorientation between neighboring grain boundaries is between 5° and 15°. The 5° criterion contributes to weeding out such substructures that can not be considered to be grains. It is worth noting that, in Figure 4a, the red lines showing a small misorientation are far less than the blue lines, indicating that the initial microstructure is mainly composed of grains with high-angle grain boundaries. In addition, few subgrain substructures are observed in Figure 4a. With the progression of plastic deformation, dislocation density is sharply increased in the case of uniaxial compression of NiTi SMA at 673 K (400 °C), as dislocation slip is the only way to sustain plastic strain in each grain. Moreover, the individual grains of a polycrystal NiTi SMA subjected to plastic deformation shall be constrained by the neighboring grains. Therefore, plastic deformation may be quite different in the different regions of the same grain. Consequently, subgrain substructures are finally formed and heterogeneously distributed in the polycrystal NiTi SMA, as shown in Figure 4b–d, where the red lines are grouped into walls.

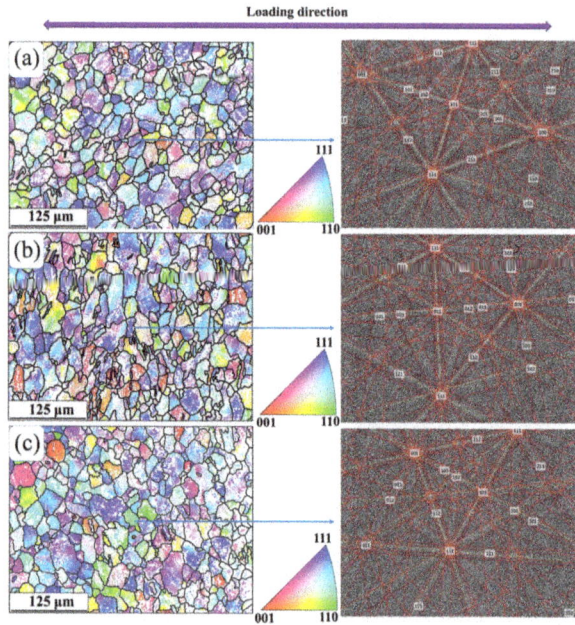

Figure 3. Microstructure evolution and corresponding Kikuchi patterns of NiTi SMA subjected to various deformation degrees: (**a**) 20%; (**b**) 30%; (**c**) 40%.

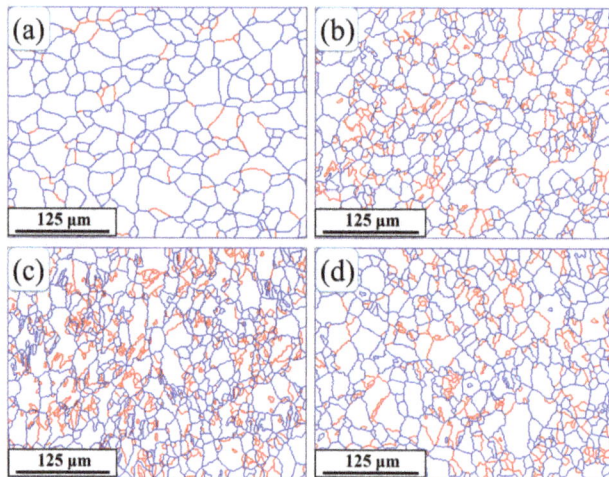

Figure 4. Maps of grain boundaries obtained from EBSD measurement in NiTi SMA samples subjected to various deformation degrees: (**a**) 0%; (**b**) 20%; (**c**) 30%; (**d**) 40%. The blue lines represent grain boundaries whose misorientation is greater than 15° and the red lines stand for grain boundaries whose misorientation ranges from 5° to 15°.

3.2. Investigation on Texture Evolution Based on CPFEM

3.2.1. Crystal Plasticity Constitutive Model

Due to the aforementioned microstructure evolution analysis, it can be noted that in the present study, dislocation slip rather than deformation twinning is found to be responsible for plastic deformation mechanism of NiTi SMA. Due to the fact that at least five independent slip systems are required for dislocations to accommodate arbitrary plastic deformation, it is necessary to introduce a secondary slip system {110}<111> into the proposed crystal plasticity finite element model in addition to {110}<100> and {010}<100> slip systems that are commonly reported [5].

Therefore, in the present study, the framework of crystal plasticity theory is based on dislocation slip. In the framework of rate-dependent single crystal plasticity, the elastic constitutive equation is specified by [17,18]:

$$\overset{\triangledown *}{\sigma} + \sigma(\mathbf{I} : \mathbf{D}^*) = \mathbf{L} : \mathbf{D}^* \tag{1}$$

where $\mathbf{I}$ is the second order identity tensor, $\mathbf{L}$ is the tensor of elastic modulus having the full set of symmetries $L_{ijkl} = L_{jikl} = L_{ijlk} = L_{klij}$, and $\mathbf{D}^*$ is the symmetric stretching rate of the lattice. The Jaumann rate $\overset{\triangledown *}{\sigma}$ is the corotational stress rate on the axes that rotate with the crystal lattice, which is related to the corotational stress rate on the axes that rotate with the material $\overset{\triangledown}{\sigma}$ by the following equation:

$$\overset{\triangledown *}{\sigma} = \overset{\triangledown}{\sigma} + \Omega^p \cdot \sigma - \sigma \ \Omega^p \tag{2}$$

where $\overset{\triangledown}{\sigma} = \dot{\sigma} - \Omega \cdot \sigma + \sigma \cdot \Omega$ and σ stands for the Cauchy stress. In addition, Ω and Ω^p are the total lattice spin tensor and plastic part of the total lattice spin tensor, respectively.

The crystal was assumed to behave as an elasto-viscoplastic solid, so the slipping shear rate $\dot{\gamma}^\alpha$ in individual α slip system is of great importance in crystal plasticity calculation. Based on the Schmid law, the slipping shear rate $\dot{\gamma}^\alpha$ can be determined by a simple rate-dependent power law relation, namely

$$\dot{\gamma}^\alpha = \dot{\gamma}_0 |\tau^\alpha / g^\alpha|^n \ \text{sign}(\tau^\alpha / g^\alpha) \tag{3}$$

where n stands for the rate dependency and if the material is rate-independent, a large value can be chosen up to 50, whereas if the material is highly rate-dependent, a typical value of 10 can be used [19]. In the present study, the value of n is chosen to be 20, indicating a certain rate dependency as reported in [20]. $\dot{\gamma}_0$ is a reference shear strain rate and is determined to be 0.001 s^{-1} in consideration of quasi-static loading rate. τ^α and g^α are resolved shear stress on the slip system α and slip resistance of this system, respectively. Furthermore, the change rate of slip resistance in each slip system is given as follows:

$$\dot{g}^\alpha = \sum_{\beta}^{n} h_{\alpha\beta} \dot{\gamma}^\beta \tag{4}$$

where $h_{\alpha\beta}$ are the slip hardening modulus, and the sum operation is performed over all the activated slip systems. Here, $h_{\alpha\alpha}$ is known as self-hardening modulus and it is derived from the hardening of slip system itself. In addition, $h_{\alpha\beta}$ ($\alpha \neq \beta$) is called latent-hardening modulus, which indicates that the hardening is caused by another slip system.

A simple hardening model is given by [21]:

$$h_{\alpha\alpha} = h(\gamma) = h_0 \sec h^2 \left| \frac{h_0 \gamma}{\tau_s - \tau_0} \right|, \ \gamma = \sum_{\alpha} \int_0^t |\dot{\gamma}^\alpha| dt$$
$$h_{\alpha\beta} = qh(\gamma) \ (\alpha \neq \beta) \tag{5}$$

where h_0 is the initial hardening modulus, τ_0 is the initial yield stress, τ_s is the saturation stress, γ is the total shear strain in all the slip systems, and q is the ratio of latent-hardening to self-hardening and $q = 1.4$ is used in the present study [8].

The aforementioned crystallographic formulations are implemented numerically into an ABAQUS standard solver through a user-defined material subroutine (UMAT), where the implicit (Euler backward) integration algorithm is adopted [18].

3.2.2. Establishment of RVE Model

In the present study, representative volume element (RVE) model based on the voxel model is used since it can effectively represent the mechanical behavior and texture evolution of polycrystalline NiTi SMA using a minimum number of grains, as shown in Figure 5. The voxel model deals with a simple computation model including very coarse three-dimensional meshes, where only one element stands for one grain [22]. Though these voxel RVE models possess rather simplified geometrical representations, they can provide very good results, which are in agreement with experimental texture measurement and mechanical response with high accuracy and computational efficiency [8,22].

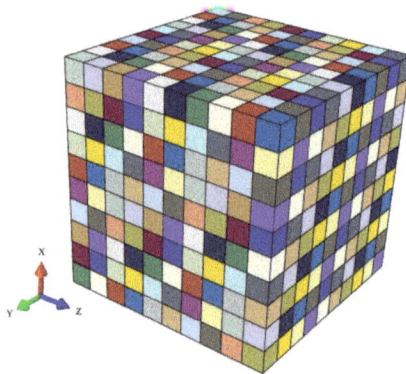

Figure 5. RVE (representative volume element model) of polycrystalline NiTi SMA based on the voxel model.

3.2.3. Parameters Calibration

Based on the three-dimensional voxel model, the parameters of the crystal plasticity finite element model proposed for NiTi SMA are calibrated from the result of the macroscale uniaxial compression experiment. Since Lv et al. [23] has reported that grain number has an influence on the overall stress of the voxel RVE model, so the starting point of the parameter calibration is to determine the optimal grain number in the RVE model, and convergence studies are carried out on five aggregates containing 64, 216, 512, 1000 and 1728 grains, respectively. Twenty independent computations are performed for each aggregate. Furthermore, prior to each computation, orientations of all the grains are regenerated randomly. In addition, the boundary conditions applied in the present study are realized by means of periodic boundary conditions in x, y, and z directions of the RVE model, which would be compressed in the x-direction at the strain rate of $0.001\ \mathrm{s}^{-1}$. Compared to boundary conditions just by causing all surfaces to be plane, though it has been confirmed that there are no large differences in the results of stress-strain curves anyway [24], the latter induces a higher constraint in the crystal plasticity finite element model.

The stress-strain responses of each computation on the basis of various grain numbers and the corresponding overall stress-strain responses are simulated, as shown in Figure 6. To obtain the macroscopic stress-strain response of NiTi SMA, homogenization based on averaging theorem over the voxel-based RVE model is adopted to make the transition from micro- to macro-variables of the RVE model. The averaged stress and strain values at each time step are defined as $\frac{1}{N}\sum\sigma_{11}$ and $\frac{1}{N}\sum\varepsilon_{11}$ (N is the total number of elements multiplied by the number of integration points at per element, σ_{11} is the component of stress tensor at each integration point and ε_{11} is the component of strain tensor at each integration point [19]. This can be realized by programming a post-processing numerical subroutine using computer language PYTHON with the help of ABAQUS interface. It can be seen from Figure 6 that with the increase of grain number in the voxel model, the scatter on the predicted stress-strain responses with random orientations decreases obviously. In addition, according to the overall stress-strain responses, the stress-strain curve based on 1000 grains is viewed as a compromise between the stress-strain curves corresponding to the selected grain numbers. Therefore, the voxel model adopted in the present study is determined to contain 1000 C3D8 elements, which stand for 1000 grains.

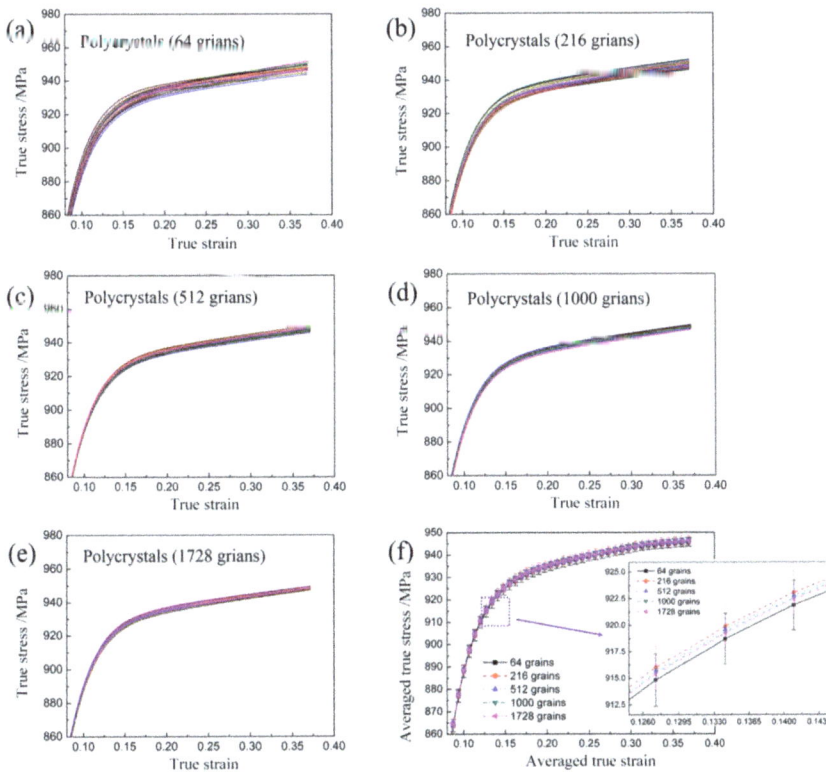

Figure 6. Simulated stress-strain curves from 20 realizations of uniaxial compression deformation of voxel model with various initial orientations: (**a**) 64 grains; (**b**) 216 grains; (**c**) 512 grains; (**d**) 1000 grains; (**e**) 1728 grains; (**f**) the averaged stress-strain curves corresponding to different grain numbers.

In general, plastic deformation is assumed to be highly sensitive to the overall texture of metal materials. Therefore, it is necessary to assign appropriate crystallographic orientations to the elements prior to finite element simulation. In the present study, 1000 grain orientations are extracted from the EBSD data by discretizing the orientation distribution function (ODF). Figure 7 shows a comparison between the measured texture from initial microstructure and the corresponding simulated texture based on 1000 extracted grain orientations. It can be seen from Figure 7a that the initial texture consists mainly of a preferred {001}<110> texture in $\varphi_2 = 45°$ section of ODF. This fact contributes to demonstrating that there exists a preferred orientation of grains, where the <110> direction is parallel to the RD direction. Such a texture is probably attributed to thermo-mechanical processing of the as-received NiTi SMA. Furthermore, the simulated ODF sections are depicted in Figure 7b. Obviously, these simulated ODF sections well reproduce the main features of the experimental ODF ones.

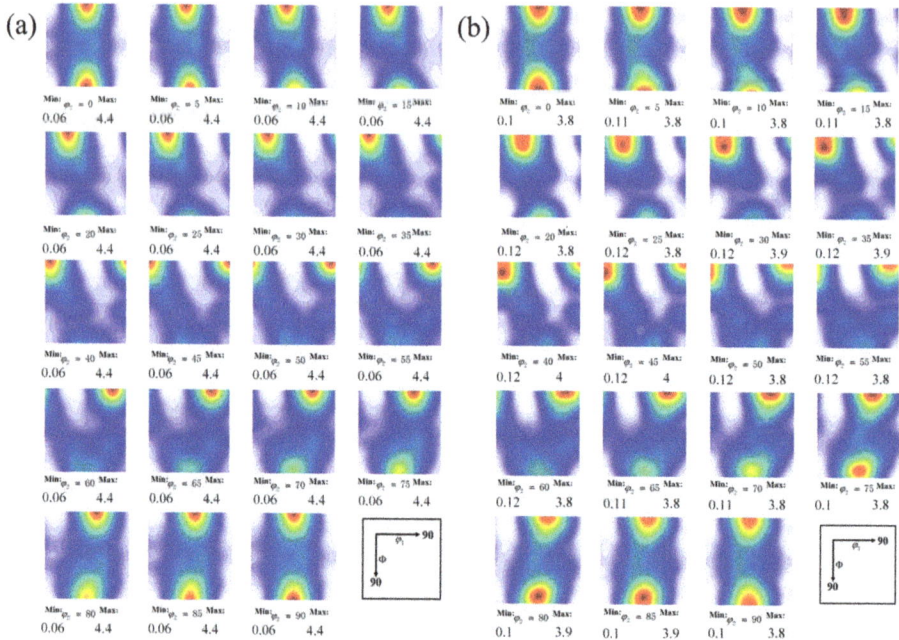

Figure 7. Orientation distribution function sections of as-received NiTi SMA used for describing initial texture: (**a**) based on EBSD experiment; (**b**) based on simulation via 1000 extracted grain orientations.

The values of anisotropically elastic parameters of single crystal NiTi SMA can be determined as $C_{11} = 130$ GPa, $C_{12} = 98$ GPa and $C_{44} = 34$ GPa [8]. Based on the constructed three-dimensional voxel-based RVE model as well as the extracted orientations from ODF measured by EBSD experiment, the remaining crystal plasticity parameters in the single-crystal constitutive equations in terms of hardening effect in different slip systems have to be determined by back-fitting the mechanical result in the constructed RVE model to the mechanical response in uniaxial compression experiment. Ezaz et al. [11] has reported that the ideal critical stress for slip systems {110}<100>, {010}<100> and {110}<111> are 2667 MPa, 9320 MPa and 5561 MPa, indicating that these three slip systems used in the present study have different hardening parameters and the slip system with a higher ideal critical stress also has higher hardening parameters, namely h_0, τ_s and τ_0 [25]. In addition, Lee et al. [26] has reported that in the case of crystal plasticity finite element simulation, the global stress-strain curve is sensitive to the adopted material parameters. Therefore, systematic variations of material parameters are important in the procedure of parameter identification. In terms of varying the material parameters, a method used frequently is "trial-error" algorithm, and it is used to optimize the numerical results by means of the optical observation on the difference between the numerical stress-strain curve and the experimental one. It is worth noting that such an optical coincidence procedure may finally result in high accuracy [24]. However, in the "trial-error" procedure, systematic variations of material parameters require a phenomenological interpretation and an understanding of the respective efforts on the mechanical response of the polycrystalline model. Therefore, such calibration is a nontrivial effort due to the number of material parameters involved. As a consequence, in the present study, the "trial-error" algorithm is never easy to find nine parameters manually. Based on the target function that depends on the user's definition, many other mathematical optimization algorithms have been proposed to facilitate the procedure of parameter identification, including the genetic algorithm (GA) method, continuous function optimization method [27–29] and their individual advantages and limitations would not be discussed in detail here. In the present study, a particle swarm optimization (PSO) algorithm is implemented for the purpose of parameter identification and the detailed introduction about PSO algorithm can be found in the literature [30]. As for the PSO algorithm, all population members generated in the first trial continuously update their problem solutions by tracking personal Best (pBest), which results in the minimal target function value by comparison with itself, and global Best (gBest), which results in the minimal target function value by comparison with all the population members in each iteration. As for combining the PSO algorithm and crystal plasticity finite element model, the coupling procedure is illustrated in Figure 8. The target function or fitness function used in the present study is identified as follows [27]:

$$F = \sqrt{\frac{1}{k}\sum_{i}^{k}\left(\sigma_i^{\exp(k)} - \sigma_i^{\sim(k)}\right)^2} \quad k = 1,2,\ldots,151 \tag{6}$$

where k is a number denoting the amplitude of 151 different true strains that range from 0.02 to 0.32 with the interval of 0.002. $\sigma_i^{\exp(k)}$ and $\sigma_i^{\sim(k)}$ are the true stresses measured in the experimental and simulated stress-strain curves for k_{th} strain amplitude, respectively. In the present study, the population size is taken to be 9, which indicates that there are nine material parameters to be determined.

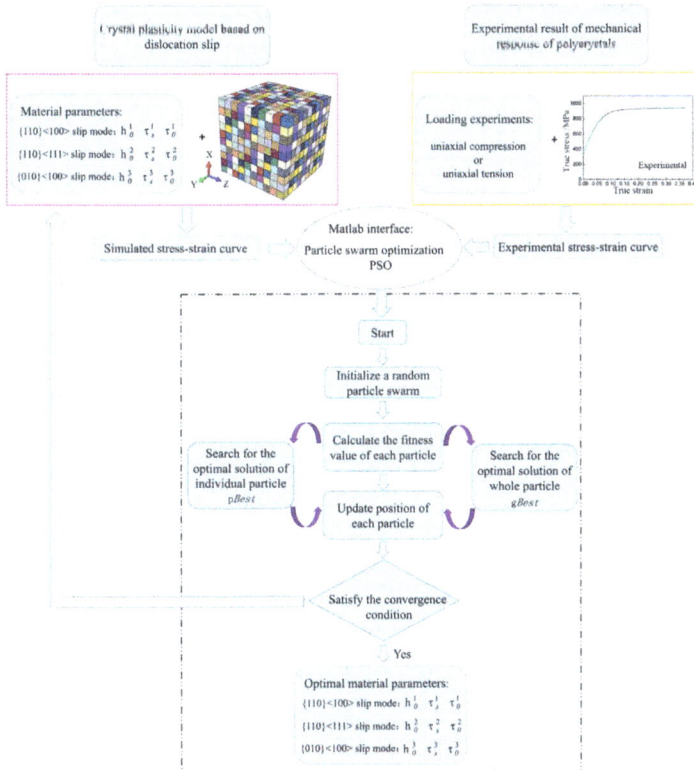

Figure 8. Flow chart of material parameter identification based on PSO (Particle swarm optimization).

Figure 9a shows the fitness-generation curve, which indicates the PSO algorithm convergence with the number of generations. It can be found from Figure 9a that, after 24 generations, the fitness value converges to a constant value of 3.95 MPa, indicating that the PSO algorithm has implemented the best optimization. The material parameters obtained by this calibration process are listed in Table 1. In addition, Figure 9b shows the experimental and simulated stress-strain curves of NiTi SMA subjected to uniaxial compression at the deformation degree of 30%. It can be found from Figure 9b that the simulated result is consistent with the experimental one and this phenomenon confirms the validity of the fitted parameters used in CPFEM.

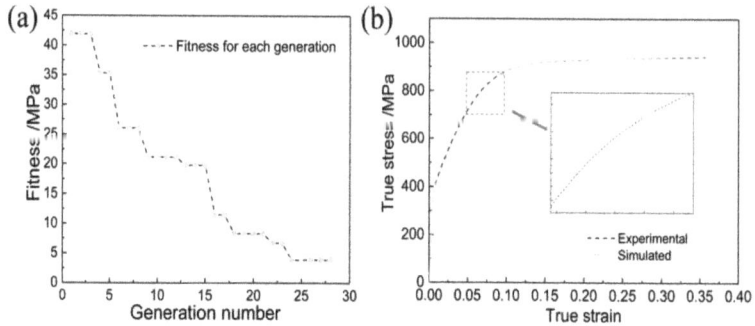

Figure 9. The convergence of PSO algorithm in the aspect of parameter identification and the validity of material parameters: (**a**) fitness function convergence as a function of generation number; (**b**) macroscopic stress-strain response of NiTi SMA subjected to uniaxial compression based on compression experiment and crystal plasticity finite element simulation.

Table 1. Material parameters of as-received NiTi SMA used for simulation.

Slip Mode	h_0	τ_s	τ_0	$\dot{\gamma}_0$	q	n
{110}<100> slip system	1283.29 MPa	354.59 MPa	134.53 MPa	0.001 s^{-1}	1.40	20
{010}<100> slip system	5191.62 MPa	505.90 MPa	489.93 MPa	0.001 s^{-1}	1.40	20
{110}<111> slip system	3574.64 MPa	390.27 MPa	239.58 MPa	0.001 s^{-1}	1.40	20

3.2.4. Texture Evolution of NiTi SMA under Uniaxial Compression

It is well known that the texture has a significant influence on the mechanical properties of NiTi SMA. Therefore, it is of great importance to predict texture evolution of NiTi SMA during plastic deformation. Based on the fitted material parameters, the constructed voxel RVE model is used to simulate the uniaxial compression process of NiTi SMA at various deformation degrees. As a consequence, the corresponding values of Euler angles at each integration point are extracted to predict texture evolution. This can be realized by programming a post-processing numerical subroutine using computer language PYTHON with the help of the ABAQUS interface. It can be generally accepted that the pole figure is an important approach to characterize the texture of metal materials during uniaxial compression or tension. Figure 10 illustrates the pole figures of NiTi SMA subjected to the various deformation degrees including 0%, 20%, 30% and 40%, which are obtained by means of EBSD experiments and CPFEM simulation. It can be seen from Figure 10 that the simulated results are in good accordance with the experimental ones, except from the legend values at various deformation degrees. The legend values show that the maximum pole density in experiment results is less than the corresponding simulated ones. This may be attributed to the limited EBSD scanning areas in EBSD experiments and the increased regions of poor identification in scanning areas. However, it can be concluded that CPFEM based on the constructed RVE model is a superior candidate for predicting texture evolution of NiTi SMA during uniaxial compression deformation. With the progression of plastic deformation, a crystallographic plane of NiTi SMA gradually rotates to be vertical to the loading direction, which lays the foundation for forming the <111> fiber texture in the case of uniaxial compression.

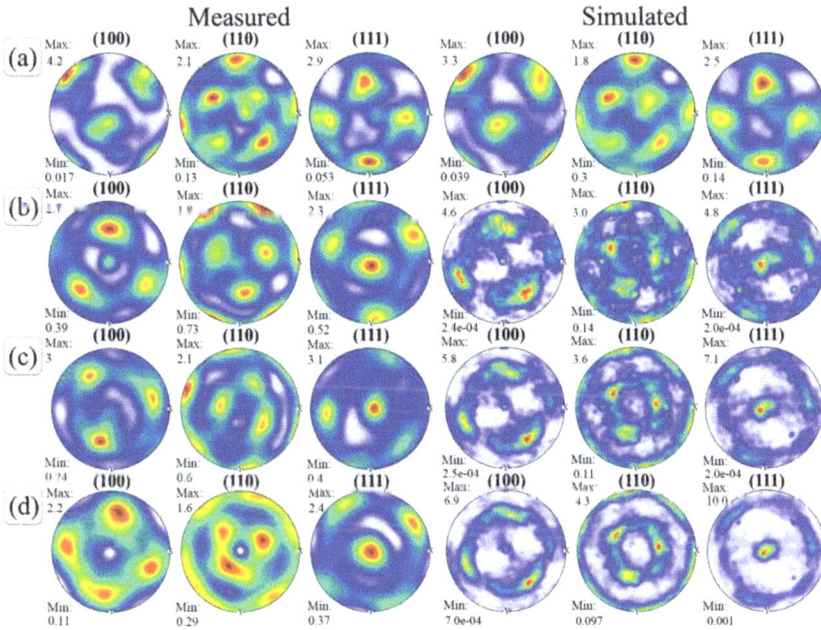

Figure 10. Pole figures showing texture evolution of NiTi SMA subjected to uniaxial compression at various deformation degrees: (**a**) 0%; (**b**) 20%; (**c**) 30%; (**d**) 40%.

4. Conclusions

(1) TEM observation demonstrates that dislocation slip rather than deformation twinning is responsible for plastic deformation mechanisms of B2 austenite NiTi SMA at 673 K (400 °C). EBSD experiment demonstrates the heterogeneous microstructure evolution during uniaxial compression, where subgrain substructures are formed and distributed within individual grains.

(2) Based on the experimental observations, {110}<100>, {010}<100> and {110}<111> slip systems are introduced into the single-crystal constitutive equations in order to accommodate plastic deformation of NiTi SMA. Particle swarm optimization (PSO) algorithm is used to identify crystal plasticity parameters from experimental results of NiTi SMA. The validity of the fitted material parameters is well confirmed based on the fact that the simulated stress-strain curve on the basis of constructed Voxel RVE model is in good agreement with the experimental result.

(3) CPFEM based on the constructed RVE model is able to accurately predict texture evolution of NiTi SMA during uniaxial compression deformation. The simulation results are in good agreement with the experimental ones. With the progression of plastic deformation, a crystallographic plane of NiTi SMA gradually rotates to be vertical to the loading direction, which lays the foundation for forming the <111> fiber texture.

Acknowledgments: The work was financially supported by National Natural Science Foundation of China (Nos. 51475101, 51305091 and 51305092).

Author Contributions: Li Hu performed the crystal plasticity finite element simulation and wrote the manuscript; Shuyong Jiang supervised the research; Yanqiu Zhang performed TEM analysis and EBSD analysis.

Conflicts of Interest: The authors declare no conflict of interest.

References

1. Vojtěch, D.; Michalcova, A.; Čapek, J.; Marek, I.; Dragounova, L. Structural and mechanical stability of the nano-crystalline Ni–Ti (50.9 at.% ni) shape memory alloy during short-term heat treatments. *Intermetallics* **2014**, *49*, 7–13. [CrossRef]
2. Sharifi, E.M.; Karimzadeh, F.; Kermanpur, A. The effect of cold rolling and annealing on microstructure and tensile properties of the nanostructured Ni 50 Ti 50 shape memory alloy. *Mater. Sci. Eng. A Struct. Mater.* **2014**, *607*, 33–37. [CrossRef]
3. Delobelle, V.; Chagnon, G.; Favier, D.; Alonso, T. Study of electropulse heat treatment of cold worked NiTi wire: From uniform to localised tensile behaviour. *J. Mater. Process. Technol.* **2016**, *227*, 244–250. [CrossRef]
4. Hu, L.; Jiang, S.H.; Zhang, Y.Q.; Zhao, Y.N.; Liu, S.W.; Zhao, C.Z. Multiple plastic deformation mechanisms of NiTi shape memory alloy based on local canning compression at various temperatures. *Intermetallics* **2016**, *70*, 45–52. [CrossRef]
5. Benafan, O.; Noebe, R.; Padula, S.; Garg, A.; Clausen, B.; Vogel, S.; Vaidyanathan, R. Temperature dependent deformation of the B2 austenite phase of a NiTi shape memory alloy. *Int. J. Plast.* **2013**, *51*, 103–121. [CrossRef]
6. Duerig, T. Some unsolved aspects of Nitinol. *Mater. Sci. Eng. A Struct. Mater.* **2006**, *438*, 69–74. [CrossRef]
7. Lin, Y.S.; Cak, M.; Paidar, V.; Vitek, V. Why is the slip direction different in different B2 alloys? *Acta Mater.* **2012**, *60*, 881–888. [CrossRef]
8. Manchiraju, S.; Anderson, P.M. Coupling between martensitic phase transformations and plasticity: A microstructure-based finite element model. *Int. J. Plast.* **2010**, *26*, 1508–1526. [CrossRef]
9. Yu, C.; Kang, G.; Kan, Q. A micromechanical constitutive model for anisotropic cyclic deformation of super-elastic niti shape memory alloy single crystals. *J. Mech. Phys. Solids* **2015**, *82*, 97–136. [CrossRef]
10. Pelton, A.; Huang, G.; Moine, P.; Sinclair, R. Effects of thermal cycling on microstructure and properties in Nitinol. *Mater. Sci. Eng. A Struct. Mater.* **2012**, *532*, 130–138. [CrossRef]
11. Ezaz, T.; Wang, J.; Sehitoglu, H.; Maier, H. Plastic deformation of NiTi shape memory alloys. *Acta Mater.* **2013**, *61*, 67–78. [CrossRef]
12. Ezaz, T.; Sehitoglu, H.; Maier, H.J. Energetics of (114) twinning in B2 NiTi under coupled shear and shuffle. *Acta Mater.* **2012**, *60*, 339–348. [CrossRef]
13. Zaki, W.; Moumni, Z. A three-dimensional model of the thermomechanical behavior of shape memory alloys. *J. Mech. Phys. Solids* **2007**, *55*, 2455–2490. [CrossRef]
14. Auricchio, F.; Reali, A.; Stefanelli, U. A three-dimensional model describing stress-induced solid phase transformation with permanent inelasticity. *Int. J. Plast.* **2007**, *23*, 207–226. [CrossRef]
15. Yu, C.; Kang, G.; Song, D.; Kan, Q. Micromechanical constitutive model considering plasticity for super-elastic NiTi shape memory alloy. *Comput. Mater. Sci.* **2012**, *56*, 1–5. [CrossRef]
16. Suwas, S.; Ray, R.K. *Crystallographic Texture of Materials*; Springer: London, UK, 2014; pp. 11–38.
17. Hill, R.; Rice, J. Constitutive analysis of elastic-plastic crystals at arbitrary strain. *J. Mech. Phys. Solids* **1972**, *20*, 401–413. [CrossRef]
18. Huang, Y.G. *A User-Material Subroutine Incorporating Single Crystal Plasticity in the Abaqus Finite Element Program*; Harvard University: Cambridge, MA, USA, 1991.
19. Abdolvand, H.; Daymond, M.R.; Mareau, C. Incorporation of twinning into a crystal plasticity finite element model: Evolution of lattice strains and texture in zircaloy-2. *Int. J. Plast.* **2011**, *27*, 1721–1738. [CrossRef]
20. Jiang, S.H.; Zhang, Y.Q.; Zhao, Y.N.; Tang, M.; Yi, W.L. Constitutive behavior of Ni-Ti shape memory alloy under hot compression. *J. Cent. South Univ. Technol.* **2013**, *20*, 24–29. [CrossRef]
21. Cailletaud, G. A micromechanical approach to inelastic behaviour of metals. *Int. J. Plast.* **1992**, *8*, 55–73. [CrossRef]

22. Diard, O.; Leclercq, S.; Rousselier, G.; Cailletaud, G. Evaluation of finite element based analysis of 3D multicrystalline aggregates plasticity: Application to crystal plasticity model identification and the study of stress and strain fields near grain boundaries. *Int. J. Plast.* **2005**, *21*, 691–722. [CrossRef]

23. Lv, L.; Zhen, L. Crystal plasticity simulation of polycrystalline aluminum and the effect of mesh refinement on mechanical responses. *Mater. Sci. Eng. A Struct. Mater.* **2011**, *528*, 6673–6679. [CrossRef]

24. Zhang, K.S.; Ju, J.W.; Li, Z.; Bai, Y.L.; Brocks, W. Micromechanics based fatigue life prediction of a polycrystalline metal applying crystal plasticity. *Mech. Mater.* **2015**, *85*, 16–37. [CrossRef]

25. Wang, Y.; Sun, X.; Wang, Y.; Hu, X.; Zbib, H.M. A mechanism-based model for deformation twinning in polycrystalline FCC steel. *Mater. Sci. Eng. A Struct. Mater.* **2014**, *607*, 206–218. [CrossRef]

26. Lee, M.G.; Wang, J.; Anderson, P.M. Texture evolution maps for upset deformation of body-centered cubic metals. *Mater. Sci. Eng. A Struct. Mater.* **2007**, *463*, 263–270. [CrossRef]

27. Xie, C.; Ghosh, S.; Groeber, M. Modeling cyclic deformation of HSLA steels using crystal plasticity. *J. Eng. Mater. Technol. Trans. ASME* **2004**, *126*, 339–352. [CrossRef]

28. Cheng, J.; Ghosh, S. A crystal plasticity FE model for deformation with twin nucleation in magnesium alloys. *Int. J. Plast.* **2015**, *67*, 148–170. [CrossRef]

29. Hasija, V.; Ghosh, S.; Mills, M.J.; Joseph, D.S. Deformation and creep modeling in polycrystalline Ti-6Al alloys. *Acta Mater.* **2003**, *51*, 4533–4549. [CrossRef]

30. Esmin, A.A.; Coelho, R.A.; Matwin, S. A review on particle swarm optimization algorithm and its variants to clustering high-dimensional data. *Artif. Intell. Rev.* **2015**, *44*, 23–45. [CrossRef]

metals

MDPI

Article

High Field X-ray Diffraction Study for Ni$_{46.4}$Mn$_{38.8}$In$_{12.8}$Co$_{2.0}$ Metamagnetic Shape Memory Film

Yoshifuru Mitsui [1,*], Keiichi Koyama [1], Makoto Ohtsuka [2], Rie Y. Umetsu [3], Ryosuke Kainuma [4] and Kazuo Watanabe [3]

[1] Graduate School of Science and Engineering, Kagoshima University, Kagoshima 890-0065, Japan; koyama@sci.kagoshima-u.ac.jp

[2] Institute of Multidisciplinary Research for Advanced Materials, Tohoku University, Sendai 980-8577, Japan; makoto.ohtsuka.d7@tohoku.ac.jp

[3] Institute for Materials Research, Tohoku University, Sendai 980-8577, Japan; rieume@imr.tohoku.ac.jp (R.Y.U.); kwata@imr.tohoku.ac.jp (K.W.)

[4] Graduate School of Engineering, Tohoku University, Sendai 980-8579, Japan; kainuma@material.tohoku.ac.jp

* Correspondence: mitsui@sci.kagoshima-u.ac.jp; Tel.: +81-99-285-8082

Received: 4 August 2017; Accepted: 7 September 2017; Published: 12 September 2017

Abstract: The transformation behaviors on metamagnetic shape memory Ni$_{46.4}$Mn$_{38.8}$In$_{12.8}$Co$_{2.0}$ film were investigated by X-ray diffraction experiments in the temperature up to 473 K and magnetic fields $\mu_0 H$ up to 5 T. The prepared film showed the parent phase with L2$_1$ structure at 473 K, and with preferred orientation along the 111 plane. The magnetic field induced reverse transformation was directly observed at $T = 366$ K, which was just around the reverse transformation starting temperature.

Keywords: high field X-ray diffraction; NiCoMnIn; metamagnetic shape memory alloys

1. Introduction

Ferromagnetic shape memory alloys (FSMAs) have been studied actively as high-performance actuator materials since a large magnetic field-induced strain of 0.2% was found in Ni$_2$MnGa alloys by Ullakko et al. [1]. That this large magnetic field induced strain in the ferromagnetic Ni$_2$MnGa single crystal is explained by the rearrangement of twin variants of martensitic phase (M-phase) [2]. To control the performance in Ni-Mn-Ga alloys (e.g., magnetic properties, martensitic transformation temperatures, etc.), they were examined by the substitution of the elements [3,4].

In 2004, Sutou et al. found that Ni-Mn-X (X = In, Sn, and Sb) alloys with Heusler-type structure showed a martensitic transformation with magnetic transition [5]. The magnetization of parent phase in Ni-Mn-X series shows large magnetization, whereas that of M-phase is small [5]. The Mössbauer spectroscopy studies on Fe^{57}-doped Ni-Mn-In and Ni-Mn-Sn systems found that the magnetism of the M-phase was paramagnetism [6,7]. The Co-doped Ni-Mn-In system was found to show a discontinuous jump in magnetization between P- and M-phase [8]. The strain of 3% was almost recovered by the application of a 7 T magnetic field, which was a so-called metamagnetic shape memory effect (MSM effect) [8]. An MSM effect was also found in Co-doped Ni-Mn-Sn alloys [9]. The crystal structure of M-phase in Ni-Co-Mn-In was reported to be the mixture of five- and seven-layered modulated monoclinic structure (10M and 14M) by electron microscopy observation [10]. Structural properties of Ni-Co-Mn-In bulk alloy were reported by using synchrotron radiation in high magnetic fields [11]. According to Reference [11], the crystal structure of M-phase was 14M. Additionally, field-induced reverse transformation under compression was observed in magnetic fields up to 5 T [11].

FSMAs films have been studied for the application as actuators [12–14]. Ni-Co-Mn-In MSM ribbons and films were also prepared by rapid solidification [15] and magnetron sputtering [16], respectively. Recently, Ni-Co-Mn-In films were examined for the application for energy harvesting [17]. According to Reference [16], the annealing temperature changes the crystal structure of M-phases. Recent reports for the Ni-Co-Mn-In films show that as-deposited film shows body-centered cubic structure, whereas the modulated structure appeared after annealing [18]. According to the phase diagram of $Ni_{50-x}Mn_{37}In_{13}Co_x$ ribbon with 30 μm thick, the reverse transformation temperature changed by ~200 K in $0 \leq x \leq 9$, and had a cusp at $x \sim 3$ [15]. Furthermore, according to Reference [18], minor changes in compositions of the film also changed the transformation behavior of the films (e.g., transformation temperature and thermal hysteresis). Therefore, the transformation behaviors and crystal structures of MSM films were sensitive to slight composition change and the annealing conditions.

Although the annealing effects, microstructure, and martensitic transformation behaviors were evaluated for MSM films and ribbons, the martensitic transformation induced by magnetic fields has not yet been confirmed by using in-situ observation techniques.

The high field X-ray diffraction (HF-XRD) technique was one of the suitable methods for investigating the structural properties in magnetic field—particularly the field-induced structural transformations. So far, the relationship between magnetic transition and structural transformation was investigated for magnetic refrigerants by using HF-XRD [19–21]. In addition, HF-XRD study has also been carried out for FSMAs and MSM alloys such as Ni_2MnGa alloys [22], Ni-Mn-Sn alloy [23], and for Ni-Co-Mn-In alloy [24]. In 2008, HF-XRD for high temperatures was developed in temperatures up to 473 K [25]. Magnetic field-induced reverse transformations in $Ni_{40}Co_{10}Mn_{34}Al_{16}$ MSM alloys were observed at 408 K using this apparatus [26].

As described above, in this study, in order to observe field-induced reverse transformation in Co-doped Ni-Mn-In film, high field X-ray diffraction experiments were performed under magnetic fields up to 5 T and temperature ranging from 293 to 473 K.

2. Materials and Methods

Co-doped Ni-Mn-In films of 1 μm thickness were deposited on a poly-vinyl alcohol (PVA) substrate using a dual magnetron sputtering apparatus (CFS-4ES, Shibaura, Yokohama, Japan). The apparatus has radio-frequency (RF) and direct current (DC) power sources. The RF power for $Ni_{45}Mn_{40}In_{15}$ target was kept at 200 W and the DC power for Co target was kept at 5 W. After separating from the PVA substrate, the films were heat-treated at 1173 K for 3.6 ks. The chemical composition of the film was determined by an inductively coupled plasma (ICP) spectrometry apparatus (Optima 3300, Perkin Elmer Inc., Waltham, MA, USA). The composition of the sample was determined to be $Ni_{46.4}Mn_{38.8}In_{12.8}Co_{2.0}$. According to the reports about Ni-Co-Mn-In films [16,18], the transformation behavior was sensitive to the annealing condition, composition of Co, and so on. In this study, as described below, the transformation temperatures of $Ni_{46.4}Mn_{38.8}In_{12.8}Co_{2.0}$ films are above room temperature (RT), which is suitable for observing the structural change by HF-XRD system at high temperature.

The martensitic transformation temperatures M_s: martensitic transformation starting temperature, M_f: martensitic transformation finishing temperature, A_s: reverse transformation starting temperature, and A_f: reverse transformation finishing temperature were determined by the magnetization measurements by using a superconducting quantum interface device (SQUID) magnetometer in magnetic fields $\mu_0 H$ up to 5 T and the temperature T ranging from 10 to 390 K. The transformation temperatures are defined using the intersection of the base line and the tangent line with the largest slope in the thermomagnetization curve.

High field X-ray diffraction measurements using Cu Kα radiation were performed for $\mu_0 H \leq 5$ T and in the temperature range from 293 to 473 K under He atmosphere. Details of the HF-XRD setup are reported in Reference [25]. $Ni_{46.4}Mn_{38.8}In_{12.8}Co_{2.0}$ films were fixed with Apiezon H Grease &I Materials Ltd., Manchester, UK) on a copper sample holder. The surface of the film was parallel to the magnetic field, and was perpendicular to the scattering vector of X-ray.

3. Results and Discussion

Figure 1 shows the thermomagnetization curve for $Ni_{46.4}Mn_{38.8}In_{12.8}Co_{2.0}$ film in $\mu_0 H = 0.05$ T, 1 T, and 5 T. In all curves, metamagnetic phase transition from M- to P-phase is clearly observed. The transformation temperatures in $\mu_0 H = 1$ T were determined to be $M_s = 373$ K, $M_f = 355$ K, $A_s = 368$ K, and $A_f = 385$ K. Meanwhile, these temperatures at 5 T were obtained to be $M_s = 373$ K, $M_f = 352$ K, $A_s = 365$ K, and $A_f = 385$ K, which were slightly lower than that in 1 T. Figure 2 shows the *M–H* curve obtained at 348, 366, and 373 K, which were $T < A_s$, $T \sim A_s$, and $A_s < T < A_f$, respectively. *M–H* data were collected after the zero-field heating from $T < M_f$. The magnetization at 348 K is very small and almost independent of magnetic fields. At 366 K, the small jump in magnetization due to metamagnetic transition was observed for $\mu_0 H \geq 4$ T. On the other hand, the curve at 373 K showed large magnetization, and magnetic transition and hysteresis were not observed.

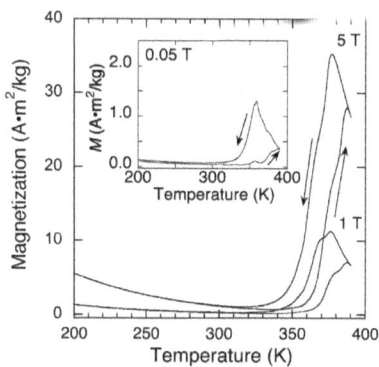

Figure 1. Thermomagnetization curves for $Ni_{46.4}Mn_{38.8}In_{12.8}Co_{2.0}$ film at $\mu_0 H = 1$ T and 5 T. The inset is the thermomagnetization curve obtained at $\mu_0 H = 0.05$ T.

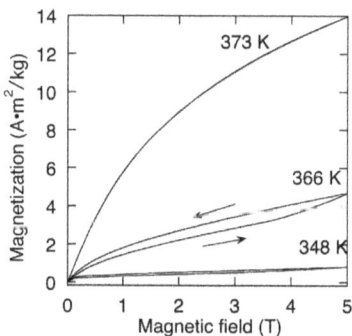

Figure 2. Isothermal magnetization curves for $Ni_{46.4}Mn_{38.8}In_{12.8}Co_{2.0}$ film at 348, 366, and 373 K. Each curve ware obtained after the zero field heating from $T > M_f$.

Figure 3 shows the XRD patterns for $Ni_{46.4}Mn_{38.8}In_{12.8}Co_{2.0}$ film at room temperature (RT) and at 473 K. The diffraction peaks at $2\theta \sim 43°$, $50°$, $74°$, and $90°$ belong to the copper sample holder. The other peaks at RT belong to the M-phase, which are indicated by the closed circles in Figure 3. The diffraction profile at 473 K is quite different from that at RT. The diffraction peaks at 473 K were indexed by $L2_1$ structure (*hklp*), which was P-phase. This profile shows the preferred orientation along (111) plane parallel to film surface. As described below, the lattice parameters of the P-phase were in good agreement with that of bulk Ni-Co-Mn-In samples. To compare the diffraction patterns of P- and

M-phase, 111$_P$, 220$_P$, and 222$_P$ seem to split during transformation, indicating the decline of crystal symmetry. In this study, it is difficult to determine the crystal structure of M-phase because the P-phase shows preferred orientation. 14M and 10M structure did not represent the obtained XRD patterns. According to the reports for bulk Ni-Co-Mn-In sample, the crystal structure of M-phase is reported to be a mixture of 14M and 10M structures [10], or 14M structures [11]. Therefore, the crystal structure of M-phase in the film is considered to be a mixture of 10M and 14M or the related modulated structure.

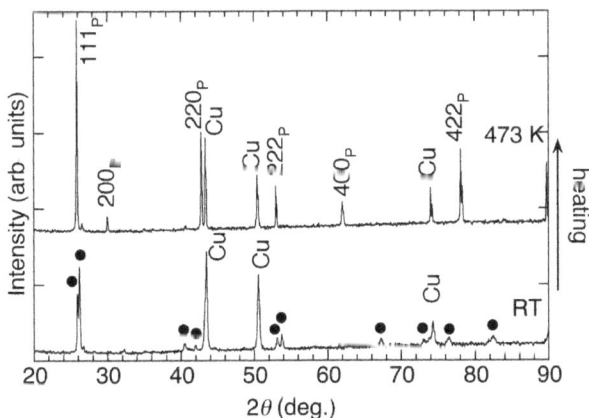

Figure 3. XRD pattern of Ni$_{46.4}$Mn$_{38.8}$In$_{12.8}$Co$_{2.0}$ film at room temperature (RT) and 473 K. hkl_Λ indicates the Miller indices of the L2$_1$ structure (parent phase). The closed circles are the diffraction peaks belonging to the martensitic phase.

Figure 4 shows the XRD patterns in the 2θ range from 52° to 55° at various temperatures in heating in a zero field (a), cooling in a zero field (b), heating in 5 T (c), and cooling in 5 T (d). In the heating process from 303 K in a zero field, the 222$_P$ diffraction developed between 363 and 373 K. For $T \geq 383$ K, only the diffraction peak of L2$_1$ structure was observed. With decreasing T from 473 K, the peak intensity of 222$_P$ diffraction began to suppress at T = 363 K, and the diffraction at 2θ ~ 53.5° appeared, and the diffraction profile did not change below 353 K. On the other hand, as seen in Figure 4c,d, the change in diffraction profile was also observed with a slightly lower temperature region than that in a zero field. These transformation behaviors were consistent with the thermomagnetization curve shown in Figure 1. Using the diffraction profiles of Figure 4a,b, the lattice parameter a of L2$_1$ structure at room temperature was evaluated. Figure 5 shows the determined lattice parameter a in a zero field. The obtained lattice parameter and temperature show linear relation. The lattice parameter a at 300 K was obtained to be 0.597 nm, which is in good agreement with previous reports for bulk sample [8].

Figures 6 and 7 show the isothermal XRD patterns in magnetic fields up to 5 T at fixed temperature of 366 K and 371 K. The measurements were carried out after zero-field heating from room temperature ($T < M_f$). The diffraction peaks of M-phase (closed circles) and P-phase (222$_P$ peak) were observed in all profiles. As seen in Figure 6, with applied magnetic field, the sharp peak belonging to the 222$_P$ diffraction was induced at 2θ ~ 53°. Although 222$_P$ diffraction was induced by the magnetic field, the reverse transformation from the M-phase to P-phase was incomplete at 5 T. With decreasing μ_0H from 5 T to 0 T, the profile did not change efficiently. On the other hand, the diffraction peaks at 371 K of P-phase became stronger than 366 K, indicating the irreversibility of the field-induced transformation. On the other hand, as seen in Figure 7, the field-induced development of 222$_P$ diffraction was not observed clearly.

Figure 8 shows the magnetic field dependence of the difference of peak intensity I_{P-M} between two peaks at 2θ ~ 53° and 2θ ~ 53.5°. As described above, because the 222$_P$ peak of P-phase and the peak of M-phase overlapped at $2\theta \sim 53°$ and the peak at 53.5° only belonged to M-phase, enhancement of $I_{P\text{-}M}$ indicated the field-induced reverse transformation. $I_{P\text{-}M}$ gradually increased to 3 T. $I_{P\text{-}M}$ showed a jump for $\mu_0 H > 3$ T, and did not recover in decreasing $\mu_0 H$. This means that the magnetic-field-induced transformation was directly observed for $\mu_0 H > 3$ T. Isothermal diffraction pattern at 371 K also showed the increase of $I_{P\text{-}M}$ with increasing H. However, the increment of $I_{P\text{-}M}$ was obtained to be 25 count from $\mu_0 H = 0$ to 5 T, which was much smaller than 366 K. Thus, it was found that the field-induced transformation of the film exhibits a narrow temperature window. This transformation behavior was qualitatively consistent with the magnetization measurements.

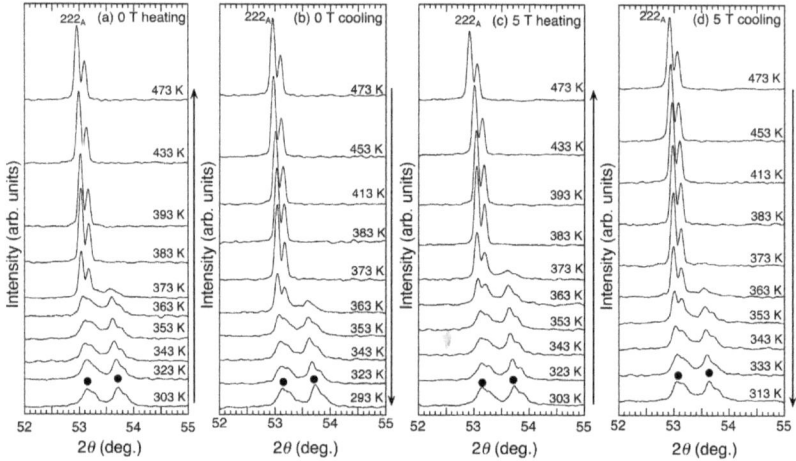

Figure 4. XRD patterns of $Ni_{46.4}Mn_{38.8}In_{12.8}Co_{2.0}$ film in (**a**) heating in a zero field, (**b**) cooling in a zero field, (**c**) heating in 5 T, and (**d**) cooling in 5 T. hkl_P indicates the Miller indices of the L2$_1$ structure (parent phase). The closed circles were the diffraction peaks derived to the martensitic phase.

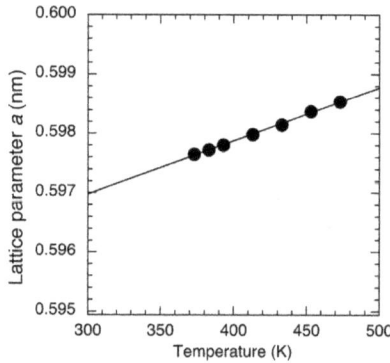

Figure 5. Lattice parameter a in the parent phase of $Ni_{46.4}Mn_{38.8}In_{12.8}Co_{2.0}$ film as a function of temperature. The solid line was obtained by the least-squares calculations.

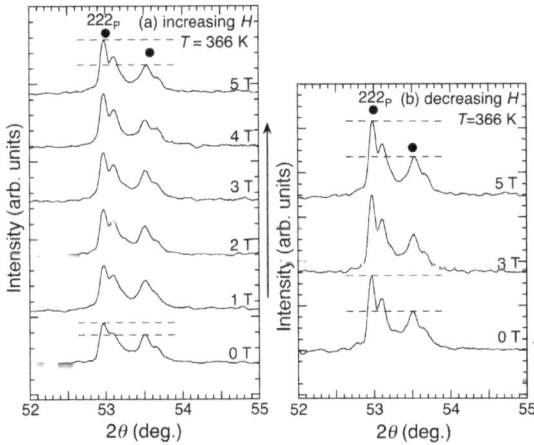

Figure 6. Isothermal XRD patterns at fixed temperature of 366 K with (**a**) increasing $\mu_0 H$ from 0 to 5 T, and (**b**) decreasing $\mu_0 H$ to 0 T. 220$_P$ indicates the Miller indices of the L2$_1$ structure (parent phase). The closed circles are the diffraction peaks belonging to the martensitic phase. The broken lines indicate the differences in intensities between the two diffraction patterns.

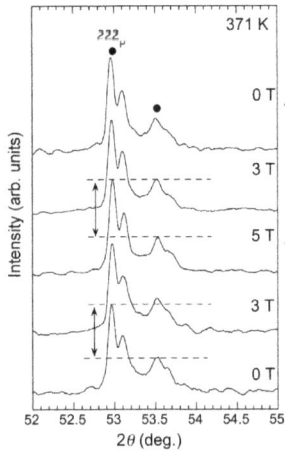

Figure 7. Isothermal XRD patterns at fixed temperature of 371 K. 220$_P$ indicates the Miller indices of the L2$_1$ structure (parent phase). The closed circles are the diffraction peaks belonging to the martensitic phase. The broken lines indicate I_{P-M}.

In this study, the application of $\mu_0 H = 5$ T was not enough to finish the reverse transformation completely at 366 K. From the *M–T* curve, the extrapolated magnetization of the P-phase was considered to be 40 A·m^2/kg at 366 K. Herein, when metamagnetic transition is seen, discontinuous change in magnetization appeared. In this study, this discontinuous change was approximated by a linear relation. If *M* increased linearly during metamagnetic transition of the films, the extrapolated line for *M–H* curves at 366 K for $\mu_0 H > 4$ T in Figure 2 reached 40 A·m^2/kg at $\mu_0 H \sim 36$ T. Thus, $\mu_0 H \sim 36$ T is required to observe the complete transformation of the film at 366 K.

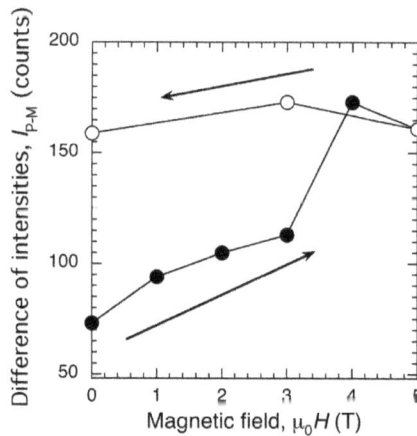

Figure 8. Differences between the peak intensity at $2\theta \sim 53°$ (martensitic + parent phases) and $53.5°$ (martensitic phase), I_{P-M}, as a function of magnetic field. The data was obtained by isothermal diffraction patterns at 366 K, which is shown in Figure 6.

4. Conclusions

The in situ observation of martensitic transformations for $Ni_{46.4}Mn_{38.8}In_{12.8}Co_{2.0}$ MSM film was performed by high field X-ray diffraction measurements in magnetic fields up to 5 T and in the temperature region from 298 to 473 K. The prepared films show the preferred orientation along (111) plane of the $L2_1$ structure at 473 K. A magnetic field-induced reverse transformation from M- to P-phase with $L2_1$ structure was observed at 366 K, which was just around A_s. The reverse transformation induced by magnetic fields was directly observed by in situ HF-XRD technique. Combining the change of the XRD patterns and the magnetization jumps, it was found that the metamagnetic transition of Ni-Co-Mn-In film is actually related to the martensitic transformation.

Acknowledgments: The work was carried out at the High Field Laboratory for Superconducting Materials, Institute for Materials Research, Tohoku University. This work was supported by Global COE Program "Materials Integration International Center of Education and Research, Tohoku University", from MEXT.

Author Contributions: Yoshifuru Mitsui performed the experiments and wrote the paper. Keiichi Koyama designed the experiments; Makoto Ohtsuka and Ryosuke Kainuma prepared the sample. Rie Y. Umetsu analyzed the data; Kazuo Watanabe supervised the HF-XRD experiments.

Conflicts of Interest: The authors declare no conflict of interest.

References

1. Ullakko, K.; Huang, J.K.; Kantner, C.; O'Handley, R.C.; Kokorin, V.V. Large magnetic-field-induced strains in Ni$_2$MnGa single crystals. *Appl. Phys. Lett.* **1996**, *69*, 1966–1968. [CrossRef]
2. Ullakko, K.; Huang, J.K.; Kokorin, V.V.; O'Handley, R.C. Magnetically controlled shape memory effect in Ni$_2$MnGa intermetallics. *Scr. Mater.* **1997**, *36*, 1133–1138. [CrossRef]
3. Vasil'ev, A.N.; Bozhko, A.D.; Khvailo, V.V.; Dikshtein, I.E.; Shavrov, V.G.; Buchelnikov, V.D.; Matsumoto, M.; Suzuki, S.; Takagi, T.; Tani, J. Structural and magnetic phase transitions in shape-memory alloys Ni$_{2+x}$Mn$_{1-x}$Ga. *Phys. Rev. B* **1999**, *59*, 1113–1120. [CrossRef]
4. Xu, X.; Nagasako, M.; Ito, W.; Umetsu, R.Y.; Kanomata, T.; Kainuma, R. Magnetic properties and phase diagram of Ni$_{50}$Mn$_{50-x}$Ga$_x$ ferromagnetic shape memory alloys. *Acta Mater.* **2013**, *61*, 6712–6723. [CrossRef]
5. Sutou, Y.; Imano, Y.; Koeda, N.; Omori, T.; Kainuma, R.; Ishida, K.; Oikawa, K. Magnetic and martensitic transformation of NiMnX (X = In, Sn, Sb) ferromagnetic shape memory alloys. *Appl. Phys. Lett.* **2004**, *85*, 4358–4360. [CrossRef]

6. Khovaylo, V.V.; Kanomata, T.; Tanaka, T.; Nakashima, M.; Amako, Y.; Kainuma, R.; Umetsu, R.Y.; Morito, H.; Miki, H. Magnetic properties of $Ni_{50}Mn_{34.8}In_{15.2}$ probed by Mössbauer spectroscopy. *Phys. Rev. B* **2009**, *80*, 144409. [CrossRef]
7. Umetsu, R.Y.; Sano, K.; Fukushima, K.; Kanomata, T.; Taniguchi, Y.; Amako, Y.; Kainuma, R. Mössbauer spectroscopy studies on magnetic properties for [57]Fe-substituted Ni-Mn-Sn metamagnetic shape memory alloys. *Metals* **2013**, *3*, 225–236. [CrossRef]
8. Kainuma, R.; Imano, Y.; Ito, W.; Sutou, Y.; Morito, H.; Okamoto, S.; Kitakami, O.; Oikawa, K.; Fujita, A.; Kanomata, T.; et al. Magnetic-field-induced shape recovery by reverse phase transformation. *Nature* **2006**, *439*, 957–960. [CrossRef] [PubMed]
9. Kainuma, R.; Imano, Y.; Ito, W.; Morito, H.; Sutou, Y.; Oikawa, K.; Fujita, A.; Ishida, K.; Okamoto, S.; Kitakami, O.; et al. Metamagnetic shape memory effect in a Heusler-type $Ni_{43}Co_7Mn_{39}Sn_{11}$ polycrystalline alloy. *Appl. Phys. Lett.* **2006**, *88*, 192513. [CrossRef]
10. Ito, W.; Imano, Y.; Kainuma, R.; Sutou, Y.; Oikawa, K.; Ishida, K. Martensitic and Magnetic transformation behaviors in Heusler-Type NiMnIn and NiCoMnIn metamagnetic shape memory alloys. *Metall. Mater. Trans. A* **2007**, *38A*, 759–766. [CrossRef]
11. Wang, Y.D.; Ren, Y.; Huang, E.W.; Nie, Z.H.; Wang, G. Direct evidence on magnetic-field-induced phase transition in a NiCoMnIn ferromagnetic shape memory alloy under a stress field. *Appl. Phys. Lett.* **2007**, *90*, 101917. [CrossRef]
12. Kohl, M.; Brugger, D.; Ohtsuka, M.; Takagi, T. A novel actuation mechanism on the basis of ferromagnetic SMA thin films. *Sens. Actuators A* **2004**, *114*, 445–450. [CrossRef]
13. Kohl, M.; Krevet, B.; Ohtsuka, M.; Brugger, D.; Liu, Y. Ferromagnetic shape memory actuators. *Mater. Trans.* **2006**, *47*, 639–644.
14. Rumpf, H.; Craciunescu, C.M.; Modrow, H.; Olimov, H.M.; Quandt, F.; Wuttig, M. Successive occurrence of ferromagnetic and shape memory properties during crystallization of NiMnGa freestanding films. *J. Magn. Magn. Mater.* **2006**, *302*, 421–428. [CrossRef]
15. Liu, J.; Scheerbaum, N.; Hinz, D.; Gutfleisch, O. Magnetostructural transformation in Ni-Mn-In-Co ribbons. *Appl. Phys. Lett.* **2008**, *92*, 162509. [CrossRef]
16. Rios, S.; Karaman, I.; Zhang, X. Crystallization and high temperature shape memory behavior of sputter-deposited NiMnCoIn thin films. *Appl. Phys. Lett.* **2010**, *96*, 173102. [CrossRef]
17. Gueltig, M.; Ossmer, H.; Ohtsuka, M.; Miki, H.; Tsuchiya, K.; Takagi, T.; Kohl, M. High frequency thermal energy harvesting using magnetic shape memory films. *Adv. Energy Mater.* **2014**, *4*, 1400751. [CrossRef]
18. Miki, H.; Tsuchiya, K.; Ohtsuka, M.; Gueltig, M.; Kohl, M.; Takagi, T. Structural and magnetic properties of magnetic shape memory alloys on Ni-Mn-Co-In self-standing films. In *Advances in Shape Memory Materials*; Springer: Basel, Switzerland, 2017; Volume 73, pp. 149–160.
19. Pecharsky, V.K.; Holm, A.P.; Gschneider, K.A., Jr.; Rink, R. Massive magnetic-field-induced structural transformation in Gd_5Ge_4 and the nature of the giant magnetocaloric effect. *Phys. Rev. Lett.* **2003**, *91*, 197204. [CrossRef] [PubMed]
20. Fujita, A.; Fukamichi, K.; Koyama, K.; Watanabe, K. X-ray diffraction study in high magnetic fields of magnetovolume effect in itinerant-electron metamagnetic $La(Fe_{0.88}Si_{0.12})_{13}$ compound. *J. Appl. Phys.* **2004**, *95*, 6687–6689. [CrossRef]
21. Koyama, K.; Kanomata, T.; Matsukawa, T.; Watanabe, K. Magnetic field effect on structural property of $MnFeP_{0.5}As_{0.5}$. *Mater. Trans.* **2005**, *46*, 1753–1756. [CrossRef]
22. Ma, Y.; Awaji, S.; Watanabe, K.; Matsumoto, M.; Kobayashi, N. Effect of high magnetic field on the two-step martensitic-phase transition in Ni_2MnGa. *Appl. Phys. Lett.* **2000**, *76*, 37–39. [CrossRef]
23. Koyama, K.; Watanabe, K.; Kanomata, T.; Kainuma, R.; Oikawa, K.; Ishida, K. Observation of field-induced reverse transformation in ferromagnetic shape memory alloy $Ni_{50}Mn_{36}Sn_{14}$. *Appl. Phys. Lett.* **2006**, *88*, 132505. [CrossRef]
24. Ito, W.; Ito, K.; Umetsu, R.Y.; Kainuma, R.; Koyama, K.; Watanabe, K.; Fujita, A.; Oikawa, K.; Ishida, K.; Kanomata, T. Kinetic arrest of martensitic transformation in the NiCoMnIn metamagnetic shape memory alloy. *Appl. Phys. Lett.* **2008**, *92*, 021908. [CrossRef]

25. Mitsui, Y.; Koyama, K.; Watanabe, K. X-ray diffraction measurements in high magnetic fields and at high temperatures. *Sci. Technol. Adv. Mater.* **2009**, *9*, 014612. [CrossRef] [PubMed]
26. Mitsui, Y.; Koyama, K.; Ito, W.; Umetsu, R.Y.; Kainuma, R.; Watanabe, K. Observation of reverse transformation in metamagnetic shape memory alloy $Ni_{40}Co_{10}Mn_{34}Al_{16}$ by high-field X-Ray diffraction measurements. *Mater. Trans.* **2010**, *51*, 1648–1650. [CrossRef]

metals

MDPI

Article

Analytical Investigation of the Cyclic Behavior of Smart Recentering T-Stub Components with Superelastic SMA Bolts

Junwon Seo [1], Jong Wan Hu [2,3,*] and Kyoung-Hwan Kim [2]

[1]   Department of Civil and Environmental Engineering, South Dakota State University, Brookings, SD 57007, USA; junwon.seo@sdstate.edu
[2]   Department of Civil and Environmental Engineering, Incheon National University, Incheon 22012, Korea; crom99@inu.ac.kr
[3]   Incheon Disaster Prevention Research Center, Incheon National University, Incheon 22012, Korea
*   Correspondence: jongp24@incheon.ac.kr; Tel.: +82-32-835-8463

Received: 16 July 2017; Accepted: 12 September 2017; Published: 21 September 2017

Abstract: Partially restrained (PR) bolted T-stub connections have been widely used in replacement of established fully restrained (FR) welded connections, which are susceptible to sudden brittle failure. These bolted T-stub connections can permit deformation, easily exceeding the allowable limit without any fracture because they are constructed with a design philosophy whereby the plastic deformation concentrates on bolt fasteners made of ductile steel materials. Thus, the PR bolted connections take advantage of excellent energy dissipation capacity in their moment and rotation behavior. However, a considerable amount of residual deformation may occur at the bolted connection subjected to excessive plastic deformation, thereby requiring additional costs to recover the original configuration. In this study, superelastic shape memory alloy (SMA) bolts, which have a recentering capability upon unloading, are fabricated so as to solve these drawbacks, and utilized by replacing conventional steel bolts in the PR bolted T-stub connection. Instead of the full-scale T-stub connection, simplified T-stub components subjected to axial force are designed on the basis of a basic equilibrium theory that transfers the bending moment from the beam to the column and can be converted into equivalent couple forces acting on the beam flange. The feasible failure modes followed by corresponding response mechanisms are taken into consideration for component design with superelastic SMA bolts. The inelastic behaviors of such T-stub components under cyclic loading are simulated by advanced three-dimensional (3D) finite element (FE) analysis. Finally, this study suggests an optimal design for smart recentering T-stub components with respect to recentering and energy dissipation after observing the FE analysis results.

Keywords: T-stub components; partially-restrained (PR) bolted connections; superelastic shape memory alloys (SMAs); prying action mechanism; failure modes; finite element (FE) analysis

1. Introduction

After the 1994 Northridge (CA, USA) and 1995 Kobe (Japan) earthquakes, partially restrained (PR) bolted connections have been utilized as an alternative to fully restrained (FR) welded connections that exhibit brittle failure within the allowable deformation limit [1–3]. In these PR bolted connections, beam members are connected by fastening connection components (i.e., end-plate, T-stub, and clip angle components) to column members with tension bolts, and thus bending moments transferred from beams are delivered to connections, including shear forces. The PR bolted connections can accommodate a rotation angle greater than the allowable limit (i.e., typically 0.03 radian for plastic deformation) before structural beam members reach full plastic moment. This is because tension bolts where plastic deformation may concentrate are fabricated with carbon steel materials, which are rich in ductility [4–6]. Contrary to the FR welded connections, the PR bolted connections cope flexibly with axial couple forces converted from bending moment, and simultaneously possess excellent energy dissipation capacity without rapid strength degradation [7–9]. When structural beam members are subjected to either initial yielding or local buckling as preliminary collapse, the typical PR bolted connections exhibit delayed deterioration to withstand additional force until the tension bolts installed arrive at ultimate strength [9–11].

Although the PR bolted connections can be substituted for the existing FE welded connections, they are restricted to modern structures, which are becoming larger and higher, and thus it is necessary to improve their performance. The tension bolts made of ductile carbon steel prevent breaking or brittle failure by permitting the concentration of plastic deformation, but immoderate residual deformation may occur at the PR bolted connection [4–6,12]. Extra cost is required to recover the original condition after strong excitation. For this reason, this study suggests new PR bolted connections that are integrated with superelastic shape memory alloy (SMA) bolts replacing conventional steel bolts in an effort to reduce the residual deformation at the connection, enhancing the recentering capability. Unless other connection members are prone to generate plastic yielding, superelastic SMA-bolted connections can recover the initial shape without any permanent displacement after the removal of the applied load.

Temperature-dependent stress and strain curves for SMA materials are presented in Figure 1. At the martensite phase transformation temperature, SMA materials are susceptible to residual deformation upon unloading, and exhibit a shape memory effect where an additional heating process is required to revert to the original shape [13]. The crystallographic conversion from martensite phase to austenite phase, which provides a recentering capability to the SMA material, takes place during this heating process [13,14]. The superelastic effect, which can automatically recover the original conditions without heating, may be observed at the austenite phase transformation temperature, which is above the martensite one, and thus superelastic SMA materials behave according to the flag-shaped hysteresis loop shown in the figure [14]. Due to the improvement of manufacturing technology, SMA materials have been able to generate the superelastic effect at room temperature for the last 30 years. In addition to medical, mechanical, and electronic instruments, superelastic SMA materials have been utilized as dampers, passive control devices, and fasteners in the civil engineering field since the early 2000s [13,14]. In particular, some representative studies on the behavioral characteristics of superelastic SMA bolted moment connections (e.g., end plate and T-stub connections) in terms of recentering and energy dissipation have been recently carried out by researchers [15–18].

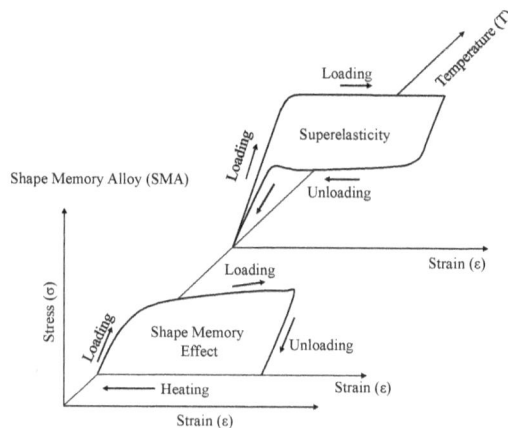

Figure 1. Temperature-dependent stress and strain curves for SMA (shape memory alloy) materials.

In this study, T-stub connections that firmly clamp structural beams to column members are designed with tension bolts made by superelastic SMAs, as shown in Figure 2. The T-stub connections can delivery bending moment (M) as well as shear force (V) from beams to columns. T-stub components including tension bolts are subjected to couple forces (P) transferred from bending moment (M), as follows:

$$M = VL, \tag{1}$$

$$P = M/d, \tag{2}$$

where L indicates the length of the beam and d denotes the depth of the beam. The simplified T-stub components instead of the full-scale T-stub connections are intended to simulate inelastic behavior under cyclically axial loads and directly evaluate the performance of superelastic SMA bolts in the T-stub component. The several T-stub components are designed with variable design parameters such as T-stub flange thickness and bolt gauge length, which can determine the extent of the prying action mechanism, and then their possible failure modes are investigated through the study of the theoretical prying action model. In lieu of experimental tests, the behaviors of such T-stub components are reproduced by advanced finite element (FE) analysis, generally used for estimating structural performance. According to the occurrence of plastic yielding in the steel T-stub member, recentering and energy dissipation capacity for T-stub components will be evaluated together with accompanying SMA bolt behavior. Finally, optimal design methodology is proposed to make the best of structural performance with respect to recentering capability and energy dissipation capacity in the T-stub component.

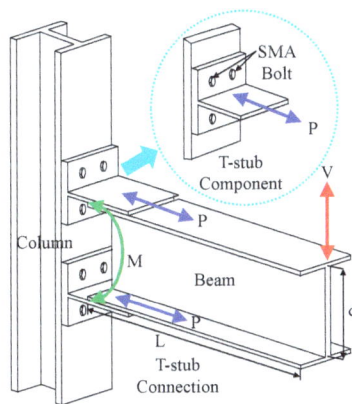

Figure 2. Schematic of the T-stub component and connection.

2. Prying Model

The prediction of the ultimate capacity for the T-stub component is quite complex in that several failure modes (e.g., yielding and fracture) correlate with each other and are tied up with uncertainties related not only to material properties but also to fabrication tolerances [7,8]. The prying action mechanism achieved by some assumption that T-stub flanges and tension bolts are considered to be beam members and spring supports, respectively, becomes a representative model to predict its response and failure mode. In this study, the prying action model specified in the AISC-LRFD (American Institute of Steel Construction-Load Resistance Factored Design) design guideline is adopted to evaluate the ultimate capacity of the T-stub component [18]. This prying action model was based upon one of the most popular models proposed by Kulak et al. [19]. The T-stub components subjected to axial force are accompanied by the yielding of the T-stub flange and the fracture of the tension bolt according to the increase of the applied load. The failure modes determined by the prying action mechanism have a significant influence on recentering and energy dissipation at the T-stub component. Thus, several T-stub components are designed by regulating the amount of prying action, which can be simultaneously affected by T-stub flange thickness and bolt gauge length.

The typical prying action mechanism occurring at the T-stub component is illustrated in Figure 3. When axial force is applied to the T-stub web, as shown in Figure 3a, T-stub flanges with relatively thin thickness are susceptible to deflection dominated by bending moment. The tension bolts can be modeled as deformable spring elements acting as flexible supports, and clamped to the column flange. As seen in this figure, prying force (Q), referred as to surplus reaction force, takes place at the edge of the T-stub flange, and increases the summation of bolt reaction forces, thereby stirring up preliminary failure. The equilibrium state defined in Equation (3) can be established by this prying action mechanism in the T-stub component. The amount of such prying force can be minimized by either increasing the thickness of the T-stub flange (t_f) or decreasing the length of the bolt gauge (g_t). When the intensity of the prying force gradually increases, the deformation caused by the deflection of the T-stub flange and the tension bolt becomes dominant [16]. This degrades the ultimate capacity of the T-stub component because the resistance strength against bending moment is relatively smaller than that against axial force. In this case, the T-stub component has a tendency to easily create the yielding of the T-stub flange as the preliminary failure. The geometric definition of the prying action model, which is appropriate to determine the ultimate capacity of the T-stub component, is presented in Figure 3b. In the prying action model suggested by Kulak et al., equivalent bolt reaction forces and maximum bending moments are assumed to act at the inside edge of the bolt shank rather than at the center of the bolt shank [16]. This prerequisite is based on the fact that more bolt head pressure is distributed into the T-stub flange inside the bolt head than outside the bolt head. It is adequate to use a' and b', defined in Equations (3) and (4), instead of a and b, with the aim of elucidating static equilibrium.

$$\Sigma B = P + Q, \tag{3}$$

$$a' = \left(a + \frac{d_b}{2} \right), \tag{4}$$

$$b' = \left(b - \frac{d_b}{2} \right), \tag{5}$$

where d_b indicates the diameter of bolt shank. The ultimate capacity of the T-stub component can be estimated on the basis of three failure modes formed by taking the relationship between T-stub flanges and tension bolts into consideration (see Figure 4). These three failure modes consist of the formation of plastic yielding in the T-stub flange (see Figure 4a), tension bolt fractures combined with flange yielding (see Figure 4b), and pure tension bolt fracture without any prying action (see Figure 4c), which correspond to Equations (6)–(8), respectively:

$$P = \frac{(1 + \delta)}{4b'} p F_y t_f^2, \tag{6}$$

$$P = \frac{\Sigma B a'}{a' + b'} + \frac{p F_y t_f^2}{4(a' + b')}, \tag{7}$$

$$P = \Sigma B, \tag{8}$$

where F_y indicates the yield stress of the T-stub flange fabricated by general carbon steel, and ΣB represents the summation of bolt reaction forces under ultimate stress. F_y and ΣB are applied to 325 MPa and 540 MPa, respectively. Moreover, p and δ represent the effective width of the T-stub flange and the ratio of the net section area to the gross section area, respectively, and thus are expressed as follows:

$$p = \frac{2W_{\text{T-stub}}}{n_{tb}}, \tag{9}$$

$$\delta = 1 - \frac{d_h}{p},$$ (10)

where n_{tb} and d_h stand for the number of the used tension bolts and the diameter of the bolt hole. $W_{T\text{-stub}}$ represents the width of the T-stub.

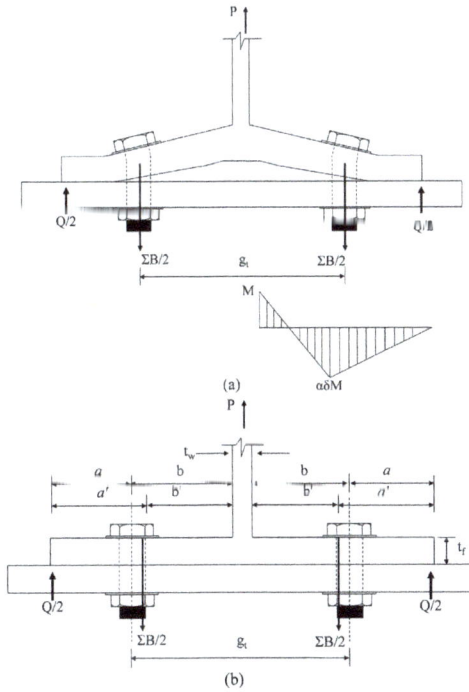

Figure 3. Typical prying action mechanism: (**a**) Prying action mechanism; (**b**) geometric definition.

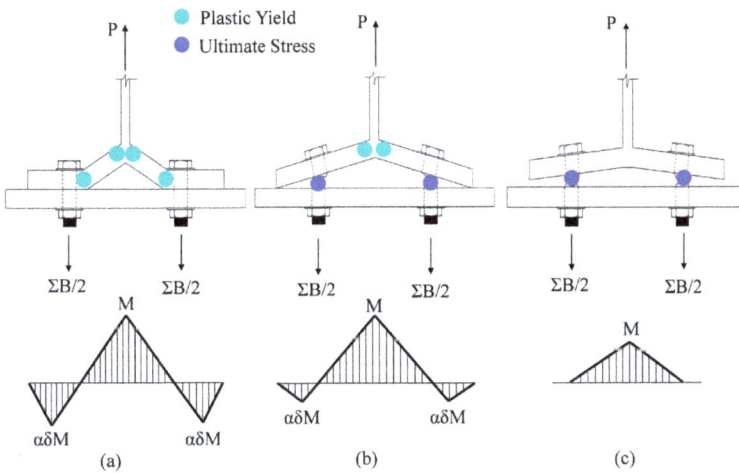

Figure 4. Three possible failure modes: (**a**) Mode 1 ($\alpha > 1$); (**b**) Mode 2 ($0 < \alpha \leq 1$); (**c**) Mode 3 ($\alpha \leq 0$).

The intensity of the prying action mechanism can be classified to use the parameter defined in Equation (11) as follows:

$$\alpha = \left(\frac{1}{\delta}\right)\left(\frac{4Pb'}{pt_f^2F_y} - 1\right). \tag{11}$$

This parameter for the level of the prying action present (α) is determined by the ratio of the moment on the centerline of the T-stub component to the moment on the centerline of the bolt shank, including the magnitude of prying force on the edge of the T-stub flange [8,19]. When the prying action parameter exceeds one ($\alpha > 1.0$), the thickness of the T-stub flange is sufficient to generate yielding caused by bending deflection. Therefore, in this case (Mode 1), the T-stub flange can be considered a fixed support beam, and the plastic hinges presented in Figure 4a take the place of the T-stub flange. Compared to the other two failure modes (Mode 2 and Mode 3), this failure mode provokes a relatively larger bending moment and prying force on the T-stub flange.

The prying action mechanism under this failure mode may serve as one of the significant points to degrade strength capacity for the entire T-stub component. The T-stub flange fabricated with carbon steel undergoes the stage of its material strain hardening after initial yielding, thereby providing extra resistance against external force until ultimate stress comes to the T-stub flange. If ultimate bolt fractures happen prior to reaching ultimate stress in the T-stub flange, tension bolt prying combined with flange yielding (Mode 2) will be preceded by initial yielding failure of the T-stub failure which is referred to as Mode 1. Once T-stub flanges are redesigned with increasing thickness and decreasing bolt gauge length, T-stub components have a high possibility to generate Mode 2 or Mode 3 failure shape. When $0 < \alpha \leq 1$, tension bolt fractures mainly produced by bending prying deflection and flange yielding take place at the same time (Mode 2). Finally, when the prying action parameter is below zero ($\alpha < 0$), the prying force becomes zero in that the T-stub flange completely separates from the column plate. Due to the absence of prying action in the tension bolt, ultimate fracture caused by pure axial force may occur after reaching ultimate stress.

On the basis of the prying action model, the general solution used for determining the capacity of the T-stub component can be plotted as the function of the T-stub flange thickness (t_f). The theoretical solution for the failure modes according to increasing flange thickness is presented in Figure 5. The line segment OB computed by Equation (6) stands for the capacity of the T-stub component, which can be determined based on the preliminary yielding failure mainly caused by prying action on the T-stub flange. The bolt fractures combined with flange yielding are displayed in the line segment BC formulated by Equation (7), while pure tension bolt fractures without any prying action are reproduced in the line segment CE computed by Equation (8). The line segment CE remains constant over flange thickness, regardless of T-stub flange capacity. In this study, the summation of bolt reaction forces (ΣB) is taken as the value of 430 kN obtained from the product of ultimate stress (540 MPa) to the total bolt shank area. In Figure 5, a' and b' are applied to 108 mm and 32 mm, respectively (Case 1 to Case 8, see Table 1). The expanded line segment AB can be computed by Equation (7) under the condition where the prying action parameter α exceeds 1.0. This line segment can be utilized for determining the capacity of the T-stub component when ultimate bolt fractures preceded by initial flange yielding occur due to bolt prying. As shown in Figure 5, the capacity of the T-stub component whose flange thickness is less than 22.5 mm can be evaluated by the line segment OB indicating initial flange yielding (Mode 1), and then finally determined by tension bolt fractures on the extended line segment AB (Mode 2) during the increase of loading capacity. In this case, tension bolts are subjected to not only axial displacement but also bending deflection caused by prying action, and thus have a tendency to more easily fail under relatively smaller external force as compared to the T-stub component belonging to the line segment CE. This implies that prying action acting on the head of the tension bolt can weaken resistance strength capacity for the T-stub component. The design of the T-stub component models in accordance with the prying action mechanism will be treated in the next section.

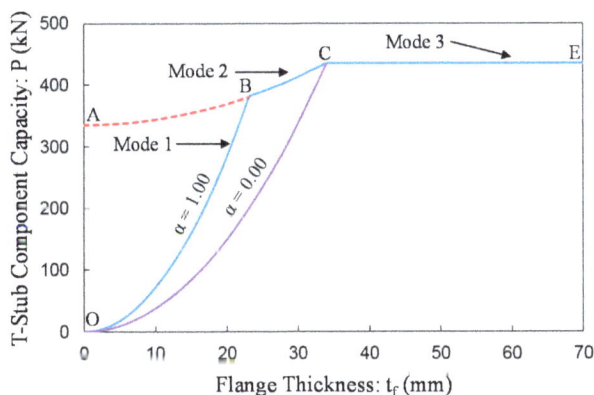

Figure 5. General solution for the failure modes according to increasing flange thickness.

3. Model Design

3.1. Specimen Design

At the behavior of the T-stub component with superelastic SMA bolts, the extent of recentering capability and energy dissipation capacity depends on the level of the prying action present. The T-stub component models presented herein were designed with several prying action levels and different failure modes, which were regulated by flange thickness and bolt gauge length. After completing the design of the T-stub component models, their behaviors were simulated by advanced nonlinear finite element (FE) analyses. Ultimately, optimal design methodology for smart recentering T-stub components with superelastic SMA bolts is intended to be proposed through the observation of the FE analysis results in an effort to make the best use of their recentering capability and energy dissipation capacity.

The component details including the dimensions of the component models are illustrated in Figure 6. The geometric sizes for individual dimensions defined in the Figure 6 are summarized in Table 1 for all T-stub component models. These component models can be classified as two groups in accordance with the ratios of H_1 to H_2 (i.e., $H_1/H_2 = 1/3$ and $H_1/H_2 = 2/3$), which can regulate the amount of prying action. The first group models ranging from Case 1 to Case 8 was designed with $H_1/H_2 = 1/3$. On the other hand, the second group models ranging from Case 9 to Case 16 was designed with $H_1/H_2 = 2/3$ indicating relatively larger prying action than the first group models. The thickness of the T-stub flange varied from 50 mm to 15 mm, and was equally divided into eight model cases. All T-stub component models were designed with 20 mm web thickness (t_w), 16 mm bolt diameter (d_h), and 140 mm bolt length. According to the thickness of the T-stub flange, the thickness of the column plate (t_c) was varied to fit the net length of the tension bolt (e.g., $t_f + t_c = 140$ mm). The superelastic SMA bolts were designed with 430 kN ultimate strength computed by the product of ultimate stress (540 MPa) to total bolt shank area.

Table 1. Geometric sizes.

Model ID	T-Stub Size											Bolt and Column Size			Failure Mode (Capacity)		
	$W_{\text{T-stub}}$	t_f	t_w	H_1	H_2	d_h	a	a'	b	b'	α	Bolt (Diameter × Length)	ΣB	t_c	Mode 1	Mode 2	Mode 3
Case-1	300 mm	50 mm	20 mm	100 mm	300 mm	16 mm	100 mm	108 mm	40 mm	32 mm	−0.60	16 × 140 mm	430 kN	90 mm	-	-	430 kN
Case-2	300 mm	45 mm	20 mm	100 mm	300 mm	16 mm	100 mm	108 mm	40 mm	32 mm	−0.48	16 × 140 mm	430 kN	95 mm	-	-	430 kN
Case-3	300 mm	40 mm	20 mm	100 mm	300 mm	16 mm	100 mm	108 mm	40 mm	32 mm	−0.31	16 × 140 mm	430 kN	100 mm	-	-	430 kN
Case-4	300 mm	35 mm	20 mm	100 mm	300 mm	16 mm	100 mm	108 mm	40 mm	32 mm	−0.06	16 × 140 mm	430 kN	105 mm	-	-	430 kN
Case-5	300 mm	30 mm	20 mm	100 mm	300 mm	16 mm	100 mm	108 mm	40 mm	32 mm	0.25	16 × 140 mm	430 kN	110 mm	-	412 kN	-
Case-6	300 mm	25 mm	20 mm	100 mm	300 mm	16 mm	100 mm	108 mm	40 mm	32 mm	0.74	16 × 140 mm	430 kN	115 mm	-	388 kN	-
Case-7	300 mm	20 mm	20 mm	100 mm	300 mm	16 mm	100 mm	108 mm	40 mm	32 mm	1.64	16 × 140 mm	430 kN	120 mm	284 kN	369 kN	-
Case-8	300 mm	15 mm	20 mm	100 mm	300 mm	16 mm	100 mm	108 mm	40 mm	32 mm	3.58	16 × 140 mm	430 kN	125 mm	160 kN	354 kN	-
Case-9	300 mm	50 mm	20 mm	200 mm	300 mm	16 mm	50 mm	58 mm	90 mm	82 mm	0.09	16 × 140 mm	430 kN	90 mm	-	394 kN	-
Case-10	300 mm	45 mm	20 mm	200 mm	300 mm	16 mm	50 mm	58 mm	90 mm	82 mm	0.22	16 × 140 mm	430 kN	95 mm	-	353 kN	-
Case-11	300 mm	40 mm	20 mm	200 mm	300 mm	16 mm	50 mm	58 mm	90 mm	82 mm	0.40	16 × 140 mm	430 kN	100 mm	-	317 kN	-
Case-12	300 mm	35 mm	20 mm	200 mm	300 mm	16 mm	50 mm	58 mm	90 mm	82 mm	0.66	16 × 140 mm	430 kN	105 mm	-	285 kN	-
Case-13	300 mm	30 mm	20 mm	200 mm	300 mm	16 mm	50 mm	58 mm	90 mm	82 mm	1.06	16 × 140 mm	430 kN	110 mm	249 kN	257 kN	-
Case-14	300 mm	25 mm	20 mm	200 mm	300 mm	16 mm	50 mm	58 mm	90 mm	82 mm	1.74	16 × 140 mm	430 kN	115 mm	173 kN	233 kN	-
Case-15	300 mm	20 mm	20 mm	200 mm	300 mm	16 mm	50 mm	58 mm	90 mm	82 mm	2.98	16 × 140 mm	430 kN	120 mm	111 kN	215 kN	-
Case-16	300 mm	15 mm	20 mm	200 mm	300 mm	16 mm	50 mm	58 mm	90 mm	82 mm	5.65	16 × 140 mm	430 kN	125 mm	62 kN	200 kN	-

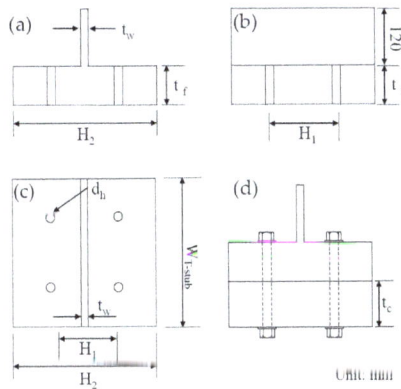

Figure 6. Component details: (**a**) Front view; (**b**) side view; (**c**) plan view; (**d**) test setup.

The design results based on the prying action mechanism for the four selected models are presented in Figure 7. The Case 3 model and the Case 7 model belong to the first group designed in accordance with $H_1/H_2 = 1/3$, but the flange thickness of the Case 3 model is twice more thick than that of the Case 7 model. In addition, the Case 11 and Case 15 models were designed with $H_1/H_2 = 2/3$. The Case 11 has 40 mm flange thickness, meaning that its flange thickness is also twice more thick than the flange thickness of the Case 15 model. Therefore, the Case 3 model and the Case 7 model lie in similar H_1/H_2 ratios comparable to the Case 11 and Case 15, respectively. The capacity of the T-stub component grows to be deteriorate when the level of the prying action present controlled by the ratio of H_1/H_2 begin to increase under the same flange thickness. It can be also found that the failure mode may shift from bolt fracture to flange yielding. The Case 3 model subjected to pure bolt fracture ($\alpha = -0.31$ for Mode 3) possesses 430 kN ultimate strength while the Case 11 model experiencing bolt fracture combined with flange yielding ($\alpha = 0.40$ for Mode 2) has smaller 317 kN ultimate strength. The Case 7 model undergoing Mode 1 failure shape ($\alpha = 1.64$) is subjected to preliminary flange yielding when force arrives at 284 kN. Thereafter, it will ultimately fail by bolt fracture combined with plastic hinge on the T-stub flange when force increases to 369 kN. Similarly, the Case 15 model undergoes Mode 2 failure shape at 215 kN force after preceded by Mode 1 failure shape at 111 kN force. The ultimate capacity and failure mode of individual T-stub component models are summarized in Table 1.

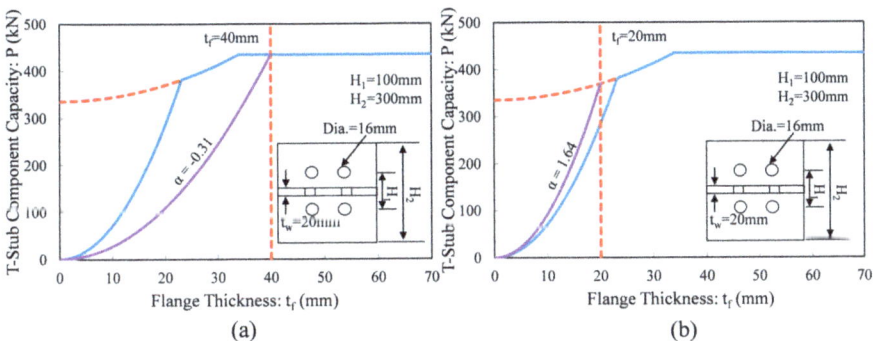

Figure 7. *Cont.*

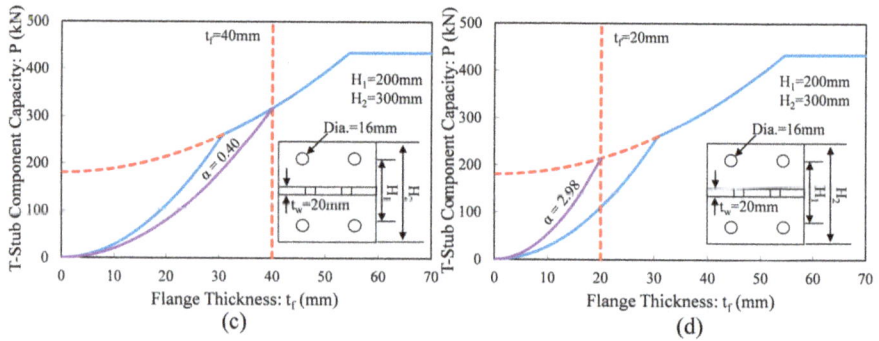

Figure 7. Design results based on prying action mechanism: (**a**) Case-3; (**b**) Case-7; (**c**) Case-11; (**d**) Case-15

3.2. 3D Finite Element Models

In lieu of experimental tests, cyclic behaviors for the T-stub components incorporating superelastic SMA tension bolts are simulated through nonlinear FE analyses, and evaluated in terms of recentering effect and energy dissipation capacity. The ABAQUS program (Version 6.12, Simulia, Pawtucket, RI, USA) was used to perform these nonlinear FE analyses [20]. The FE models were fabricated with 16 T-stub component specimens presented in Table 1, as considering not only geometric nonlinearity but also nonlinear material property during analyses. The FE models composed of element mesh, loading, and boundary conditions (BCs) are illustrated in Figure 8. These FE models are made up of 3D eight-node solid elements (C3D8) with nonlinear material properties. All 3D eight-node solid elements used in the Case 3 model were made in the form of the 10 mm × 10 mm × 10 mm cube (16,300 element numbers). The number of the used elements varies according to the model cases. The individual parts were divided by structural meshes in order to align the solid elements in an effective manner, and thus the analytical prediction becomes more accurate. The separate step was generated with an intention to impose initial adjustment displacement acting as pretension force on the middle surface of the tension bolt. Instead of modeling the entire column, the BCs applied to the column plate can be substituted to save time and cost.

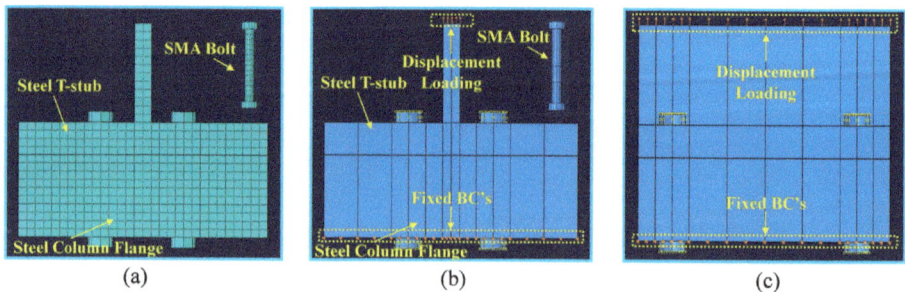

Figure 8. 3D FE (Finite element) models (Case 3 models): (**a**) Element mesh; (**b**) loading and BCs (boundary conditions) of front view; (**c**) loading and BCs (boundary conditions) of side view.

After applying initial adjustment displacement, the second step was independently generated to impose displacement-controlled cyclic loading on the edge of the T-stub web as shown in Figure 8a,b. Figure 9 shows the used displacement loading history, where the amplitude of the cycle ascends to 4mm maximum displacement as time goes on. This displacement loading history was reproduced by utilizing the amplitude function provided in the ABAQUS program, and assigned to the released BCs (loading points) imposed on the edge of the T-stub web. The displacement-controlled loading history can be converted into reaction force by the summation of the forces at the loading faces. During compression, referred to as minus displacement loading, compression bearing occurs at the contact surface between T-stub flange and column plate, and reaction force including stiffness increase very fast. When applying compression to the T-sub component, the summation of bolt reaction forces is negligible. Accordingly, the amplitude of compression displacement, which is much smaller than that of tension displacement, should be applied to FE models. The stress contour fields were measured at the specific displacement loading points such as S1, S2, S3, and S4.

Besides geometric nonlinearity, material nonlinearity was considered during FE modelling for the purpose of performing accurate analyses. Including column plates, T-stub members were constructed with typical Gr.50 carbon steel materials which include 325 MPa yield stress, 200 GPa elastic modulus, and 1.5% strain hardening ratio. The behavior of this steel material was reproduced by utilizing the isotropic hardening material model, which displays stress relaxation, Bauschinger effect, and ratchetting response [20]. The nonlinear material properties were assigned to individual parts composed of 3D solid elements. The default material model used to numerically simulate the behavior of the superelastic SMA material is absent in the ABAQUS program, and thus the user-defined material (UMAT) model was employed to FE modeling [21]. The material input properties required to operate the UMAT model were obtained from uni-axial pull-out tests performed by DesRoches et al. [14]. The stress and strain curves for superelastic SMA materials are shown in Figure 10. The material input properties needed to reproduce the behavior of the superelastic SMA during FE analyses were taken as 40 GPa for elastic modulus, 0.33 for Poisson's ratio, 440 MPa for martensite start stress, 540 MPa for martensite finish stress, 250 MPa for austenite start stress, 140 MPa for austenite finish stress, 0.045 radian for slip strain, and 25 °C for transformation reference temperature. The simulated stress and strain curve was modeled as a series of straight lines, which reflect the path of each phase transformation. In general, the simulated stress and strain curve exhibits good agreement with the experimental stress and strain curve.

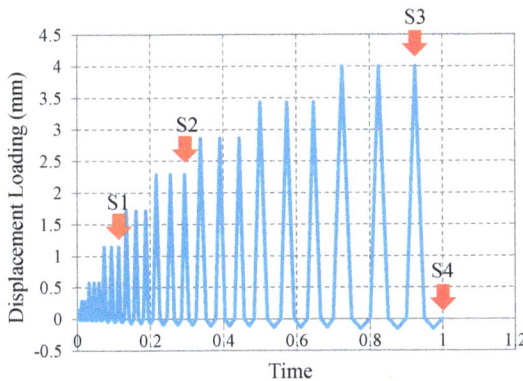

Figure 9. Displacement loading history.

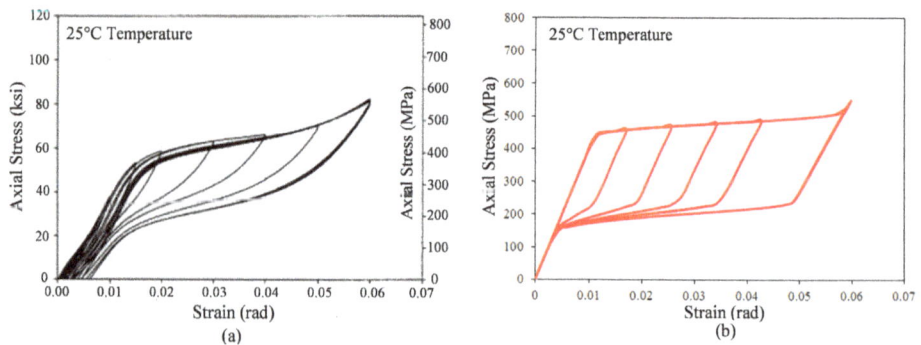

Figure 10. Material properties of SMA materials: (**a**) Experimental curve; (**b**) simulated curve.

4. Analysis Results

4.1. Behavior of T-Stub Components

The behavior of the T-stub component models constructed with variable flange thickness and bolt gauge length was reproduced through advanced nonlinear FE analyses, and then their performance was appraised based on the observation of the analysis results according to design parameters. The intensity of the prying action mechanism as well as the magnitude of prying force will be evaluated by examining the deflection of the T-stub flange, which is measured at the specific loading points. The force and displacement curves simulated by FE analyses are presented in Figure 11. These curves are obtained by imposing the displacement-controlled loading history (see Figure 9) on the edge of the T-stub web. When applying 4 mm maximum displacement, residual displacements upon unlading (Δ_{res}) are also examined to assess recentering capability according to each model case. The Case 1 model as one of the representative cases with Mode 3 failure shape exhibits the flag-shaped hysteresis loop which completely coincides with the behavior of the superelastic SMA materials, and arrives at 400 kN maximum force under 4 mm displacement. Overall, the four models classified as the first group with Mode 3 failure shape (Case 1 to Case 4) completely recovered their original configuration without residual displacement, meaning that T-stub flanges maintain an elastic state throughout FE analyses and that plastic deformation concentrates on the superelastic SMA bolts. The yielding of the hysteresis loop generally occurs at approximately 380 kN force under below 1.7 mm. It is affirmed that these four models are able to permit 7 mm displacement to reach 430 kN force corresponding to ultimate strength capacity for superelastic SMA bolts, as considering their post-yield stiffness in the hysteresis loop.

As the parameter α is close to the unit, the T-stub component models that fail by Mode 2 failure shape begin to create residual displacement upon unloading. Furthermore, ultimate strength measured at 4 mm starts to decrease below 400 kN force. Although the T-stub component models whose the parameter α is close to zero arrive at 400 kN force under 4 mm displacement, they are not expected to reach the ultimate strength of the tension bolts (430 kN) due to flange yielding and bolt prying. There is a tendency to augment the amount of residual displacement and to reduce the capacity of the T-stub component when the parameter α regulated by flange thickness and bolt gauge length starts to increase. The T-stub component models which fail by Mode 1 failure shape exhibit strength capacity well below 400 kN force under 4 mm displacement, and engender considerable amount of residual displacement. Strength degradation mainly happens due to plastic hinge caused by the prying action mechanism on the T-stub flange. The Case 16 model which has the largest prying action present just generates 189 kN force at 4 mm displacement and as much as 4 mm residual displacement upon unloading. In addition to residual displacement, the lowering of the energy dissipation capacity represented by the area of the hysteresis loop can be attributed to the large amount of the prying action mechanism. As compared with the first group models with $H_1/H_2 = 1/3$, the second group models with relatively larger $H_1/H_2 = 1/3$ are prone to strength degradation and residual displacement under the same flange thickness because of more severe prying action.

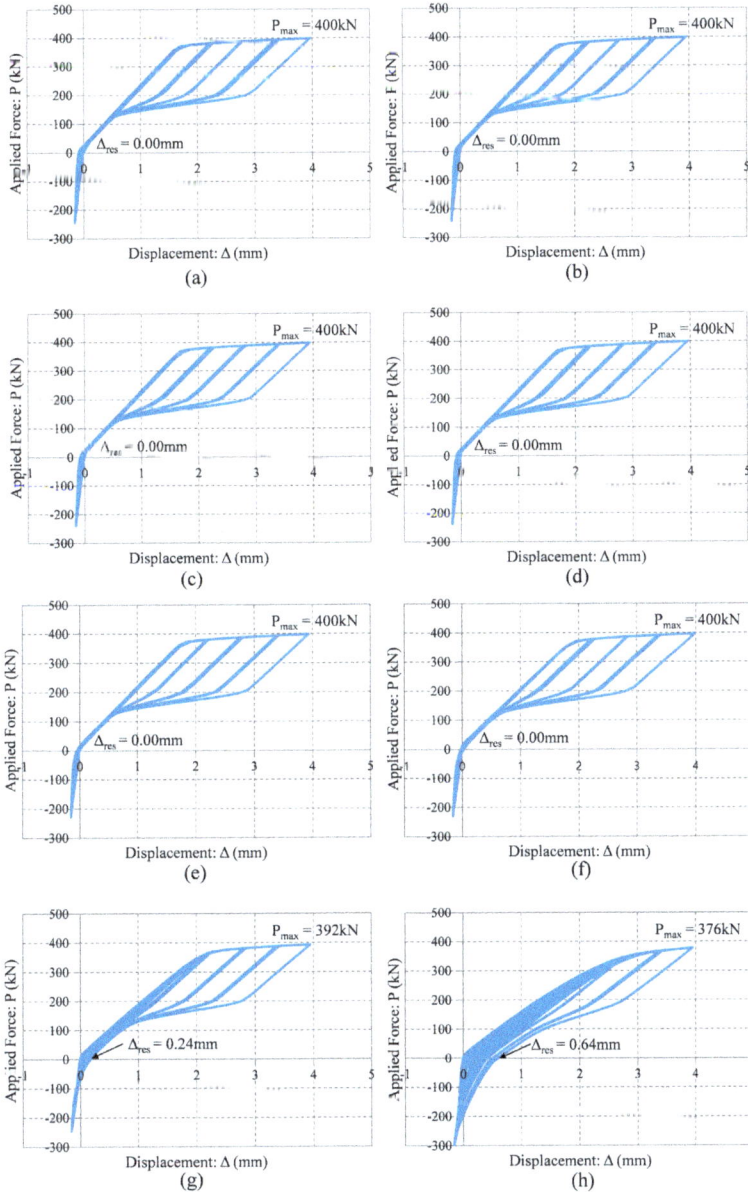

Figure 11. *Cont.*

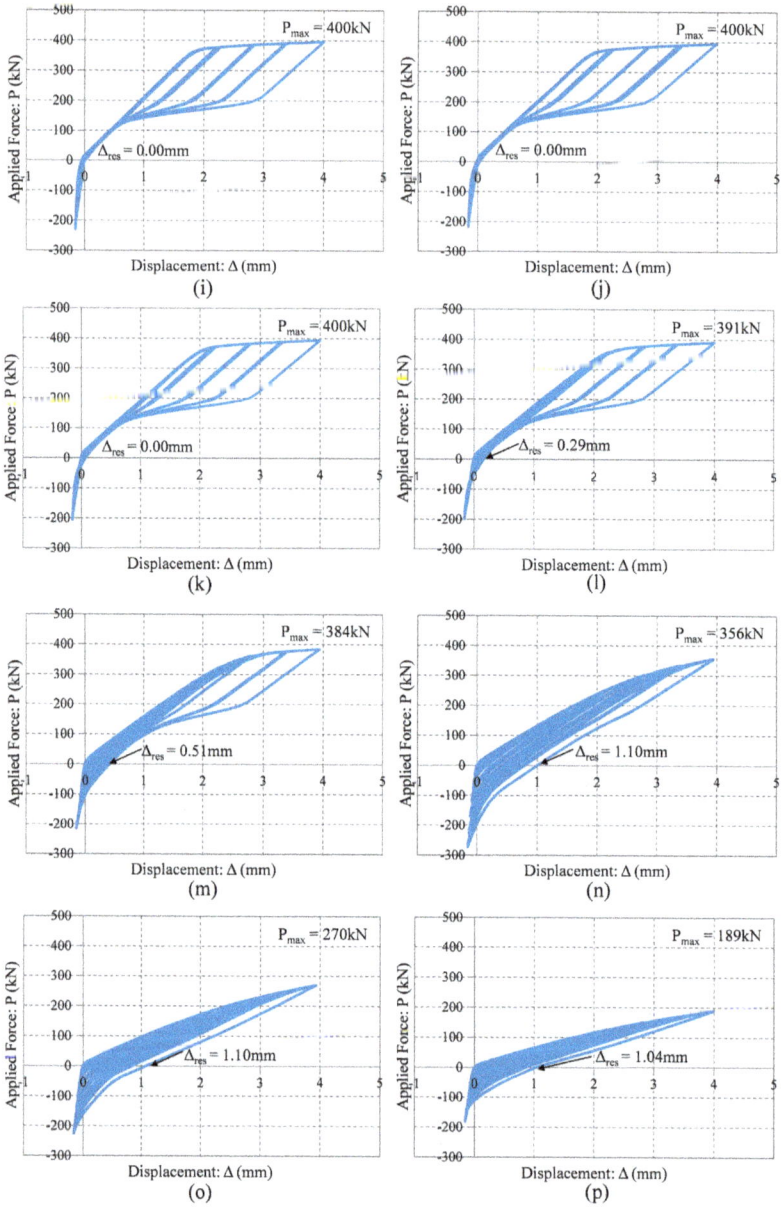

Figure 11. Total displacement and force curves for the 16 model cases: (**a**) Case-1 ($\alpha = -0.60$); (**b**) Case-2 ($\alpha = -0.48$); (**c**) Case-3 ($\alpha = -0.31$); (**d**) Case-4 ($\alpha = -0.06$); (**e**) Case-5 ($\alpha = 0.25$); (**f**) Case-6 ($\alpha = 0.74$); (**g**) Case-7 ($\alpha = 1.64$); (**h**) Case-8 ($\alpha = 3.58$); (**i**) Case-9 ($\alpha = 0.09$); (**j**) Case-10 ($\alpha = 0.22$); (**k**) Case-11 ($\alpha = 0.40$); (**l**) Case-12 ($\alpha = 0.66$); (**m**) Case-13 ($\alpha = 1.06$); (**n**) Case-14 ($\alpha = 1.74$); (**o**) Case-15 ($\alpha = 2.98$); (**p**) Case-16 ($\alpha = 5.65$).

4.2. Stress Filed Contours

It is also necessary to check both deformed configurations and stress field contours so as to elucidate the response mechanism acting on the T-stub component. According to the individual displacement loading steps (S1 to S4 defined in Figure 9), stress field contours distributed over the T-stub components and deformed configurations with five times the deformation scale factor (DSF) are presented in Figure 12. The Case 3 model completely separates the T-stub flange from the column plate, and relatively thick flange thickness enables this model to maintain original configuration with nearly zero deflection. The stress field contours distributed over the Case 3 model show that the T-stub member remain elastic during all displacement loading steps. Instead, stress concentration takes place at the tension bolts subjected to axial force without bolt prying. The Case 7 model and the Case 11 model separate the T-stub flange from the column plate, but deformed like an arch with the distribution of slightly higher stress as compared to the Case 3 model. The T-stub flange of the Case 15 model subjected to Mode 1 failure shape is still attached to the column plate when even applied to 4 mm displacement loading. This model is perfectly bent as a bow by severe prying action, and plastic hinges are found at the stress filed contour distributed over the T-stub flange. Although the Case 7 model and the Case 15 model completely recover to original shape, residual stress greater than their yield stress is observed at the T-stub flange. Both models are prone to generating bending deflection arising due to the prying action mechanism, and susceptible to severe damage even under relatively small loading. It can be shown that the intensity of prying action on the T-stub flange leads to directly decreasing the capacity of the T-stub component. As we expect, superelastic SMA bolts show nearly zero residual stress at their original position (S4).

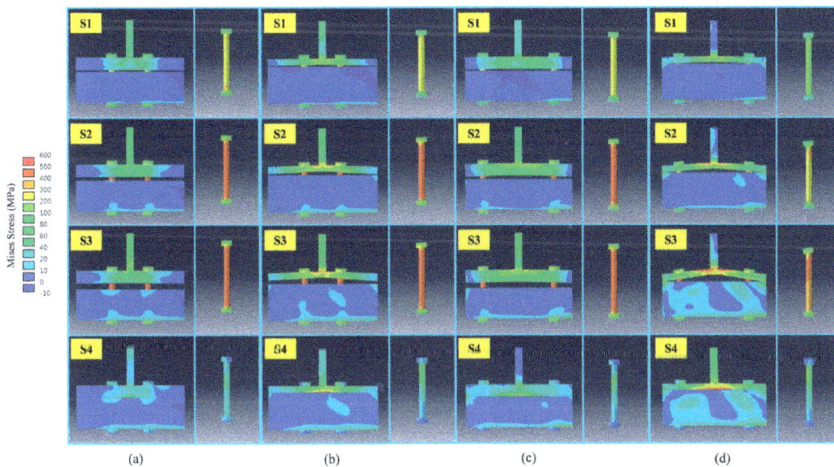

Figure 12. Von Mises stress contours and deformation configurations (DSF = 5.0; DSF: Deformation scale factor): (**a**) Case-3; (**b**) Case-7; (**c**) Case-11; (**d**) Case-15.

4.3. Recentering Capabilities of Shape Memory Alloy (SMA) Bolts

In this section, the behavior of the superelastic SMA bolts is required to be examined with an intention to verify the adequacy of FE modeling and to clarify the response of the T-stub component, including energy dissipation capacity. The T-stub members presented herein are connected to the column plates by using four superelastic SMA bolts. The average bolt reaction force versus uplift displacement curves for four model cases selected in this study are presented in Figure 13. The T-stub component models with Mode 3 failure shape indicating relatively smaller prying action response display good energy dissipation at the behavior of the superelastic SMA bolts. For instance, the Case

3 model has the largest area of the hysteresis loop standing for energy dissipation capacity while the Case 15 model dissipate nearly zero amount of kinetic energy within the elastic range. In the Case 3 model, the maximum bolt uplift is able to reach almost 4 mm displacement, equal to the maximum displacement load because the superelastic SMA bolts majorly undergo axial force without bolt prying. Small discrepancy only results from the deformation of the T-stub web. On the other hand, the maximum bolt uplift of the Case 15 model only reaches 1.5 mm under maximum displacement loading, indicating that most of the deformable contributions may be attributed to the deformation of the T-stub flange arising due to prying. For these reasons, the Case 3 model that can produce the largest bolt uplift among other models makes the best use of recentering and energy dissipation, which are supplied by the superelastic SMA bolts.

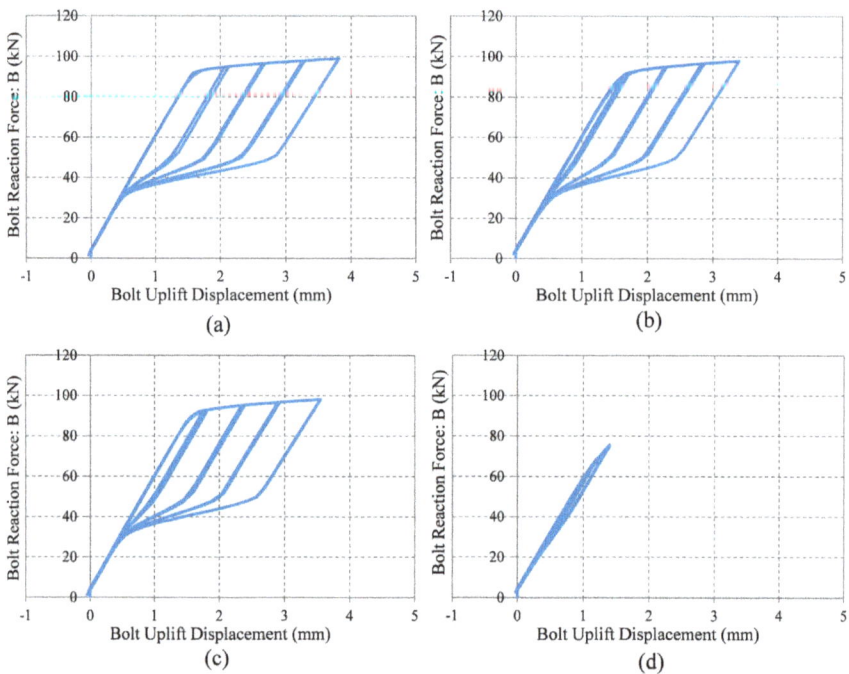

Figure 13. Average bolt reaction force and uplift displacement curves for four model cases: (a) Case-3; (b) Case-7; (c) Case-11; (d) Case-15.

For another performance examination, the total bolt reaction force versus applied force curves are presented in Figure 14. The force versus displacement curves for the T-stub component under the last cycle of the displacement loading history are presented in Figure 15. The energy dissipation capacity of the T-stub component models can be evaluated by computing the area of these curves. The zero prying lines indicating $P = \Sigma B$ are also plotted as the red dotted lines. The total bolt reaction forces ($4B$) begin at the non-zero value because of initial bolt pretension, and then gradually increase as the forces applied to the T-stub web ascend. In the beginning of the displacement loading history, prying force (Q) acting on the edge of the T-stub flange increases owing to the initial bolt pretension [22]. However, except for the Case 15 model, the effect of prying force gradually dwindles away to nothing as the displacement loading history goes on. Finally, applied force coincides with total bolt reaction force, thereby meeting the equilibrium state. The Case 15 model can preserve prying force even under the last cycle of the displacement loading history, and thus meets the equilibrium conditions defined in Equation (3). When compression force is imposed on the T-stub component, the T-stub flange is subjected to bearing compression. This bearing compression leads to the applied force increasing very quickly, but has no influence on the bolt reaction force.

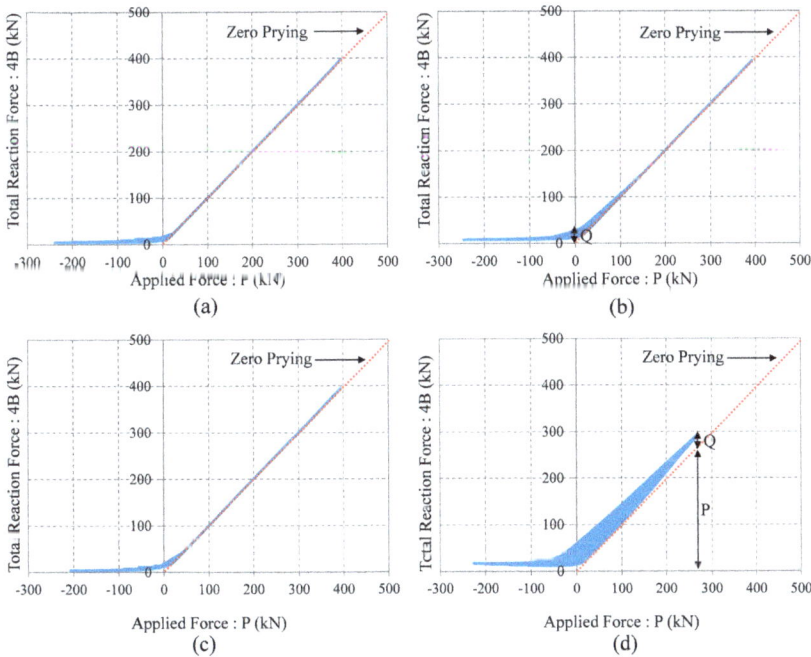

Figure 14. Prying action results: (**a**) Case-3; (**b**) Case-7; (**c**) Case-11; (**d**) Case-15.

4.4. Energy Dissipation Capacity of T-Stub Components

As shown in Figure 15, the Case 3 model possesses the best energy dissipation capacity, while the Case 15 model possesses the least energy dissipation capacity at the hysteresis loop. In particular, the behavior of the Case 3 model is very similar to the behavior of the superelastic SMA bolts combined in parallel, which is characterized by the flag-shaped hysteresis loop. In addition to recentering, the kinetic energy dissipated in the T-stub component mainly results from the response of the superelastic SMA bolts rather than the metallic yielding of the T-stub flange. Accordingly, optimal design methodology that makes the best use of recentering capability and energy dissipation capacity at the smart recentering T-stub component can be achieved by concentrating plastic deformation on the superelastic SMA bolts, and simultaneously other component members shall be designed to maintain the elastic condition.

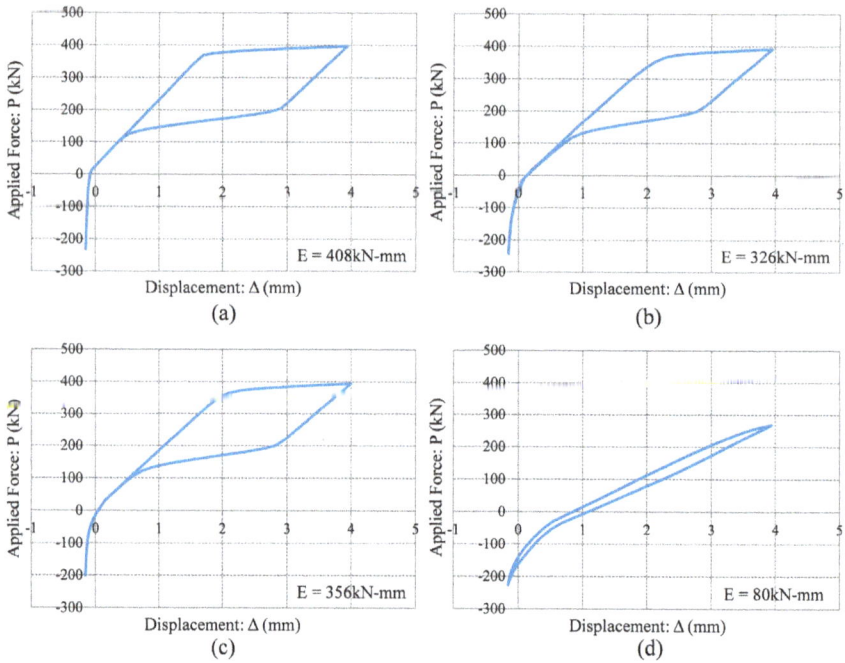

Figure 15. Energy dissipation capacity for four model cases during the last loading cycle: (**a**) Case-3; (**b**) Case-7; (**c**) Case-11; (**d**) Case-15.

5. Concluding Remarks

In this study, T-stub components that are subjected to axial couple forces converted from bending moment usually acting on the bolted PR connection are designed, and then new smart recentering T-stub components utilizing superelastic SMA bolts are proposed with an aim to enhance their recentering capability. The T-stub component models are constructed with different design parameters such as flange thickness and bolt gauge length. Instead of conducting the experimental tests, the behavior of such T-stub components is simulated by the FE analyses. Including the capacity of the T-stub component, recentering capability and energy dissipation capacity can be determined by the prying action mechanism regulated by the design parameters used. The T-stub component models designed with thinner flange thickness and longer bolt gauge length have a relatively smaller strength capacity due to the occurrence of flange yielding as the preliminary failure mode, and do not generate adequate deformation of the tension bolts. However, the T-stub models with thick flange thickness and short bolt gauge length lead to ultimate bolt fractures arising due to axial force instead of bolt prying, and thus accommodate enough resistance against external force. These models can effectively reduce the residual displacement by appropriately utilizing recentering and energy dissipation provided by the superelastic SMA bolts, and can behave as a flag-shaped hysteresis loop. Therefore, smart recentering T-sub components equipped with superelastic SMA bolts should be constructed based on the design concept that plastic deformation only concentrates on the tension bolts in order to maximize their performance with respect to recentering and energy dissipation.

Acknowledgments: This research was supported by a grant (17CTAP-C129811-01) from the Land Transport Technology Promotion Research Project Program, funded by the Ministry of Land, Infrastructure, and Transport of the Korean government.

Metals **2017**, *7*, 386

Author Contributions: Jong Wan Hu and Kyoung-Hwan Kim conducted finite element analysis, and checked finite element analysis results. Jong Wan Hu and Junwon Seo wrote the manuscript.

Conflicts of Interest: The authors declare no conflict of interest.

References

1. Leon, R.T.; Hajjar, J.F.; Gustafson, M.A. Seismic response of composite moment-resisting connections I: Performance. *ASCE J. Struct. Eng.* **1998**, *124*, 868–876. [CrossRef]
2. Green, T.P.; Leon, R.T.; Rassati, G.A. Bidirectional tests on partially restrained, composite beam-column connections. *ASCE J. Struct. Eng.* **2004**, *130*, 320–327. [CrossRef]
3. Hu, J.W.; Kang, Y.S.; Choi, D.H.; Park, T. Seismic design, performance, and behavior of composite-moment frames with steel beam-to-concrete filled tube column connections. *KSSC Int. J. Steel Struct.* **2010**, *10*, 177–191. [CrossRef]
4. Hu, J.W.; Leon, R.T. Analyses and evaluations for composite-moment frames with SMA PR-CFT connections. *Nonlinear Dyn.* **2011**, *65*, 433–455. [CrossRef]
5. Hu, J.W.; Kim, D.; Choi, E. Numerical investigation on the cyclic behavior of smart recentering clip-angle connections with superelastic shape memory alloy fasteners. *Proc. Inst. Mech. Eng. Part C J. Mech. Eng. Sci.* **2013**, *227*, 1315–1327. [CrossRef]
6. Hu, J.W.; Choi, E.; Leon, R.T. Design, analysis, and application of innovative composite PR connections between steel beams and CFT columns. *Smart Mater. Struct.* **2011**, *20*, 025019. [CrossRef]
7. Swanson, J.A.; Leon, R.T. Bolted steel connections: Tests on T-stub components. *ASCE J. Struct. Eng.* **2000**, *126*, 50–56. [CrossRef]
8. Swanson, J.A.; Leon, R.T. Stiffness modeling of bolted T-stub connection components. *ASCE J. Struct. Eng.* **2001**, *127*, 498–505. [CrossRef]
9. Hu, J.W. Seismic Performance Evaluations and Analyses for Composite Moment Frames with Smart SMA PR-CFT Connections. Ph.D. Thesis, Georgia Institute of Technology, Atlanta, GA, USA, April 2008.
10. Leon, R.T. *Seismic Performance of Bolted and Riveted Connections. Background Reports: Metallurgy, Fracture Mechanics, Welding, Moment Connections, and Frame System Behavior*; FEMA Publication No. 288; Federal Emergency Management Association (FEMA): Washington, DC, USA, 1997.
11. Leon, R.T. Analysis and design problems for PR composite frames subjected to seismic loads. *Eng. Struct.* **1998**, *20*, 364–371. [CrossRef]
12. Hu, J.W. Seismic analysis and parametric study of SDOF lead-rubber bearing (LRB) isolation systems with recentering shape memory alloy (SMA) bending bars. *J. Mech. Sci. Technol.* **2016**, *30*, 2987–2999. [CrossRef]
13. Song, G.; Ma, N.; Li, H. Applications of shape memory alloys in civil structures. *Eng. Struct.* **2006**, *28*, 1266–1274. [CrossRef]
14. DesRoches, R.; McCormick, J.; Delemont, M. Cyclic properties of superelastic shape memory alloy wires and bars. *ASCE J. Struct. Eng.* **2004**, *130*, 38–46. [CrossRef]
15. Amin, M.; Ghassemieh, M. Shape memory alloy-based moment connections with superior self-centering properties. *Smart Mater. Struct.* **2016**, *25*, 075028.
16. Yang, G.; Lee, D.H. Method for increasing the energy dissipation capacity of T-stub connections. *Int. J. Steel Struct.* **2015**, *15*, 595–603. [CrossRef]
17. Abolmaali, A.; Treadway, J.; Aswath, P.; Lu, F.K.; McCarthy, E. Hysteresis behavior of T-stub connections with superelastic shape memory fasteners. *J. Construct. Steel Res.* **2006**, *62*, 831–838. [CrossRef]
18. American Institute of Steel Construction (AISC). *Manual of Steel Construction: Load and Resistance Factor Design (LRFD)*, 3rd ed.; American Institute of Steel Construction (AISC): Chicago, IL, USA, 2001.
19. Kulak, G.L.; Fisher, J.W.; Struik, J.H.A. *Guide to Design Criteria for Bolted and Riveted Joint*, 2nd ed.; John Wiley & Sons: Hoboken, NJ, USA, 1987.
20. *Abaqus*, version 6.12; Standard User's Manual; Simulia: Pawtucket, RI, USA, 2012.

Metals **2017**, *7*, 386

21. Auricchio, F.; Sacco, E. A one-dimensional model for superelastic shape-memory alloys with different properties between martensite and austenite. *Int. J. Non-Linear Mech.* **1997**, *32*, 1101–1114. [CrossRef]
22. Croccolo, D.; De Agostinis, M.; Fini, S.; Olmi, G. Tribological properties of bolts depending on different screw coatings and lubrications: An experimental study. *Tribol. Int.* **2017**, *107*, 199–205. [CrossRef]

metals

MDPI

Article

Damping Characteristics of Inherent and Intrinsic Internal Friction of Cu-Zn-Al Shape Memory Alloys

Shyi-Kaan Wu [1], Wei-Jyun Chan [2] and Shih-Hang Chang [2,*]

[1] Department of Materials Science and Engineering, National Taiwan University, Taipei 106, Taiwan; skw@ntu.edu.tw
[2] Department of Chemical and Materials Engineering, National I-Lan University, I-Lan 260, Taiwan; r0423009@ms.niu.edu.tw
* Correspondence: shchang@niu.edu.tw; Tel.: +886-2-2363-7846

Received: 28 August 2017; Accepted: 22 September 2017; Published: 28 September 2017

Abstract: Damping properties of the inherent and intrinsic internal friction peaks ($IF_{PT} + IF_I$) of Cu-xZn-11Al (x = 7.0, 7.5, 8.0, 8.5, and 9.0 wt. %) shape memory alloys (SMAs) were investigated by using dynamic mechanical analysis. The Cu-7.5Zn-11Al, Cu-8.0Zn-11Al, and Cu-8.5Zn-11Al SMAs with $(IF_{PT} + IF_I)_{\beta_3(L2_1) \to \gamma_3'(2H)}$ peaks exhibit higher damping capacity than the Cu-7.0Zn-11Al SMA with a $(IF_{PT} + IF_I)_{\beta_3(L2_1) \to \gamma_3'(2H)}$ peak, because the γ_3' martensite phase possesses a 2H type structure with abundant movable twin boundaries, while the β_3' phase possesses an 18R structure with stacking faults. The Cu-9.0Zn-11Al SMA also possesses a $(IF_{PT} + IF_I)_{\beta_3(L2_1) \to \gamma_3'(2H)}$ peak but exhibits low damping capacity because the formation of γ phase precipitates inhibits martensitic transformation. The Cu-8.0Zn-11Al SMA was found to be a promising candidate for practical high-damping applications because of its high ($IF_{PT} + IF_I$) peak with tan $\delta > 0.05$ around room temperature.

Keywords: shape memory alloys (SMAs); martensitic transformation; internal friction; dynamic mechanical analysis

---

1. Introduction

Shape memory alloys (SMAs) have been widely investigated for a broad range of applications because of their unique shape memory effect and superelasticity [1]. Numerous studies have shown that SMAs also exhibit a high damping capacity during martensitic transformation, and are effective for energy dissipation applications [2–5]. The damping capacity of SMAs is typically determined using an inverted torsion pendulum or dynamic mechanical analysis (DMA). During damping measurements, SMAs normally exhibit a significant internal friction peak (IF peak) at the martensitic transformation temperature, and the damping capacity is closely related to experimental parameters, including temperature rate, frequency, and applied strain amplitude [2,6].

The IF peak of SMAs typically comprises three individual terms (i.e., IF = $IF_{Tr} + IF_{PT} + IF_I$) [6–8]. IF_{Tr} denotes transient internal friction, which appears only at low frequencies and a non-zero temperature change rate. IF_{PT} is the inherent internal friction corresponding to phase transformation, which is independent of temperature rate. IF_I is the intrinsic internal friction of the austenitic or martensitic phase, and it depends strongly on microstructural properties such as dislocations, vacancies, and twin boundaries. It has been reported that IF_I is also temperature rate dependent, since time-dependent pinning affects the intrinsic damping and depends on the concentration of mobile pinning points throughout the heat treatment procedure during thermal cycling and deformation of Cu-based alloys [9–12]. The damping capacities of IF_{PT} and IF_I are usually more important than that of IF_{Tr} because most high-damping applications of SMAs are realized at a steady temperature.

Chang and Wu [13–22] have systematically studied the inherent and intrinsic internal friction ($IF_{PT} + IF_I$) peaks for various SMAs by applying DMA using the isothermal method. According to their results, TiNi-based SMAs exhibit acceptable damping capacities during martensitic transformations with tan δ > 0.02. Cu-Al-Ni and Ni_2MnGa SMAs show (IF_{PT} + IF_I) peaks above room temperature. However, the damping capacity of the (IF_{PT} + IF_I) peaks for these SMAs was not as good as expected. The martensitic transformation temperatures of Cu-Zn-Al SMAs can be controlled by carefully adjusting their chemical composition [23], suggesting that Cu-Zn-Al SMAs have the potential to exhibit significant (IF_{PT} + IF_I) peaks above room temperature. Numerous studies have reported the transformation behaviors, crystal structures, and mechanical properties of Cu-Zn-Al SMAs [23–35]. Besides, several works in the literature have also investigated the internal friction properties of Cu-Zn-Al SMAs [36–40]. To date, the damping properties of (IF_{PT} + IF_I) peaks for Cu-Zn-Al SMA have not been investigated, to the best of our knowledge. Therefore, the aim of this study was to investigate the inherent and intrinsic internal friction properties of Cu-Zn-Al SMAs with regard to their damping properties.

2. Materials and Methods

Polycrystalline samples of Cu-xZn-11Al (x = 7.0, 7.5, 8.0, 8.5, and 9.0 wt. %) SMAs were prepared from pure copper (purity 99.9 wt. %), zinc (purity 99.9 wt. %), and aluminum (purity 99.9 wt. %). The raw materials were melted at 1100 °C in an evacuated quartz tube for 6 h and then slowly cooled in the furnace to room temperature to form Cu-xZn-11Al SMA ingots. The ingots were solution-treated at 850 °C for 12 h, followed by quenching in ice water. Each ingot was cut into bulks with dimensions of 30.0 mm × 6.0 mm × 3.0 mm for the DMA tests. The crystallographic features of the solution-treated Cu-xZn-11Al SMAs were determined using a Rigaku Ultima IV X-ray diffraction (XRD) instrument with Cu Kα radiation (λ = 0.154 nm) at room temperature. Microstructural observations of Cu-xZn-11Al SMAs were performed with a Tescan 5136MM scanning electron microscope (SEM). The chemical compositions of Cu-xZn-11Al SMAs were determined with an Oxford Instruments x-act energy-dispersive X-ray spectroscope (EDS). According to the EDS results, the determined chemical compositions for Cu-xZn-11Al with x = 7.0, 7.5, 8.0, 8.5, and 9.0 SMAs were Cu-7.14Zn-11.25Al, Cu-7.57Zn-11.29Al, Cu-8.18Zn-11.40Al, Cu-8.69Zn-10.91Al, and Cu-8.98Zn-10.83Al, respectively, suggesting that the determined chemical composition of each specimen was close to that of the expected composition. The martensitic transformation temperatures and the transformation enthalpy (ΔH) values for Cu-xZn-11Al SMAs were determined using a TA Q10 differential scanning calorimeter (DSC) under a constant cooling/heating rate of 10 °C·min^{-1}. The damping capacity (tan δ) for Cu-xZn-11Al SMAs was determined using TA 2980 DMA equipment with a single cantilever clamp and a liquid nitrogen cooling apparatus. The parameters for the DMA tests were a temperature rate of 3 °C·min^{-1}, frequency of 1 Hz, and strain amplitude of 1.0×10^{-4}. The inherent and intrinsic internal friction (IF_{PT} and IF_I) of Cu-xZn-11Al SMAs were also investigated by DMA, but tested under a temperature rate of 1 °C·min^{-1}, frequency of 10 Hz, and strain amplitude of 1.0×10^{-4}.

3. Results and Discussion

3.1. XRD and SEM Results

Figure 1 presents the XRD results of Cu-xZn-11Al (x = 7.0, 7.5, 8.0, 8.5, and 9.0) SMAs. As shown in the figure, the Cu-7.0Zn-11Al SMA exhibits diffraction peaks at 2θ = 39.0°, 41.1°, 43.0°, 44.7°, 46.3°, and 47.8°, which correspond to the $(12\bar{2})$, (201), (0018), $(12\bar{8})$, (1210), and $(20\bar{1}0)$ diffraction planes, respectively, of the 18R structure [24]. Furthermore, the Cu-7.5Zn-11Al, Cu-8,0Zn-11Al, and Cu-8.5Zn-11Al SMAs exhibit diffraction peaks at 2θ = 40.0°, 42.7°, and 45.3°, which correspond to the (200), (002), and (201) diffraction planes, respectively, of the 2H structure [34]. The Cu-9.0Zn-11Al SMA shows only a sharp diffraction peak at 2θ = 43.2°, which corresponds to the (220) diffraction plane of the $L2_1$ structure. Therefore, we can conclude that the Cu-7.0Zn-11Al SMA is in the β'_3(18R) martensite phase at room temperature. On the other hand, the Cu-7.5Zn-11Al, Cu-8,0Zn-11Al, and

Cu-8.5Zn-11Al SMAs are in the γ_3'(2H) martensite phase, whereas the Cu-9.0Zn-11Al SMA is in the β_3(L2$_1$) parent phase at room temperature.

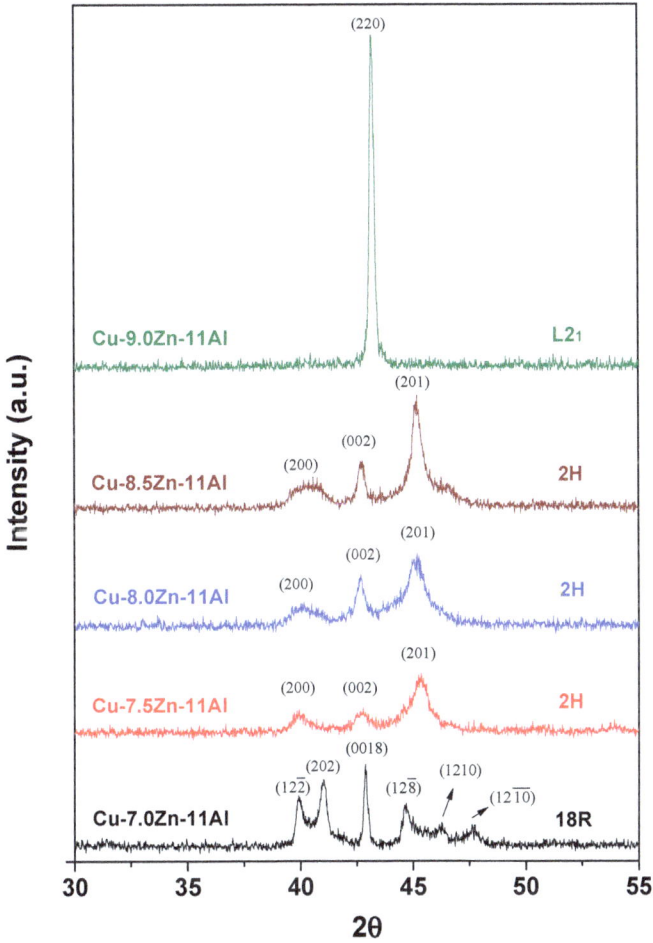

Figure 1. XRD patterns for Cu-xZn-11Al SMAs with various Zn contents.

Figure 2a–e shows the SEM images for the Cu-7.0Zn-11Al, Cu-7.5Zn-11Al, Cu-8.0Zn-11Al, Cu-8.5Zn-11Al, and Cu-9.0Zn-11Al SMAs, respectively. As shown in Figure 2a, the Cu-7.0Zn-11Al SMA exhibits typical self-accommodating, zig-zag groups of β_3' martensite variants, indicating that it possesses an 18R structure at room temperature [33]. Figure 2b illustrates that the γ_3'(2H) martensite structure is dominant in the Cu-7.5Zn-11Al SMA. Figure 2c,d demonstrate that the Cu-8.0Zn-11Al and Cu-8.5Zn-11Al SMAs exhibit a γ_3'(2H) martensite phase at room temperature, where the γ_3'(2H) martensite plates become broader and more significant with the increase in Zn content. Figure 2e reveals that the Cu-9.0Zn-11Al SMA does not show obvious martensite variants because it adopts the β_3(L2$_1$) parent phase at room temperature. However, abundant γ phase precipitates appear along the grain boundaries of the alloy. This feature has also been reported by Condó et al. [31], wherein the γ phase normally formed when the electron/atom (*e*/*a*) ratios of Cu-Zn-Al SMAs were above 1.53.

Figure 2f depicts the magnification of the precipitates presented in Figure 2e and shows that the γ phase precipitates possess a typical crisscross structure with a size of approximately 20 μm.

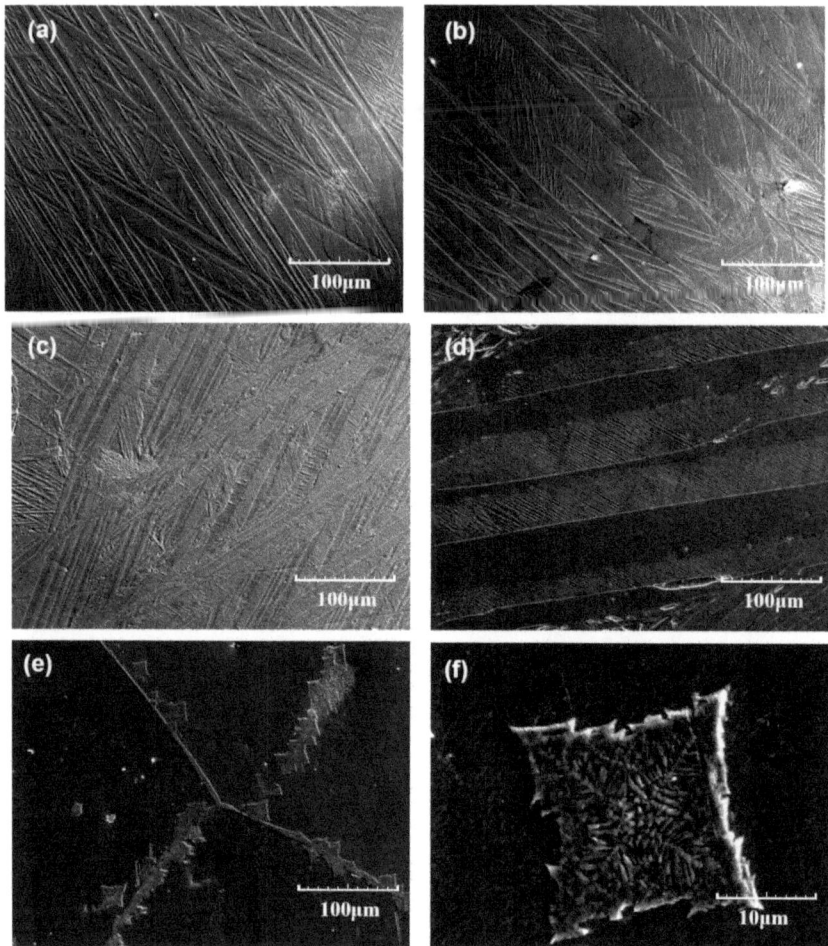

Figure 2. SEM images of (**a**) Cu-7.0Zn-11Al, (**b**) Cu-7.5Zn-11Al, (**c**) Cu-8.0Zn-11Al, (**d**) Cu-8.5Zn-11Al, and (**e**) Cu-9.0Zn-11Al SMAs under the same magnification. (**f**) A magnified SEM image of (**e**).

3.2. DSC Results

Figure 3 shows the DSC curves of Cu-xZn-11Al (x = 7.0, 7.5, 8.0, 8.5, and 9.0) SMAs. As shown in Figure 3, each Cu-xZn-11Al SMA exhibits a single martensitic transformation peak in both cooling and heating curves. According to the XRD and SEM results shown in Figures 1 and 2, one can conclude that the Cu-7.0Zn-11Al SMA possesses a $\beta_3(L2_1) \rightarrow \beta'_3(18R)$ martensitic transformation in cooling and a $\beta'_3(18R) \rightarrow (L2_1)$ transformation in heating. On the other hand, the Cu-7.5Zn-11Al, Cu-8.0Zn-11Al, and Cu-8.5Zn-11Al SMAs all exhibit a $\beta_3(L2_1) \rightarrow \gamma'_3(2H)$ martensitic transformation in cooling and a $\gamma'_3(2H) \rightarrow \beta_3(L2_1)$ transformation in heating. Although the Cu-9.0Zn-11Al SMA is in the $\beta_3(L2_1)$ parent phase at room temperature, according to the report by Ahlers and Pelegrina [27], the Cu-9.0Zn-11Al SMA should also exhibit a $\beta_3(L2_1) \leftrightarrow \gamma'_3(2H)$ martensitic transformation, for its e/a value was calculated to be 1.528. Figure 3 also shows that the martensite start (Ms) temperature of the Cu-xZn-11Al SMAs decreases significantly from 104.0 °C to −21.2 °C when Zn content is increased from x = 7.0 to 9.0. This is consistent with the study reported by Ahlers [23], in which the Ms temperature of the Cu-Zn-Al SMA depended strongly on its chemical composition. On the other hand, the ΔH values of the specimens shown in Figure 3 are not significantly different, as all are close to 6 J/g.

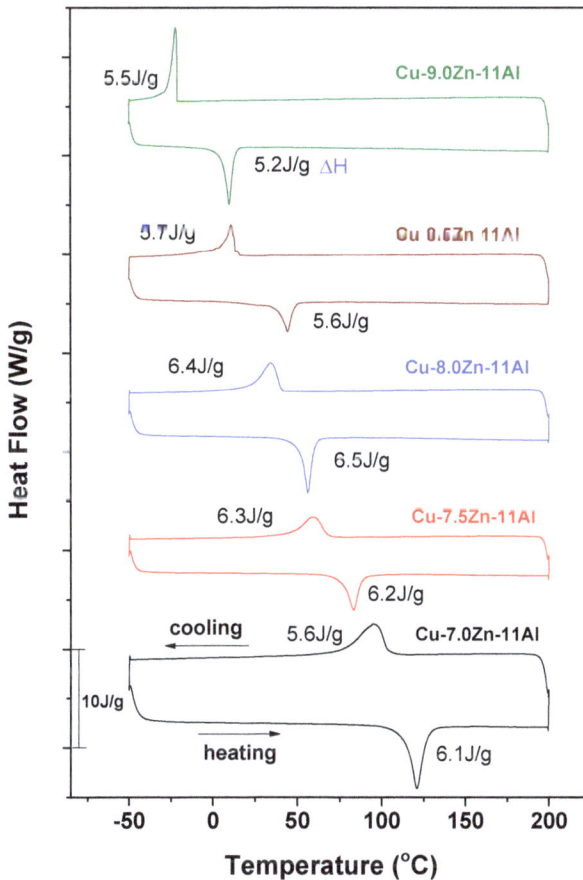

Figure 3. DSC curves for Cu-xZn-11Al SMAs with various Zn contents.

3.3. DMA Results

Figure 4 shows the DMA curves of Cu-xZn-11Al (x = 7.0, 7.5, 8.0, 8.5, and 9.0) SMAs measured at a controlled temperature rate of 3 °C·min^{-1}, frequency of 1 Hz, and strain amplitude of 1.0×10^{-4}. Only the DMA cooling curves are shown in Figure 4, for clarity. The Cu-7.0Zn-11Al SMA possesses a $\beta_3(L2_1) \rightarrow \beta_3'(18R)$ IF peak with tan δ = 0.065 at approximately 95.0 °C. Compared to the Cu-7.0Zn-11Al SMA, the Cu-7.5Zn-11Al, Cu-8.0Zn-11Al, and Cu-8.5Zn-11Al SMAs exhibit a more significant $\beta_3(L2_1) \rightarrow \gamma_3'(2H)$ IF peak with higher tan δ values, above 0.12. However, the IF peak temperatures for the Cu-7.5Zn-11Al, Cu-8.0Zn-11Al and Cu-8.5Zn-11Al SMAs were determined to be 59.8, 30.1, and −9.6 °C, respectively, which are much lower than that of the Cu-7.0Zn-11Al SMA. The Cu-9.0Zn-11Al SMA also possesses a $\beta_3(L2_1) \rightarrow \gamma_3'(2H)$ IF peak at approximately −11.4 °C; however, its tan δ value is only 0.082.

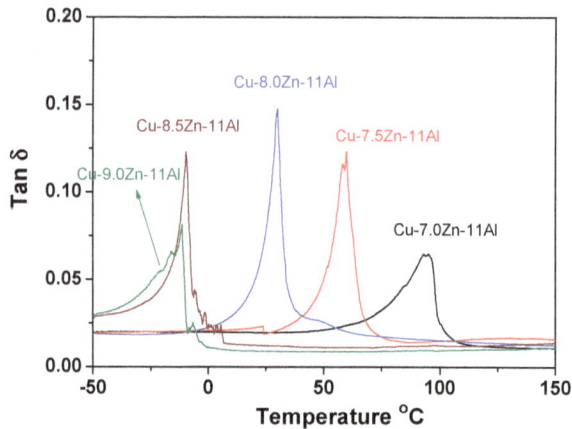

Figure 4. DMA tan δ curves measured at 1 Hz and 3 °C·min^{-1} cooling rate for Cu-xZn-11Al SMAs with various Zn contents.

3.4. IF$_{PT}$ and IF$_I$ Measurements

To investigate the inherent internal friction characteristics of Cu-xZn-11Al (x = 7.0, 7.5, 8.0, 8.5, and 9.0) SMAs, each specimen should be determined by DMA, but also assessed by the isothermal method reported previously [13]. A typical isothermal test-procedure can be described as follows: The SMA is initially cooled from the high temperature parent phase at a constant cooling rate and then maintained isothermally at a set temperature for a sufficient time interval to ensure that the IF$_{Tr}$ term decays completely, leaving only the IF$_{PT}$ and IF$_I$ terms. Then, the SMA should be heated to a sufficiently high temperature to ensure that the SMA is completely in the parent phase state. Subsequently, the SMA is cooled to another set temperature and kept at a constant temperature to determine the IF$_{PT}$ and IF$_I$ values at that temperature. The aforementioned isothermal method can effectively and accurately determine the IF$_{PT}$ and IF$_I$ values of most SMAs [13–22]. However, this method is not suitable for the Cu-xZn-11Al SMAs in this study, because the repeated thermal cycling may influence the martensitic transformation properties of Cu-xZn-11Al SMAs, as demonstrated in Figure 5, which shows the DSC curve of the Cu-7.5Zn-11Al SMA for 10 repeated heating and cooling cycles. As per this figure, the Ms temperature of the Cu-7.5Zn-11Al SMA gradually decreased from 76.4 to 71.5 °C over the course of 10 repeated thermal cycles. In addition, the ΔH value of the Cu-7.5Zn-11Al SMA decreased from 6.3 J/g to 5.4 J/g. The decreasing Ms temperature and ΔH value can be attributed to the introduction of defects and dislocations during repeated thermal cycling, depressing the martensitic transformations of the Cu-7.5Zn-11Al SMA. Similar results were also observed in a previous study on the Ti$_{51}$Ni$_{39}$Cu$_{10}$ SMA [16].

To address this issue, the IF$_{PT}$ and IF$_I$ values for Cu-xZn-11Al SMAs were also determined by DMA, but the DMA was conducted at a high frequency where the IF$_{Tr}$ term can be neglected [41]. In addition, Nespoli et al. [42] demonstrated that the IF values of SMAs determined by DMA at a 1 °C·min^{-1} cooling rate and 10 Hz frequency are very close to the IF$_{PT}$ and IF$_I$ determined under isothermal conditions. Accordingly, in this study, we used identical experimental parameters (1 °C·min^{-1} cooling rate and 10 Hz frequency) to determine the IF$_{PT}$ and IF$_I$ of Cu-xZn-11Al SMAs, and the results are presented in Figure 6. From Figure 6, it can be seen that the Cu-7.0Zn-11Al SMA possesses an inherent and intrinsic internal friction peak during the $\beta_3(L2_1) \rightarrow \beta_3'(18R)$ martensitic transformation $(IF_{PT} + IF_I)_{\beta_3(L2_1) \rightarrow \beta_3'(18R)}$ with tan δ = 0.026 at approximately 95.1 °C. Compared to the Cu-7.0Zn-11Al SMA, the Cu-7.5Zn-11Al SMA exhibited a higher $(IF_{PT} + IF_I)_{\beta_3(L2_1) \rightarrow \gamma_3'(2H)}$ peak with a higher tan δ value of 0.040, but at a lower temperature of approximately 56.5 °C. Figure 6 also shows that both the Cu-8.0Zn-11Al and Cu-8.5Zn-11Al SMAs exhibit a high $(IF_{PT} + IF_I)_{\beta_3(L2_1) \rightarrow \gamma_3'(2H)}$

peak with a tan δ value above 0.05 at approximately 32.2 and −6.8 °C, respectively. However, the Cu-9.0Zn-11Al SMA possesses a small $(\mathrm{IF_{PT}} + \mathrm{IF_I})_{\beta_3(L2_1) \to \gamma_3'(2H)}$ peak with tan δ = 0.030 at a low temperature of approximately −22.9 °C. In contrast to Figure 4, Figure 6 shows that the tan δ value of the $(\mathrm{IF_{PT}} + \mathrm{IF_I})$ peak for each specimen measured at 10 Hz is much lower than that of the corresponding IF peak measured at 1 Hz, suggesting that the $\mathrm{IF_{Tr}}$ term disappears when Cu-xZn-11Al SMAs are measured at 10 Hz. Accordingly, we calculated the contribution of the $(\mathrm{IF_{PT}} + \mathrm{IF_I})$ peak to the overall IF peak for each Cu-xZn-11Al SMAs, which was approximately 35% for all SMAs.

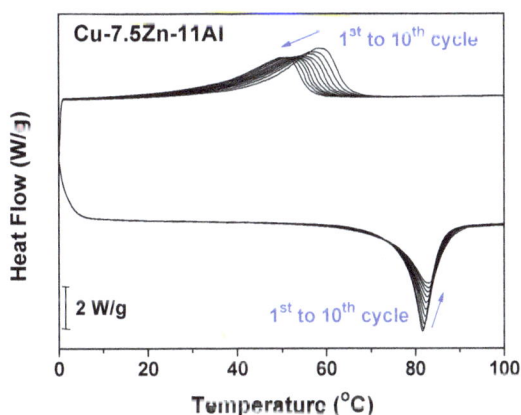

Figure 5. DSC curve of Cu-7.5Zn-11Al SMA determined for 10 repeating heating and cooling cycles.

Figure 6 also shows that the $(\mathrm{IF_{PT}} + \mathrm{IF_I})_{\beta_3(L2_1) \to \gamma_3'(2H)}$ peaks for the Cu-7.5Zn-11Al, Cu-8.0Zn-11Al, and Cu-8.5Zn-11Al SMAs (tan δ > 0.04) are much higher than that of $(\mathrm{IF_{PT}} + \mathrm{IF_I})_{\beta_3(L2_1) \to \beta_3'(18R)}$ for the Cu-7.0Zn-11Al SMA (tan δ = 0.026). In addition, the transformed γ_3' martensite phases of the Cu-7.5Zn-11Al, Cu-8.0Zn-11Al, and Cu-8.5Zn-11Al SMAs possess a 2H type structure with abundant internal twin boundaries, which are easily moved to dissipate energy during damping [34]. On the other hand, the transformed β_3' phase for the Cu-7.0Zn-11Al SMA only possesses an 18R structure with stacking faults, instead of movable twin boundaries [30]. Figure 6 reveals that the Cu-9.0Zn-11Al SMA also possesses an $(\mathrm{IF_{PT}} + \mathrm{IF_I})_{\beta_3(L2_1) \to \gamma_3'(2H)}$ peak during martensitic transformation, while exhibiting a lower tan δ value (0.030) compared to the other $(\mathrm{IF_{PT}} + \mathrm{IF_I})_{\beta_3(L2_1) \to \gamma_3'(2H)}$ peaks for Cu-xZn-11Al SMAs with lower Zn contents. This can be explained by the fact that abundant γ phase precipitates form in the Cu-9.0Zn-11Al SMA (Figure 2). These undesirable γ phase precipitates restrict the mobility of the parent phase/martensite interfaces, leading to a small $(\mathrm{IF_{PT}} + \mathrm{IF_I})_{\beta_3(L2_1) \to \gamma_3'(2H)}$ peak. Therefore, one can conclude that the Cu-8.0Zn-11Al SMA is more suitable for practical high-damping applications because of its high $(\mathrm{IF_{PT}} + \mathrm{IF_I})_{\beta_3(L2_1) \to \gamma_3'(2H)}$ peak with a tan δ value above 0.05 at around room temperature. However, according to the SEM results shown in Figure 2, the $\gamma_3'(2H)$ martensite plates in Cu-xZn-11Al SMAs become broader with the increase in Zn content. Increasing the width of the martensite band normally decreases the number of twin boundaries, suggests that Cu-xZn-11Al SMAs with higher Zn content should exhibit lower damping capacity. Nevertheless, this is not seen in the DMA results shown in Figures 4 and 6. The reason for this unexpected DMA results is not clear yet, further follow-up studies will be carried out.

3.5. Comparison of the IF$_{PT}$ and IF$_I$ of the Cu-8.0Zn-11Al SMA with Other SMAs

According to our previous studies, $\mathrm{Ti_{50}Ni_{50}}$ [13], $\mathrm{Ti_{50}Ni_{40}Cu_{10}}$ [21], and $\mathrm{Ti_{50}Ni_{47}Fe_3}$ [22] SMAs all have acceptable damping capacity, exemplified by their $(\mathrm{IF_{PT}} + \mathrm{IF_I})$ peaks with tan δ > 0.02. However, their low martensitic transformation temperatures seriously restrict the use of these SMAs for practical high-damping applications. Although the Cu-14.0Al-4Ni SMA [20] exhibits an $(IF_{PT} + IF_I)$ peak at approximately 70 °C, its damping capacity is extremely low. The group III Ni_2MnGa SMAs [18] exhibited good inherent internal friction, where tan δ > 0.02, over a wide temperature range from -100 to 100 °C. However, the undesirable brittle nature of the Ni_2MnGa SMAs limited their workability and their use in high-damping applications. In this study, the Cu-8.0Zn-11Al SMA was shown to exhibit a high $(IF_{PT} + IF_I)_{\beta_3(L2_1) \rightarrow \gamma_3'(2H)}$ peak with a tan δ value above 0.053 at 32.2 °C. Except for the much higher $(IF_{PT} + IF_I)$ peaks above room temperature, as compared to other SMAs, the Cu-Zn-Al SMAs also have better workability, lower cost, and acceptable mechanical properties, and desirable Ms temperatures can be obtained by adjusting the chemical composition of the alloys. Consequently, Cu-Zn-Al SMAs are promising high-damping materials under isothermal conditions.

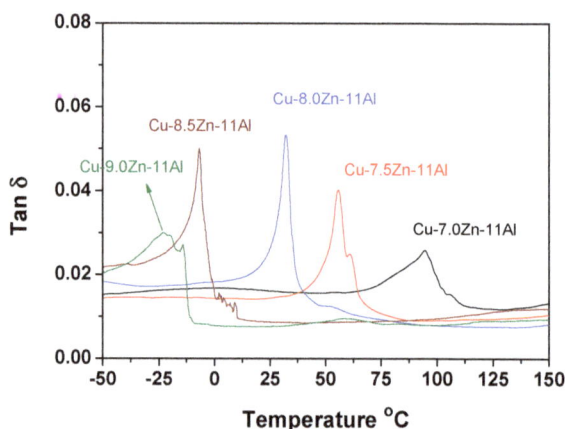

Figure 6. DMA tan δ curves measured at 10 Hz and 1 °C·min^{-1} cooling rate for Cu-xZn-11Al SMAs with various Zn contents.

4. Conclusions

Cu-xZn-11Al (x = 7.0, 7.5, 8.0, 8.5, and 9.0 wt. %) SMAs can exhibit a wide martensitic transformation temperature range from 104.0 to -21.2 °C by adjusting their chemical compositions. Cu-xZn-11Al SMAs with a higher ΔH value exhibit a higher IF peak because of the larger amount of martensite being transformed during martensitic transformation. The $(IF_{PT} + IF_I)_{\beta_3(L2_1) \rightarrow \gamma_3'(2H)}$ peaks for Cu-7.5Zn-11Al, Cu-8.0Zn-11Al, and Cu-8.5Zn-11Al SMAs are much higher than the $(IF_{PT} + IF_I)_{\beta_3(L2_1) \rightarrow \beta_3'(18R)}$ peak for the Cu-7.0Zn-11Al SMA because the transformed γ_3' martensite phase possesses a 2H type structure with abundant movable twin boundaries, while the transformed β_3' phase possesses an 18R structure with stacking faults. The Cu-9.0Zn-11Al SMA also possesses an $(IF_{PT} + IF_I)_{\beta_3(L2_1) \rightarrow \gamma_3'(2H)}$ peak during martensitic transformation; however, the abundant γ phase precipitates inhibit the movement of parent phase/martensite interfaces during damping, resulting in a lower tan δ value. The Cu-xZn-11Al SMAs are promising for practical high-damping applications under isothermal conditions because they possess good workability, low cost, acceptable mechanical properties, and the high damping capacities of the $(IF_{PT} + IF_I)$ peaks around and above room temperature. Among them, Cu-8.0Zn-11Al SMA has a high $(IF_{PT} + IF_I)$ peak with tan δ > 0.05 appearing at $\approx$25 °C.

Acknowledgments: The authors gratefully acknowledge the financial support for this research provided by the Ministry of Science and Technology (MOST), Taiwan, under Grants MOST104-2221-E197-004-MY3 (S.-H. Chang) and MOST105-2221-E002-043-MY2 (S.-K. Wu).

Author Contributions: Wei-Jyun Chan contributed to the experimental procedures, results, and discussion sections of this paper. Shyi-Kaan Wu contributed to the results and discussion sections, and he is the principal investigator (PI) of the Grant MOST105-2221-E002-043-MY2. Shih-Hang Chang also contributed to the results and discussions sections, and he is the principal investigator of the Grant MOST 104-2221-E197-004-MY3. These grants are also mentioned in the Acknowledgement of this paper.

Conflicts of Interest: The authors declare no conflict of interest.

References

1. Otsuka, K.; Ren, X. Physical metallurgy of Ti-Ni-based shape memory alloys. *Prog. Mater. Sci.* **2005**, *50*, 511–678. [CrossRef]
2. Mercier, O.; Melton, K.N.; De Préville, Y. Low-frequency internal friction peaks associated with the martensitic phase transformation of NiTi. *Acta Metall.* **1979**, *27*, 1467–1475. [CrossRef]
3. Wu, S.K.; Lin, H.C.; Chou, T.S. A study of electrical resistivity, internal friction and shear modulus on an aged Ti$_{49}$Ni$_{51}$ alloy. *Acta Metall. Mater.* **1990**, *38*, 95–102. [CrossRef]
1. Caluani, B.; Biscarini, A.; Campanella, R.; Trotta, L.; Mazzolai, G.; Tuissi, A.; Mazzolai, F.M. Mechanical spectroscopy and twin boundary properties in a Ni$_{50.8}$Ti$_{49.2}$ alloy. *Acta Mater.* **1999**, *47*, 1965–1976. [CrossRef]
5. Fan, G.; Zhou, Y.; Otsuka, K.; Ren, X.; Nakamura, K.; Ohba, T.; Suzuki, T.; Yoshida, I.; Yin, F. Effects of frequency, composition, hydrogen and twin boundary density on the internal friction of Ti$_{50}$Ni$_{50-x}$Cu$_x$ shape memory alloys. *Acta Mater.* **2006**, *54*, 5221–5229. [CrossRef]
6. Bidaux, J.E.; Schaller, R.; Benoit, W. Study of the h.c.p.-f.c.c. phase transition in cobalt by acoustic measurements. *Acta Metall.* **1989**, *37*, 803–811. [CrossRef]
7. Van Humbeeck, J.; Stoiber, J.; Delaey, L.; Gotthardt, R. The high damping capacity of shape memory alloys. *Z. Metallkunde* **1995**, *86*, 176–183.
8. Dejonghe, W.; De Batist, R.; Delaey, L. Factors affecting the internal friction peak due to thermoelastic martensitic transformation. *Scr. Metall.* **1976**, *10*, 1125–1128. [CrossRef]
9. Kustov, S.; Golyandin, S.; Sapozhnikov, K.; Morin, M. Application of acoustic technique to determine the temperature range of quenched-in defect mobility in Cu-Al-Be β′$_1$ martensitic phase. *Scr. Mater.* **2000**, *43*, 905–911. [CrossRef]
10. Sapozhnikov, K.; Golyandin, S.; Kustov, S.; Van Humbeeck, J.; De Batist, R. Motion of dislocations and interfaces during deformation of martensitic Cu-Al-Ni crystals. *Acta Mater.* **2000**, *48*, 1141–1151. [CrossRef]
11. Sapozhnikov, K.; Golyandin, S.; Kustov, S.; Van Humbeeck, J.; Schaller, R.; De Batist, R. Transient internal friction during thermal cycling of Cu-Al-Ni single crystals in β′$_1$ martensitic phase. *Scr. Mater.* **2002**, *47*, 459–465. [CrossRef]
12. Kustov, S.; Golyandin, S.; Sapozhnikov, K.; Cesari, E.; Van Humbeeck, J.; De Batist, R. Influence of martensite stabilization on the low-temperature non-linear anelasticity in Cu-Zn-Al shape memory alloys. *Acta Mater.* **2002**, *50*, 3023–3044. [CrossRef]
13. Chang, S.H.; Wu, S.K. Inherent internal friction of B2 → R and R → B19′ martensitic transformations in equiatomic TiNi shape memory alloy. *Scr. Mater.* **2006**, *55*, 311–314. [CrossRef]
14. Chang, S.H.; Wu, S.K. Internal friction of R-phase and B19′ martensite in equiatomic TiNi shape memory alloy under isothermal conditions. *J. Alloys Compd.* **2007**, *437*, 120–126. [CrossRef]
15. Chang, S.H.; Wu, S.K. Internal friction of B2 → B19′ martensitic transformation of Ti$_{50}$Ni$_{50}$ shape memory alloy under isothermal conditions. *Mater. Sci. Eng. A* **2007**, *454–455*, 379–383. [CrossRef]
16. Chang, S.H.; Wu, S.K. Inherent internal friction of Ti$_{51}$Ni$_{39}$Cu$_{10}$ shape memory alloy. *Mater. Trans.* **2007**, *48*, 2143–2147. [CrossRef]
17. Chang, S.H.; Wu, S.K. Isothermal effect on internal friction of Ti$_{50}$Ni$_{50}$ alloy measured by step cooling method in dynamic mechanical analyzer. *J. Alloys Compd.* **2008**, *459*, 155–159. [CrossRef]
18. Chang, S.H.; Wu, S.K. Low-frequency damping properties of near-stoichiometric Ni$_2$MnGa shape memory alloys under isothermal conditions. *Scr. Mater.* **2008**, *59*, 1039–1042. [CrossRef]
19. Chang, S.H.; Wu, S.K. Determining transformation temperatures of equiatomic TiNi shape memory alloy by dynamic mechanical analysis test. *J. Alloys Compd.* **2013**, *577*, s241–s244. [CrossRef]
20. Chang, S.H. Influence of chemical composition on the damping characteristics of Cu-Al-Ni shape memory alloys. *Mater. Chem. Phys.* **2011**, *125*, 358–363. [CrossRef]
21. Chang, S.H.; Hsiao, S.H. Inherent internal friction of Ti$_{50}$Ni$_{50-x}$Cu$_x$ shape memory alloys measured under isothermal conditions. *J. Alloys Compd.* **2013**, *586*, 69–73. [CrossRef]

22. Chang, S.H.; Chien, C., Wu, S.K. Damping characteristics of the inherent and intrinsic internal friction of $Ti_{50}Ni_{50-x}Fe_x$ (x = 2, 3, and 4) shape memory alloys. *Mater. Trans.* **2015**, *57*, 351–356. [CrossRef]
23. Ahlers, M. Martensite and equilibrium phases in Cu-Zn and Cu-Zn-Al alloys. *Prog. Mater. Sci.* **1986**, *30*, 135–186. [CrossRef]
24. Suzuki, T.; Kojima, R.; Fujii, Y.; Nagasawa, A. Reverse transformation behaviour of the stabilized martensite in Cu-Zn-Al alloy. *Acta Metall.* **1989**, *37*, 163–168. [CrossRef]
25. Tolley, A.; Rios Jara, D.; Lovey, F.C. 18R to 2H transformations in Cu-Zn-Al alloys. *Acta Metall.* **1989**, *37*, 1099–1108. [CrossRef]
26. Pelegrina, J.L.; Ahlers, M. The martensitic phases and their stability in Cu-Zn and Cu-Zn-Al alloys—I. The transformation between the high temperature β phase and the 18R martensite. *Acta Metall. Mater.* **1992**, *40*, 3205–3211. [CrossRef]
27. Ahlers, M.; Pelegrina, J.L. The martensitic phases and their stability in Cu-Zn and Cu-Zn-Al alloys—II. The transformation between the close packed martensitic phases. *Acta Metall. Mater.* **1992**, *40*, 3213–3220. [CrossRef]
28. Pelegrina, J.L.; Ahlers, M. The martensitic phases and their stability in Cu-Zn and Cu-Zn-Al alloys—III. The transformation between the high temperature phase and the 2H martensite. *Acta Metall. Mater.* **1992**, *40*, 3221–3227. [CrossRef]
29. Nakata, Y.; Yamamoto, O.; Shimizu, K. Effect of Aging in Cu-Zn-Al Shape Memory Alloys. *Mater. Trans. Jpn. Inst. Met.* **1993**, *34*, 429–437.
30. Saule, F.; Ahlers, M. Stability, stabilization and lattice parameters in Cu-Zn-Al martensites. *Acta Metall. Mater.* **1995**, *43*, 2373–2384. [CrossRef]
31. Condó, A.M.; Arneodo Larochette, P.; Tolley, A. Gamma phase precipitation processes in quenched beta phase Cu-Zn-Al alloys at an electron concentration of 1.53. *Mater. Sci. Eng. A* **2002**, *328*, 190–195. [CrossRef]
32. Arneodo Larochette, P.; Condó, A.M.; Pelegrina, J.L.; Ahlers, M. On the stability of the martensitic phases in Cu-Zn-Al, and its relationship with the equilibrium phases. *Mater. Sci. Eng. A* **2006**, *438–440*, 747–750. [CrossRef]
33. Asanović, V.; Delijić, K.; Jauković, N. A study of transformations of β-phase in Cu-Zn-Al shape memory alloys. *Scr. Mater.* **2008**, *58*, 599–601. [CrossRef]
34. Haberkorn, N.; Condó, A.M.; Espinoza, C.; Jaureguizahar, S.; Guimpel, J.; Lovey, F.C. Bulk-like behavior in the temperature driven martensitic transformation of Cu-Zn-Al thin films with 2H structure. *J. Alloys Compd.* **2014**, *591*, 263–267. [CrossRef]
35. Domenichini, P.; Condó, A.M.; Soldera, F.; Sirena, M.; Haberkorn, N. Influence of the microstructure on the resulting 18R martensitic transformation of polycrystalline Cu-Al-Zn thin films obtained by sputtering and reactive annealing. *Mater. Charact.* **2016**, *114*, 289–295. [CrossRef]
36. Zhao, Z.Q.; Chen, F.X.; Yang, D.Z. The internal friction associated with bainitic transformation, and bainitic transformation in a Cu-Zn-Al shape memory alloy. *J. Phys. Condens. Matter* **1989**, *1*, 1395–1404. [CrossRef]
37. Zhao, Z.Q.; Chen, F.X.; Li, S.Z.; Yang, D.Z. The internal friction associated with martensitic transformation in a CuZnAl shape memory alloy. *Scr. Metall. Mater.* **1991**, *25*, 669–672. [CrossRef]
38. Stoiber, J.; Bidaux, E.; Gotthardt, R. The movement of single $\beta_1 \to \beta_1'$ interfaces in Cu-Zn-Al as studied by a new technique of internal friction measurement. *Acta Metall. Mater.* **1994**, *42*, 4059–4070. [CrossRef]
39. Xiao, T.; Johari, G.P.; Mai, C. Time dependence of internal friction and shape change in Cu-Zn-Al shape memory alloys. *Metall. Trans. A* **1993**, *24*, 2743–2749. [CrossRef]
40. Cimpoeşu, N.; Stanciu, S.; Mayer, M.; Ioniţă, I.; Hanu Cimpoeşu, R. Effect of stress on damping capacity of a shape memory alloy CuZnAl. *J. Optoelectron. Adv. Mater.* **2010**, *12*, 386–391.
41. Perez-Saez, R.B.; Recarte, V.; No, M.L.; San Juan, J. Anelastic contributions and transformed volume fraction during thermoelastic martensitic transformations. *Phys. Rev. B* **1998**, *57*, 5684–5692. [CrossRef]
42. Nespoli, A.; Villa, E.; Passaretti, F. Quantitative evaluation of internal friction components of NiTiCu-Y shape memory alloys. *Thermochim. Acta* **2016**, *641*, 85–89. [CrossRef]

Article

Temperature Dependences of the Electrical Resistivity on the Heusler Alloy System Ni₂MnGa₁₋ₓFeₓ

Yoshiya Adachi [1,*], Yuki Ogi [1], Noriaki Kobayashi [2], Yuki Hayasaka [3], Takeshi Kanomata [4], Rie Y Umetsu [5], Xiao Xu [6] and Ryosuke Kainuma [6]

[1]   Graduate School of Science and Engineering, Yamagata University, Yonezawa 992-8510, Japan; ogilab1016@gmail.com
[2]   Faculty of Engineering, Yamagata University, Yonezawa 992-8510, Japan; noriaki.k728@gmail.com
[3]   Faculty of Engineering, Tohoku Gakuin University, Tagajo 985-8537, Japan; foxcity.yuki@gmail.com
[4]   Research Institute for Engineering and Technology, Tohoku Gakuin University, Tagajo 985-8537, Japan; kanomata@mail.tohoku-gakuin.ac.jp
[5]   Institute for Materials Research, Tohoku University, Sendai 980-8577, Japan; rieume@imr.tohoku.ac.jp
[6]   Department of Materials Science, Graduate School of Engineering, Tohoku University, Sendai 980-8579, Japan; xu@material.tohoku.ac.jp (X.X.); kainuma@material.tohoku.ac.jp (R.K.)
*   Correspondence: adachy@yz.yamagata-u.ac.jp; Tel.: +81-238-26-3381

Received: 10 August 2017; Accepted: 28 September 2017; Published: 3 October 2017

Abstract: Temperature dependences of the electrical resistivity have been measured on the Heusler alloy system Ni₂MnGa₁₋ₓFeₓ. The phase diagram of Ni₂MnGa₁₋ₓFeₓ was constructed on the basis of the experimental results. The structural and magnetic transition temperatures are consistent with those previously determined by magnetic measurements. The changes of the electrical resistivity at the martensitic transition temperature, $\Delta\rho$, were studied as a function of Fe concentration x. The $\Delta\rho$ abruptly increased in the concentration range between $x = 0.15$ and 0.20. The magnetostructural transitions were observed at $x = 0.275, 0.30$, and 0.35.

Keywords: ferromagnetic shape memory alloy; martensitic transition; electrical resistivity; Heusler alloy; Fe-doped Ni₂MnGa

1. Introduction

Recently, Ni-Mn based ferromagnetic shape memory alloys (FSMAs) with full Heusler-type structure have attracted much attention because of their potential applications in smart materials. These Heusler alloys exhibit a giant field-induced shape memory effect, large magnetoresistance, and large magnetocaloric effect [1–6]. Among FSMAs with a Heusler-type ($L2_1$-type) structure, the stoichiometric compound Ni₂MnGa has been the most studied. Ni₂MnGa orders ferromagnetically with the Curie temperature $T_C = 365$ K [7]. On cooling, the premartensitic phase appears below $T_p = 260$ K. With further decrease of temperature, Ni₂MnGa undergoes a first-order martensitic transition at $T_M = 200$ K [7]. Recently, Singh et al. performed a high-resolution synchrotron X-ray powder diffraction study for Ni₂MnGa and discussed the incommensurate nature of the modulate structures of the premartensitic (intermediate) and martensitic phases [8,9]. The ferromagnetic state remains below T_M. The spontaneous magnetization of Ni₂MnGa just below T_M is larger than that just above T_M.

The T_C, T_p, and T_M of Ni₂MnGa can be tuned in a wide range by doping with a fourth element. For Ni₂Mn₁₋ₓCuₓGa ($0 \leq x \leq 0.4$) [10], T_M increases with increasing the concentration x, while T_C decreases with x. With further increase of x, the magnetostructural transitions between the paramagnetic austenite (Para-A) and the ferromagnetic martensite (Ferro-M) phase occur in limited concentration range. The characteristics of the phase diagram of Ni₂Mn₁₋ₓCuₓGa ($0 \leq x \leq 0.4$) are closely similar to those of $Ni_{2+x}Mn_{1-x}Ga$ ($0 \leq x \leq 0.36$) [11,12]. The effect of the substitution of Fe and Co atoms in Ni-Mn-Ga alloy was studied [13,14]. Recently, Hayasaka et al. determined the phase diagram in the temperature-concentration plane of $Ni_2MnGa_{1-x}Fe_x$ ($0 \leq x \leq 0.40$) [15]. The characteristics of the determined phase diagram of $Ni_2MnGa_{1-x}Fe_x$ ($0 \leq x \leq 0.40$) are very similar to those of $Ni_{2+x}Mn_{1-x}Ga$ ($0 \leq x \leq 0.36$) [11,12] and $Ni_2Mn_{1-x}Cu_xGa$ ($0 \leq x \leq 0.4$) [10], where the magnetostructural transition between Para-A and Ferro-M occurs. However the microscopic understanding of the robust phase diagrams observed in $Ni_{2+x}Mn_{1-x}Ga$, $Ni_2Mn_{1-x}Cu_xGa$, and $Ni_2MnGa_{1-x}Fe_x$ is not clear in this stage. Furthermore, there is only a small amount of information about the electric properties of the Cu and Fe element doped Ni_2MnGa. In this paper, the electric properties are examined experimentally to gain deeper insight into the electronic properties of $Ni_2MnGa_{1-x}Fe_x$ alloys.

2. Experimental Procedures

The experiments were made on the same $Ni_2MnGa_{1-x}Fe_x$ ($0 \leq x \leq 0.40$) alloys that were used in our previous studies [15]. Namely, the polycrystalline $Ni_2MnGa_{1-x}Fe_x$ ($0 \leq x \leq 0.40$) alloys were prepared by the repeated arc melting of the appropriate quantities of constituent elements, 99.99% Ni, 99.99% Mn, 99.99% Fe, and 99.9999% Ga in an argon atmosphere. The samples with $x = 0.30$ and 0.35 were prepared by the melting of appropriate quantities of the constituent elements with high purity in an induction furnace (DIAVAC LIMITED, Yachiyo, Japan). The reaction products were sealed in evacuated silica tubes, heated at 850 °C for 3 days and at 600 °C for 1 day, and then quenched into water. The crystal structure was investigated by X-ray diffraction (Rigaku, Tokyo, Japan) measurements at room temperature using Cu-Kα radiation. The lattice parameters for the samples were the same as the previous report [15].

The measurements of the electrical resistivity ρ were carried out by a conventional DC (direct current) four-probe method in the temperature range from 80 K to 450 K. The samples were cut out using a diamond disk saw (BUEHLER, Lake Bluff, IL, USA) into the size of about $1.0 \times 1.0 \times 10$ mm^3. The thermal process in the measurements started from 80 K, heated up to 450 K, and cooled down again to 80 K.

3. Results and Discussion

The temperature dependence of the electrical resistivity ρ of the sample with $x = 0.05$ is given in Figure 1a. The obvious slope change near 377 K is indicative of the ferromagnetic ordering. Below the Curie temperature T_C, the ρ shows a steep decrease with decreasing temperature. This can be attributed to the disappearance of electron scattering on magnetic fluctuations. The behavior of ρ around T_C is a common feature for the Heusler alloys with ferromagnetic ordering. Assuming that the break point on the ρ vs. T curves corresponds to the Curie temperature of the sample with $x = 0.05$, the value of $T_C = 377$ K is very close to that determined from the initial permeability μ vs. T curve [15]. A prominent jump-like feature of ρ appears at around 250 K, indicating the occurrence of the martensitic transition. The martensitic transition temperature T_M was defined by the equation: $T_M = (T_{Ms} + T_{Af})/2$, where T_{Ms} and T_{Af} are the martensitic transition starting temperature and the reverse martensitic transition finishing temperature, respectively. The values of T_{Ms} and T_{Af} were defined as the cross points of the linear extrapolation lines of the ρ vs. T curves from both higher and lower temperature ranges. As shown in Figure 1a, a temperature hysteresis is formed around T_M between 240 K and 260 K, confirming that the martensitic transition is first-order. On the other hand, such a temperature hysteresis behavior is absent for the ferromagnetic transition around 377 K. With further decrease of temperature from T_M, ρ of the sample with $x = 0.05$ represents a typical metallic behavior. The inset in Figure 1a shows the temperature dependence of $d\rho/dT$. A noticeable slope change in $\rho(T)$ around 260 K marks the onset of the premartensitic transition. The premartensitic transition temperature T_p is estimated to be 264 K, as shown in the inset in Figure 1a. The values of T_M and T_p are in good agreement with those determined from the magnetic measurements earlier [15]. For the sample with $x = 0.025$, anomalies on $d\rho/dT$ vs. T curves are observed around T_p (see the inset in Figure 1b. However, as shown in the insets of Figure 1c–f, no anomalies on dρ/dT vs. T curves are observed in the temperature range between T_C and T_M, indicating that the premartensitic phase disappears in the concentration range of $x \geq 0.10$. As seen in Figure 1a–f, the martensitic transition temperature increases with increase of Fe concentration x. On the other hand, T_C increases slightly with x.

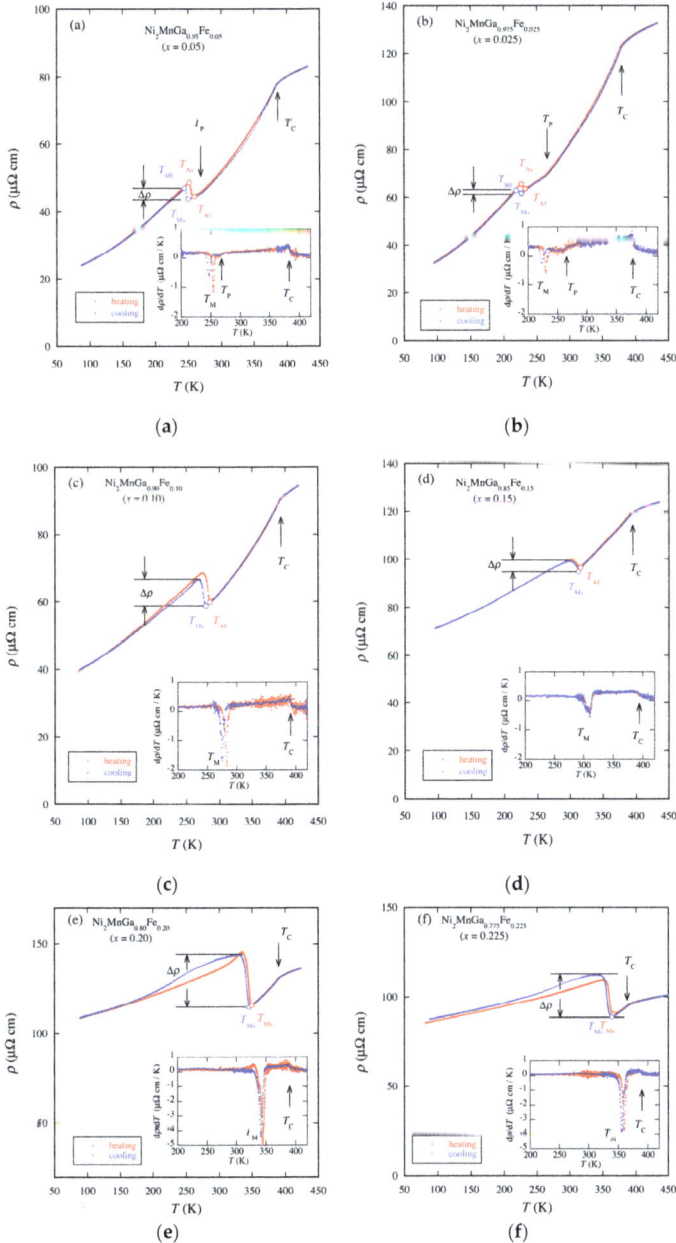

Figure 1. *Cont.*

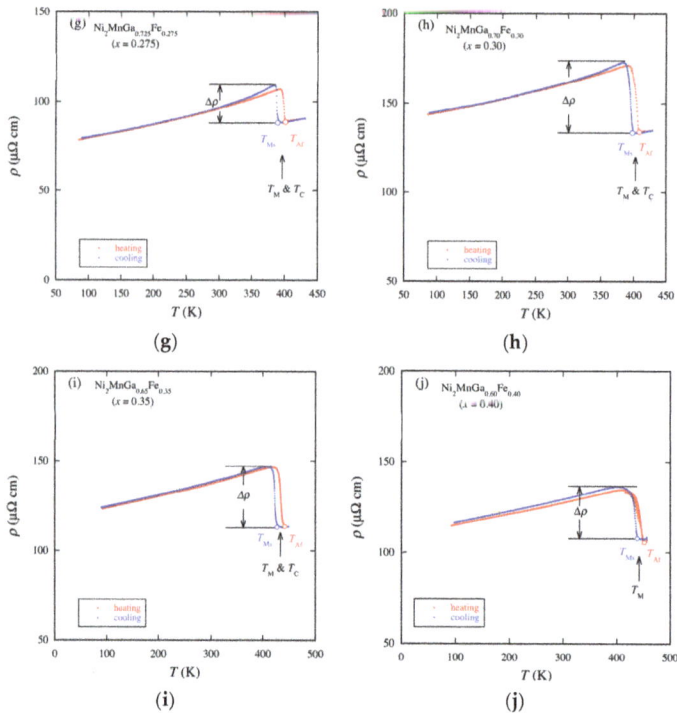

Figure 1. Temperature dependences of the electrical resistivity ρ for $Ni_2MnGa_{1-x}Fe_x$ (x = 0.025–0.40). The inset shows the temperature dependence of $d\rho/dT$. (**a**) $Ni_2MnGa_{0.95}Fe_{0.05}$ (x = 0.05); (**b**) $Ni_2MnGa_{0.975}Fe_{0.025}$ (x = 0.025); (**c**) $Ni_2MnGa_{0.90}Fe_{0.10}$ (x = 0.10); (**d**) $Ni_2MnGa_{0.85}Fe_{0.15}$ (x = 0.15); (**e**) $Ni_2MnGa_{0.80}Fe_{0.20}$ (x = 0.20); (**f**) $Ni_2MnGa_{0.775}Fe_{0.225}$ (x = 0.225); (**g**) $Ni_2MnGa_{0.725}Fe_{0.275}$ (x = 0.275); (**h**) $Ni_2MnGa_{0.70}Fe_{0.30}$ (x = 0.30); (**i**) $Ni_2MnGa_{0.65}Fe_{0.35}$ (x = 0.35); (**j**) $Ni_2MnGa_{0.60}Fe_{0.40}$ (x = 0.40).

Figure 1g–j show the temperature dependence of ρ for the samples with x = 0.275, 0.30, 0.35, and 0.40. According to the phase diagram of $Ni_2MnGa_{1-x}Fe_x$ ($0 \leq x \leq 0.40$) reported by Hayasaka et al. [15], the samples with $0.27 \leq x \leq 0.37$ underwent the magnetostructural transition from the Ferro-M state to the Para-A state. For the sample with x = 0.40, the Ferro-M to the paramagnetic martensite (Para-M) transition appeared. As shown in Figure 1h, the ρ increases abruptly around 402 K with deceasing temperature, indicating that the magnetic transition between the ferromagnetic phase and the paramagnetic phase was first-order. We did not observe any anomaly below T_C on the ρ vs. T curves, so T_M is considered to coincide with T_C. The T_C ($=T_M$) was estimated to be 402 K for the sample with x = 0.30, where T_C was defined to be $T_C = T_M = (T_{Ms} + T_{Af})/2$. Similar ρ vs. T curves are observed for the samples with x = 0.275 and 0.35 (see Figure 1g,i). As seen in Figure 1a–j, the jump of ρ at T_M, $\Delta\rho$, of the samples with $x \geq 0.20$ are considerably larger than that of samples with $0.025 \leq x \leq 0.15$.

Figure 2 shows the concentration dependence of $\Delta\rho$ for $Ni_2MnGa_{1-x}Fe_x$. As seen in Figure 2, $\Delta\rho$ shows a tendency to increase with increasing x, but abruptly increases in the concentration range between x = 0.15 and 0.20.

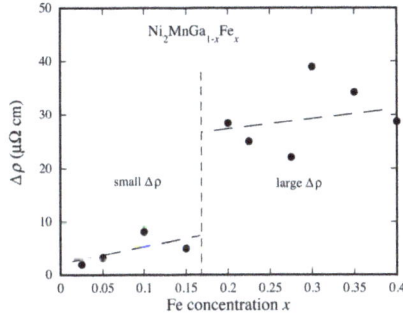

Figure 2. The concentration dependence of $\Delta\rho$ for $Ni_2MnGa_{1-x}Fe_x$ ($x = 0.025$–0.40). The black dots are $\Delta\rho$ obtained from the experimental data as shown in Figure 1, with dashed lines being guides to the eye. The vertical dotted lines distinguish the electric properties of $Ni_2MnGa_{1-x}Fe_x$ to two regions.

Figure 3 shows the phase transition temperatures T_M, T_p and T_C determined from the ρ vs. T curves in this study. The closed triangle in the figure represents the phase transition temperature determined from the magnetic measurement [15]. As shown in Figure 3, the phase transition temperatures determined in this study are in good agreement with those reported earlier [15]. The phase diagram of Figure 3 is very similar to those of $Ni_{2+x}Mn_{1-x}Ga$ ($0 \leq x \leq 0.36$) [11,12] and $Ni_2Mn_{1-y}Cu_yGa$ ($0 \leq x \leq 0.4$) [10] and $Ni_2MnGa_{1-x}Cu_x$ ($0 \leq x \leq 0.25$) [16], as mentioned above. In order to understand the phase diagram of $Ni_2Mn_{1-x}Cu_xGa$ ($0 \leq x < 0.4$), the phenomenological Landau-type free energy as a function of the martensitic distortion and magnetization was constructed and analyzed [10]. Satisfactory agreements between the experiments and theory were obtained, except for the appearance of the premartensitic phase. The analysis showed that the biquadratic coupling term of the martensitic distortion and magnetization plays an important role in the interplay between the martensitic phase and ferromagnetic phase.

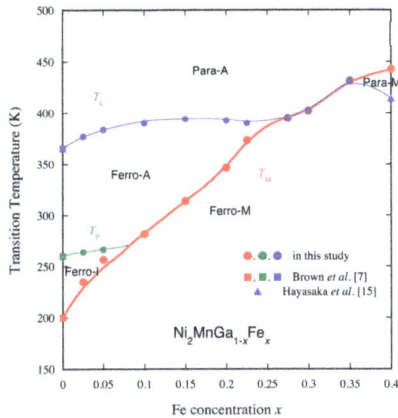

Figure 3. Phase diagram for $Ni_2MnGa_{1-x}Fe_x$ ($0 \leq x \leq 0.40$). Para-A and Para M represent the paramagnetic austenite phase and the paramagnetic martensite phase, respectively. Ferro-A and Ferro-M represent the ferromagnetic austenite phase and the ferromagnetic martensite phase, respectively. Ferro-I means the ferromagnetic premartensite (intermediate) phase.

The total electrical resistivity is given for usual ferromagnetic materials as follows,

$$\rho(T) = \rho_0 + aT + bT^2, \tag{1}$$

where ρ_0 is the residual part, the second term and the third term are the electrical resistivity parts due to electron–phonon scattering and magnetic origin, respectively, with a and b being fitting parameteters. Of course, Equation (1) is a phenomenological fit to the experimental data, and we do not claim that magnetic scattering is purely quadratic in temperature.

Many authors fitted Equation (1) to their experimental data for many Heusler alloys. [17–23]. Table 1 gives the ρ_0, a, and b values, the validity range of fit, and T_C values of $Ni_2MnGa_{1-x}Fe_x$. The concentration dependence of ρ_0 determined by fitting the ρ vs. T data ($T < T_M$) to Equation (1) is shown in Figure 4. As seen in Figure 4, the ρ_0 value roughly increases with the concentration x, but abruptly changes in the concentration range between $x = 0.15$ and 0.20, as well as the behavior of $\Delta\rho$. It can be explained as the impurity effect that the ρ_0 increases with increasing x. The values of b for the samples with $x \leq 0.30$ are two orders of magnitude smaller than those of typical weak itinerant electron ferromagnets Ni_3Al, $ZrZn_2$ and Sc_3In [24–27]. A general theory of electrical resistivity in an itinerant electron system has been proposed by Ueda and Moriya on the basis of spin fluctuations [28]. Their work predicts a strong T^2 dependence of electrical resistivity due to spin fluctuations.

Table 1. The ρ_0 ($\mu\Omega$ cm), a ($\mu\Omega$ cm K^{-1}), b ($\mu\Omega$ cm K^{-2}) values according to the fit $\rho(T) = \rho_0 + aT + bT^2$ in the temperature range below T_M (i.e., in the Ferro-M (ferromagnetic martensite) phase). The range of validity of the fit, T_M and T_C are given in Table 1.

Alloys	ρ_0	$a \times 10^{-2}$	$b \times 10^{-4}$	Range (K)	T_M (K)	T_C (K)
$Ni_2MnGa_{0.975}Fe_{0.025}$ ($x = 0.025$)	22.2	4.73	6.37	100~220	235	383
$Ni_2MnGa_{0.95}Fe_{0.05}$ ($x = 0.05$)	18.2	3.34	3.54	100~230	257	377
$Ni_2MnGa_{0.90}Fe_{0.10}$ ($x = 0.10$)	31.1	8.61	1.84	100~260	282	390
$Ni_2MnGa_{0.85}Fe_{0.15}$ ($x = 0.15$)	60.8	9.18	1.41	100~270	314	393
$Ni_2MnGa_{0.80}Fe_{0.20}$ ($x = 0.20$)	101.6	6.47	1.67	100~270	347	392
$Ni_2MnGa_{0.775}Fe_{0.225}$ ($x = 0.225$)	80.8	5.25	0.84	100~320	356	390
$Ni_2MnGa_{0.725}Fe_{0.275}$ ($x = 0.275$)	75.0	4.91	0.86	100~350	395	395
$Ni_2MnGa_{0.70}Fe_{0.30}$ ($x = 0.30$)	139	4.42	1.04	100~370	402	402
$Ni_2MnGa_{0.65}Fe_{0.35}$ ($x = 0.35$)	118	5.06	0.43	100~390	431	431
$Ni_2MnGa_{0.60}Fe_{0.40}$ ($x = 0.40$)	110	5.27	0.26	100~390	442	414 [1]

[1] Quoted from Ref. [15].

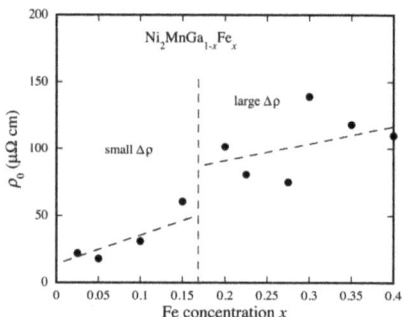

Figure 4. The concentration dependence of the residual electrical resistivity ρ_0 for $Ni_2MnGa_{1-x}Fe_x$. The black dots are ρ_0 obtained from the analyzed data as shown in Table 1, with dashed lines being guides to the eye. The vertical dotted lines distinguish the electric properties of $Ni_2MnGa_{1-x}Fe_x$ to two regions.

Metals **2017**, 7, 413

In this study, as shown in Figures 2 and 4, we found the borderline which distinguishes the electric properties of $Ni_2MnGa_{1-x}Fe_x$ ($0 \le x \le 0.40$) alloys. The Fe concentration variation on the temperature dependence of resistivity for $Ni_2MnGa_{1-x}Fe_x$ ($0 \le x \le 0.40$) alloys may be caused by the change of the charge carrier concentration between the martensite and austenite phases or the change of the mechanism of the martensitic transition. Now, it is not clear for the origin of appearance of the borderline. This may be related to the electron scattering at the twin and domain boundaries which appear in the martensitic phase. An investigation of microstructure observation will also be necessary.

4. Conclusions

Temperature dependences of the electrical resistivity have been measured on the Heusler alloy system $Ni_2MnGa_{1-x}Fe_x$. The phase diagram of $Ni_2MnGa_{1-x}Fe_x$ was constructed on the basis of the experimental results. The magnetostructural transitions were observed at the Fe concentrations $x = 0.275$, 0.30, and 0.35 in $Ni_2MnGa_{1-x}Fe_x$ alloys as well as a previous report [15]. The changes of the electrical resistivity at the martensitic transition temperature were studied as the function of Fe concentration x.

Acknowledgments: The authors would like to express our sincere thanks to Toetsu Shishido and Kazuo. Obara of Institute for Materials Research, Tohoku University (Sendai, Japan) for their help on the sample preparation. This study was partly supported by a Grant-in-Aid for Scientific Research, provided by the Japan Society for the Promotion of Science.

Author Contributions: Yoshiya Adachi and Takeshi Kanomata conceived and designed the experiments; Yuki Ogi, Noriaki Kobayashi, Yuki Hayasaka, Rie Y Umetsu and Xiao Xu performed the experiments; Xiao Xu and Ryosuke Kainuma analyzed the data; Yoshiya Adachi, Xiao Xu and Takeshi Kanomata wrote the paper.

Conflicts of Interest: The authors declare no conflict of interest.

References

1. Brown, P.J.; Kanomata, T.; Matsumoto, M.; Neumann, K.-U.; Ziebeck, K.R.A. *Magnetism and Structure in Functional Materials*; Planes, A., Mañosa, L., Saxena, A., Eds.; Springer: Berlin/Heiderberg, Germany, 2005; pp. 113–140.
2. Kainuma, R.; Imano, Y.; Ito, W.; Sutou, Y.; Morito, H.; Okamoto, S.; Kitakami, O.; Oikawa, K.; Fujita, A.; Kanomata, T.; et al. Magnetic-field-induced shape recovery by reverse phase transformation. *Nature* **2006**, *439*, 957–960. [CrossRef] [PubMed]
3. Planes, A.; Mañosa, L.; Acet, M. Magnetocaloric effect and its relation to shape-memory properties in ferromagnetic Heusler alloys. *J. Phys. Condens. Matter* **2009**, *21*, 233201. [CrossRef] [PubMed]
4. Chernenko, V.A. (Ed.) *Advances in Magnetic Shape Memory Materials*; Trans. Tech. Publications Ltd.: Zurich, Switzerland, 2011.
5. Acet, M.; Mañosa, L.I.; Planes, A. Magnetic-Field-Induced Effects in Martensitic Heusler-Based Magnetic Shape Memory Alloys. In *Handbook of Magnetic Materials*; Elsevier Science: Amsterdam, The Netherlands, 2011.
6. Yu, G.H.; Xu, Y.L.; Liu, Z.H.; Qiu, H.M.; Zhu, Z.Y.; Huang, X.P.; Pan, L.Q. Recent progress in Heusler-type magnetic shape memory alloys. *Rare Met.* **2015**, *34*, 527–539. [CrossRef]
7. Brown, P.J.; Crangle, J.; Kanomata, T.; Matsumoto, M.; Neumann, K.U.; Ouladdiaf, B.; Ziebeck, K.R.A. The crystal structure and phase transitions of the magnetic shape memory compound Ni_2MnGa. *J. Phys. Condens. Matter* **2002**, *14*, 10159–10171. [CrossRef]
8. Singh, S.; Bednarcik, J.; Barman, S.R.; Felser, C.; Pandey, D. Premartensite to martensite transition and its implications for the origin of modulation in Ni_2MnGa ferromagnetic shape-memory alloy. *Phys. Rev. B* **2015**, *92*, 054112. [CrossRef]
9. Singh, S.; Petricek, V.; Rajput, P.; Hill, A.H.; Suard, E.; Barman, S.R.; Pandey, D. High-resolution synchrotron X-ray powder diffraction study of the incommensurate modulation in the martensite phase of Ni_2MnGa: Evidence for nearly 7M modulation and phason broadening. *Phys. Rev. B* **2014**, *90*, 014109. [CrossRef]

10. Kataoka, M.; Endo, K.; Kudo, N.; Kanomata, T.; Nishihara, H.; Shishido, T.; Umetsu, R.Y.; Nagasako, M.; Kainuma, R. Martensitic transition, ferromagnetic transition, and their interplay in the shape memory alloys Ni$_2$Mn$_{1-x}$Cu$_x$Ga. *Phys. Rev. B* **2010**, *82*, 214423. [CrossRef]

11. Vasil'ev, A.N.; Bozhko, A.D.; Khovailo, V.V.; Dikshtein, I.E.; Shavrov, V.G.; Buchelnikov, V.D.; Matsumoto, M.; Suzuki, S.; Takagi, T.; Tani, J. Structural and magnetic phase transitions in shape-memory alloys Ni$_{2+x}$Mn$_{1-x}$Ga. *Phys. Rev. B* **1999**, *59*, 1113–1120. [CrossRef]

12. Khovaylo, V.V.; Buchelnikov, V.D.; Kainuma, R.; Koledov, V.V.; Ohtsuka, M.; Shavrov, V.G.; Takagi, T.; Taskaev, S.V.; Vasiliev, A.N. Phase transitions in Ni$_{2+x}$Mn$_{1-x}$Ga with a high Ni excess. *Phys. Rev. B* **2005**, *72*, 224408. [CrossRef]

13. Soto, D.; Hernández, V.A.; Flores-Zúñiga, H.; Moya, X.; Mañosa, L.; Planes, A.; Aksoy, S.; Acet, M.; Krenke, T. Phase diagram of Fe-doped Ni-Mn-Ga ferromagnetic shape-memory alloys. *Phys. Rev. B* **2008**, *77*, 184103. [CrossRef]

14. Soto-Parra, D.E.; Moya, X.; Mañosa, L.; Planes, A.; Flores-Zúñiga, H.; Alvarado-Hernández, F.; Ochoa-Gamboa, R.A.; Matutes-Aquino, J.A.; Ríos-Jara, D. Fe and Co selective substitution in Ni$_2$MnGa: Effect of magnetism on relative phase stability. *Philos. Mag.* **2010**, *90*, 2771–2792. [CrossRef]

15. Hayasaka, Y.; Aoto, S.; Date, H.; Kanomata, T.; Xu, X.; Umetsu, R.Y.; Nagasako, M.; Omori, T.; Kainuma, R.; Sakon, T.; et al. Magnetic phase diagram of ferromagnetic shape memory alloys Ni$_2$MnGa$_{1-x}$Fe$_x$. *J. Alloys Compd.* **2014**, *591*, 280–285. [CrossRef]

16. Endo, K.; Kanomata, T.; Kimura, A.; Kataoka, M.; Nishihara, H.; Umetsu, R.Y.; Obara, K.; Shishido, T.; Nagasako, M.; Kainuma, R.; et al. Magnetic phase diagram of the ferromagnetic shape memory alloys Ni$_2$MnGa$_{1-x}$Cu$_x$. *Mater. Sci. Forum* **2011**, *684*, 165–176. [CrossRef]

17. Kunzler, J.V.; Grandi, T.A.; Schreiner, W.H.; Pureur, P.; Brandão, D.E. Electrical resistivity measurements on the Cu$_2$MnAl heusler alloy. *J. Phys. Chem. Solids* **1979**, *40*, 427–429. [CrossRef]

18. Kunzler, J.V.; Grandi, T.A.; Schreiner, W.H.; Pureur, P.; Brandão, D.E. Spin-disorder resistivity in the Cu$_2$Mn(Al$_{1-x}$Sn$_x$) Heusler alloys. *J. Phys. Chem. Solids* **1980**, *41*, 1023–1026. [CrossRef]

19. Schreiner, W.H.; Brandão, D.E.; Ogiba, F.; Kunzler, J.V. Electrical resistivity of Heusler alloys. *J. Phys. Chem. Solids* **1982**, *43*, 777–780. [CrossRef]

20. Zhang, M.; Liu, G.; Cui, Y.; Hu, H.; Liu, Z.; Chen, J.; Wu, G.; Sui, Y.; Qian, Z.; Zhang, X. Magnetism and transport properties of melt-spun ribbon Cu$_2$MnAl Heusler alloy. *J. Magn. Magn. Mater.* **2004**, *278*, 328–333. [CrossRef]

21. Marchenkova, E.B.; Kourov, N.I.; Marchenkov, V.V.; Pushin, V.G.; Korolev, A.V.; Weber, H.W. Low temperature kinetic properties and structure of Ni$_{50+x}$Mn$_{25-x+y}$Ga$_{25-y}$ alloys with shape memory. *J. Phys. Conf. Ser.* **2009**, *150*, 022054. [CrossRef]

22. Ingale, B.; Gopalan, R.; Chandrasekaran, V.; Ram, S. Structural, magnetic, and magnetotransport studies in bulk Ni$_{55.2}$Mn$_{18.1}$Ga$_{26.7}$ alloy. *J. Appl. Phys.* **2009**, *105*, 023903. [CrossRef]

23. Kourov, N.I.; Marchenkov, V.V.; Pushin, V.G.; Belozerova, K.A. Electrical properties of ferromagnetic Ni$_2$MnGa and Co$_2$CrGa Heusler alloys. *J. Exp. Theor. Phys.* **2013**, *117*, 121–125. [CrossRef]

24. Fluitman, J.H.J.; Boom, R.; De Chatel, P.F.; Schinkel, C.J.; Tilanus, J.L.L.; De Vries, B.R. Possible explanations for the low temperature resistivities of Ni$_3$Al and Ni$_3$Ga alloys in terms of spin density fluctuation theories. *J. Phys. F Met. Phys.* **1973**, *3*, 109–117. [CrossRef]

25. Yoshizawa, M.; Seki, H.; Ikeda, K.; Okuno, K.; Saito, M.; Shigematsu, K. Magnetic field effects on electrical resistivity and ferromagnetism in Ni$_3$Al alloys. *J. Phys. Soc. Jpn.* **1992**, *61*, 3313–3321. [CrossRef]

26. Ogawa, S. Electrical resistivity of weak itinerant ferromagnet ZrZn$_2$. *J. Phys. Soc. Jpn.* **1976**, *40*, 1007–1009. [CrossRef]

27. Ikeda, K.; Gschneider, K.A., Jr.; Kobayashi, N.; Noto, K. Magnetoresistance of Sc$_3$In. *J. Magn. Magn. Mater.* **1984**, *42*, 1–11. [CrossRef]

28. Ueda, K.; Moriya, T. Contribution of spin fluctuations to the electrical thermal resistivities of weakly and nearly ferromagnetic metals. *J. Phys. Soc. Jpn.* **1975**, *39*, 605–615. [CrossRef]

metals

MDPI

Article

Evidence of Change in the Density of States during the Martensitic Phase Transformation of Ni-Mn-In Metamagnetic Shape Memory Alloys

Rie Y Umetsu [1,*], Xiao Xu [2], Wataru Ito [3] and Ryosuke Kainuma [2]

[1] Institute for Materials Research, Tohoku University, Sendai 980-8577, Japan
[2] Department of Materials Science, Graduate School of Engineering, Tohoku University, Sendai 980-8579,
     Japan; xu@material.tohoku.ac.jp (X.X.); kainuma@material.tohoku.ac.jp (R.K.)
[3] National Institute of Technology, Sendai College, Natori 981-1239, Japan; ito@sendai-nct.ac.jp
*   Correspondence: rieume@imr.tohoku.ac.jp; Tel.: +81-22-215-2199

Received: 11 August 2017; Accepted: 27 September 2017; Published: 4 October 2017

Abstract: Specific heat measurements were performed at low temperatures for $Ni_{50}Mn_{50-x}In_x$ alloys to determine their Debye temperatures (θ_D) and electronic specific heat coefficients (γ). For $x \leq 15$, where the ground state is the martensite (M) phase, θ_D decreases linearly and γ increases slightly with increasing In content. For $x \geq 16.2$, where the ground state is the ferromagnetic parent (P) phase, γ increases with decreasing In content. Extrapolations of the composition dependences of θ_D and γ in both the phases suggest that these values change discontinuously during the martensitic phase transformation. The value of θ_D in the M phase is larger than that in the P phase. The behavior is in accordance with the fact that the volume of the M phase is more compressive than that of the P phase. On the other hand, γ is slightly larger in the P phase, in good agreement with the reported density of states around the Fermi energy obtained by the first-principle calculations.

Keywords: electronic specific heat coefficient; density of states; shape memory alloy; martensitic phase transformation

1. Introduction

Since a unique martensitic phase transformation in off-stoichiometric Heusler NiMnZ (Z = In, Sn, and Sb) alloys were first reported by Sutou et al. [1], NiMnIn- and NiMnSn-based alloys have attracted widespread interest as high-performance multiferroic materials. The alloys show many interesting properties, such as metamagnetic shape memory effect [2,3] inverse magnetocaloric effect [4–6], giant magnetoresistance effect [7,8], and giant magnetothermal conductivity [9]. These interesting physical properties are related to drastic changes in magnetic properties between the ferromagnetic parent (P) phase and the martensite (M) phase with weak magnetism. Such large changes of the various physical properties should be due to the change in the electronic state during the martensitic phase transformation. First-principles density-functional calculations and hard X-ray photoelectron spectroscopy (HAXPES) experiments were performed to investigate the electronic states in the P and M phases of Ni-Mn-In alloys [10]. In these calculations, composition dependence of the partial density of states (DOS) in the P phase for Ni-Mn-In was investigated using the periodic supercells, in addition to that in the M phase, where tetragonal distortion was introduced. The valence-band photoelectron spectra of the $Ni_{50}Mn_{34}In_{16}$ alloy in warming and cooling processes showed that the peak near the Fermi energy (E_F) in the P phase disappeared in the M phase temperature range, suggesting the formation of the pseudo-gap. The minority-spin Ni $3d-e_g$ state has also been concluded to play an important role in stabilizing the M phase in off-stoichiometric composition [10].

Herein, we performed specific heat measurements in Ni-Mn-In alloy system to investigate the electronic specific heat coefficient (γ), which will provide information on the DOS around E_F, and the Debye temperature (θ_D). The obtained results may provide indirect information on the electronic state, in contrast to the photoelectron spectroscopy observations that provide direct information. However, it has been reported that the value of γ for $L1_0$-type NiMn alloy and related materials (i.e., PdMn and PtMn) where the existence of the pseudo-gap around E_F characterizes their unique electronic states agrees well with the value obtained by theoretical calculations [11–13]. Systematic study of the specific heat measurements in a wide composition region for Ni-Mn-In will help understand the behavior of the metamagnetic shape memory alloys in this system.

2. Experimental Procedure

$Ni_{50}Mn_{50-x}In_x$ ($0 \leq x \leq 25$) alloys were fabricated by induction melting in an Ar atmosphere. The specimens were sealed in a quartz capsule and were annealed at 1173 K for 1 day before quenching in water. The microstructure and composition were confirmed by the electron probe microanalyzer. The crystal structure was investigated by powder X-ray diffraction and transmission electron microscope observations. The related results were reported in the previous paper [14]. The magnetic measurements were carried out on a superconducting quantum interference device (SQUID; Quantum Design Ltd., San Diego, CA, USA) magnetometer. Specific heat measurements were carried out by the relaxation method using a physical properties measurement system (PPMS produced by Quantum Design Ltd., San Diego, CA, USA) at temperatures below 20 K. The absolute value of the specific heat was checked by measuring one of the standard pure Cr.

3. Results and Discussion

3.1. Magnetic Measurements

Figure 1a,b show magnetization (*M–H*) curves obtained at 5 K and thermomagnetization (*M–T*) curves obtained under a magnetic field of 10 kOe for $Ni_{50}Mn_{50-x}In_x$ alloy specimens having $x \leq 15$, respectively. Here, the ground state of the specimens is the martensite (M) phase. The specimen with $x = 0$ (NiMn) has been reported to be collinear-type antiferromagnetic with the Néel temperature higher than the martensitic transformation temperature [15]. Therefore, the magnetic state is deduced to be antiferromagnetic for low In concentrations, and the antiferromagnetic exchange interaction in the system decreases with increasing In content [16,17]. The NiMn has an $L1_0$-type tetragonal structure with lattice parameters of $a = 0.374$ and $c = 0.352$ nm, and it transforms to B2-type cubic one at the martensitic transformation temperature [15]. With increasing the In concentration, the crystal structure of M phase varies to 14 M and 10 M stacking monoclinic structures [14,18,19]. The straight line in the *M–H* curves (Figure 1a) is characteristic of the antiferromagnetic properties. The slope increases with increasing In content, and the *M–H* curve indicates small hysteresis at $x = 13$. In the *M–H* curve for $x = 15$ (inset of Figure 1a), large hysteresis is observed, and magnetization tends to saturation. This variation arises in the magnetization curves because ferromagnetic exchange interaction is introduced with increasing In content. AC magnetization measurements have suggested that the ground state for $x = 15$ is the blocking state, showing frequency dependence in both the real and imaginary parts of the susceptibility [20]. The variation of magnetic property from the antiferromagnetic state to the blocking state is also confirmed by the *M–T* curves for $Ni_{50}Mn_{50-x}In_x$ alloys with $x \leq 15$. In the measurements for obtaining the *M–T* curves, the specimens were cooled to low temperatures in zero magnetic field, and the magnetization was measured in warming process and cooling process under the same applied magnetic field. Magnetization at lower temperatures gradually increases with increasing In content, and magnetic field cooling effect is observed for $x = 10$, 13, and 15. The large magnetization change observed at ~300 K in the *M–T* curve for $x = 15$ (inset of Figure 1b) is due to the martensitic phase transformation, and the magnetic property at temperatures just below the transition temperature has been concluded to be paramagnetic based on Mössbauer spectroscopy [21].

These results along with previous reports on AC magnetization measurements [20], suggest that antiferromagnetic long-range ordering might have disappeared somewhere in the composition region.

$M–H$ curves obtained at 5 K and $M–T$ curves obtained under a magnetic field of 500 Oe for specimens with $x \geq 16.2$ in $Ni_{50}Mn_{50-x}In_x$ alloys are shown in Figure 2a,b, respectively. Here, the ground state of the system is the ferromagnetic parent (P) phase. That is, no martensitic transformation occurs down to low temperatures in these composition regions. The crystal structure is basically the $L2_1$-type structure. The lattice parameter has been reported to be $a = 0.6071$ nm for the $x = 25$ at room temperature and to decrease linearly with increasing the Mn composition [22]. The Figure 2b shows only warming process in $M–T$ curves because there is almost no magnetic field cooling effect, in contrast to that for $x \leq 15$ in Figure 1b. The saturated magnetization increases with decreasing In content from the stoichiometric composition of $Ni_{50}Mn_{25}In_{25}$ (=Ni_2MnIn). On the other hand, concentration dependence of the Curie temperature (T_C) does not show the systematic variation as is shown by saturation magnetization. Figure 2b shows that the value of T_C is ~320 K, independent of In content.

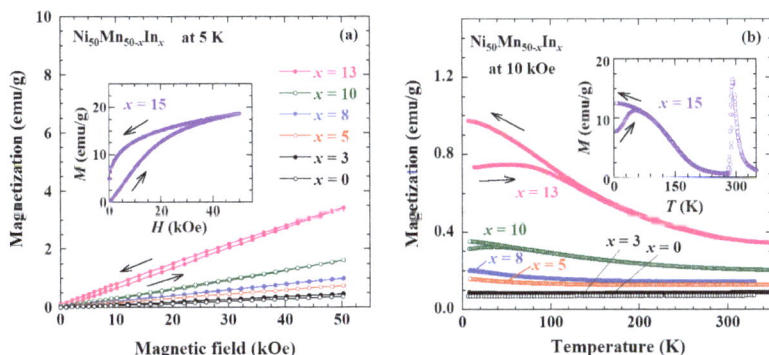

Figure 1. (a) Magnetization curves at 5 K and (b) thermomagnetization curves measured at 10 kOe for the specimens with $x \leq 15$ in $Ni_{50}Mn_{50-x}In_x$ alloys. The arrows in the figures mean applying and removing magnetic fields in (a), and the warming and cooling processes in (b). Here, the ground state of the specimens is the martensite phase. The insets are those for $x = 15$.

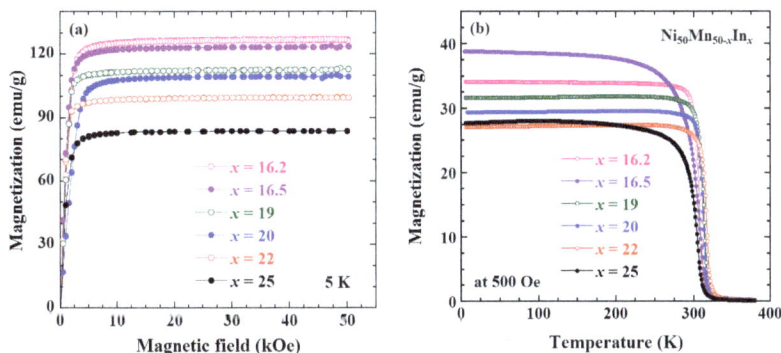

Figure 2. (a) Magnetization curves at 5 K and (b) thermomagnetization curves measured at 500 Oe in the warming process for the specimens with $x \geq 16.2$ in $Ni_{50}Mn_{50-x}In_x$ alloys. Here, the ground state of the system is the ferromagnetic parent phase and no martensitic transformation occurs down to low temperatures.

Figure 3 indicates concentration dependences of saturation magnetization (I_s) and T_C for Ni$_{50}$Mn$_{50-x}$In$_x$ alloys with $x \geq 16.2$, together with reported experimental values [22] and the theoretically calculated one for the stoichiometric composition [23]. Here, I_s is determined by linear extrapolation of the M^2 versus H/M curve to $H/M = 0$ (Arrott plot), from the data in Figure 2a. I_s increases almost linearly with decreasing In content, or in other words, with increasing Mn content. For the stoichiometric composition of Ni$_{50}$Mn$_{25}$In$_{25}$ (=Ni$_2$MnIn), the Ni atoms locate at the $8c$ site, and the Mn and In atoms at the $4a$ and $4b$ sites in the Wyckoff position, respectively. In the present series of the specimens, excess Mn substituted for In at the $4b$ site. Therefore, the increasing of I_s with increasing the Mn means that the magnetic moment of Mn atoms at the $4b$ site couples ferromagnetically with that of the Mn atoms at the ordinary site ($4a$ site). A solid straight line is drawn, assuming that the magnitude of the magnetic moment of Mn at the $4b$ site is the same as that of Mn at the $4a$ site, and the experimental values are in good agreement with the expected line [24]. Here, the values of the magnetic moments (m) are used as $m_{Mn} = 3.719$, $m_{Ni} = 0.277$ and $m_{In} = -0.066$ μ_B, which are obtained from the first principle calculation [23]. The T_C values for all the specimens with $x \geq 16.2$ are similar (the variation is <15 K) and they are independent to the composition. This behavior suggests that the T_C is governed by the exchange interaction between Mn atoms at the $4a$ site, and the interaction by the Mn atoms at the $4b$ site does not affect the total exchange interaction in the system. This behavior of T_C was studied earlier over a wide concentration region by Miyamoto et al. [25], who reported that T_C decreases drastically with increases In concentration in Ni$_{50}$Mn$_{50-y}$In$_y$ alloys with $y > 25$. Here, excess In substituted the Mn atoms at the $4a$ site, therefore, the decrease in T_C is probably caused by the lack of Mn–Mn exchange interactions at the $4a$ site, if the magnetic moment of Mn atoms are assumed not to change.

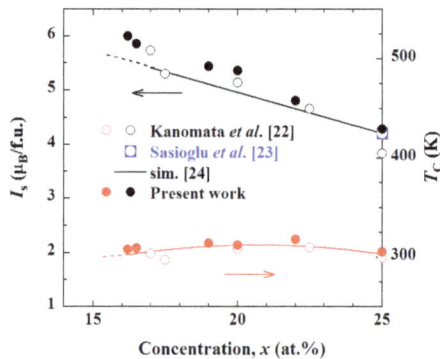

Figure 3. Concentration dependences of the spontaneous magnetization (I_s) evaluated from the magnetization curves in Figure 2a, along with reported experimental and theoretically calculated values [22,23], and of the Curie temperature (T_C) for Ni$_{50}$Mn$_{50-x}$In$_x$ alloys with $x \geq 16.2$. The solid straight line for I_s is simulated assuming that the excess Mn atoms have same magnetic moment as that of Mn at the ordinary sites and couples ferromagnetically [24]. The solid line for T_C is guide to the eye.

3.2. Specific Heat Measurements

Figure 4a,b show the relationship between specific heat (C) and temperature (T) in C/T–T^2 plots for $x \leq 15$ and $x \geq 16.2$, respectively, the insets show C–T plots for each specimen. The specific heat is generally expressed as the summation of some contributions like electronic, lattice, and magnetic contributions. When the experiments are performed at low temperatures, the magnetic excitation has little contribution to the specific heat and the other two components become dominant as follows [26]:

$$C = \gamma T + \beta T^3, \tag{1}$$

where, the first and second terms are the electronic and lattice contributions, respectively, in which γ is the electronic specific heat coefficient and β is the lattice coefficient. In the classical model, γ is thought to correspond to the total DOS. In the C/T–T^2 plot, when a linear relationship is obtained, γ and β are given by the intercept and the slope, respectively. Furthermore, θ_D is given by the relation

$$\theta_D = \sqrt[3]{12\pi^4 R/5\beta}, \tag{2}$$

where R is the gas constant [26]. As shown in Figure 4a,b, a linear relationship is obtained, and the behaviors of these coefficients in terms of composition are different in each figure (i.e., in the M phase and in the P phase). In Figure 4a, for specimens with $x \leq 15$, in which the ground state is the M phase, γ increases with increasing In content. The slope also increases, corresponding to the decrease of θ_D. On the other hand, in Figure 4b for specimens with $x \geq 16.2$, in which the ground state is the ferromagnetic P phase, γ changes slightly and slope is increased with increasing In content.

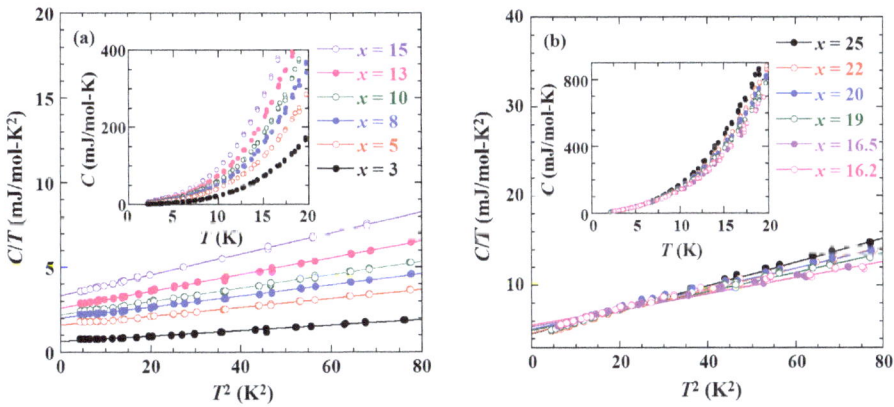

Figure 4. (a) C/T–T^2 plot for the specific heat (C) as a function of temperature (T) for (a) $x \leq 15$ and (b) $x \geq 16.2$ in $Ni_{50}Mn_{50-x}In_x$ alloys. Insets are C–T plots for each specimen.

Concentration dependences of γ and θ_D in $Ni_{50}Mn_{50-x}In_x$ alloys are shown in Figure 5, together with the reported values of $\gamma = 0.31$ mJ/mol·K^2 and $\theta_D = 422$ K for $x = 0$ [11]. The θ_D value decreases linearly with increasing In content in the M phase region. It has been reported that the crystal structure changes continuously from $L1_0$-type tetragonal structure for $x = 0$ to monoclinic multi-layered stacking structure with increasing In content [18]. Magnetic property is collinear-type antiferromagnetic with high Néel temperature for $x = 0$ [15]. This property changes to complicated magnetic properties, in which long-range magnetic ordering is absent and blocking behavior is observed [18,21]. Since neither crystal structure nor magnetic properties change sharply with changing In content, the decrease in θ_D may be caused by the substitution effect of the heavy element In. In the M phase range, γ increases gradually with increasing In content. The electronic state of $x = 0$ is characterized by the formation of a clear pseud-gap around E_F and it has been reported that the total DOS is significantly low and the unique electronic state may correlate with the high antiferromagnetic stability [11]. Gradual increase of γ in M phase region would be due to the loss in antiferromagnetic stability.

In the P phase region, both γ and θ_D increase with decreasing In content. Although the variation of θ_D is similar to that observed in the M phase, the values do not coincide in the middle composition region. From the extrapolations in both the phases (the dotted lines for θ_D), it seems that θ_D changes discontinuously around $x = 15$, the difference corresponds to the change in θ_D during the martensitic phase transformation. Overall, θ_D in the M phase is larger than that in the P phase, in accordance with the experimental fact that the volume of the M phase is more compressive than that of the P phase [27]. The variation of γ agrees well with the value obtained by first-principle density-functional calculations [10]. DOS in the P phase for the stoichiometric and off-stoichiometric compositions have been calculated, and γ values converted from total DOS are also plotted in Figure 5. DOS in the M phase was also calculated by applying the lattice distortion. The gradual increase of γ in the P phase is in accordance with that obtained by calculations. The theoretical calculations have shown that the minor peak corresponding to the Ni-3d e_g state appears at the energy level of -0.1 eV from E_F in the minority band for the off-stoichiometric composition, and the minority peak shifts closer to the E_F with increasing In content. Therefore, the shift in the minor peak will increase the total DOS because the DOS in the majority band is independent of the In content. Furthermore, the experimental value of γ at $x = 13$ is almost the same as the calculated one. Based on the extrapolations of the concentration dependence of γ in both the phases, a change in DOS is expected in the transformation. In other words, DOS is reduced when the P phase transforms to the M phase. This behavior has also been explained by the theoretical calculations that the minority-spin Ni-3d x^2-y^2 splits on introducing the lattice distortion and a pseudo-gap is formed at E_F [10]. The other calculation also suggested that the splitting of the Ni-3d e_g states around E_F plays an important role in occurrence of the martensitic transformation in the off-stoichiometric Ni-Mn-In alloy [28].

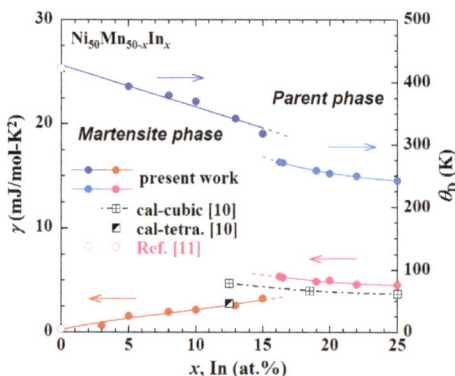

Figure 5. Concentration dependences of the electronic specific heat coefficient (γ) and the Debye temperature (θ_D) for $Ni_{50}Mn_{50-x}In_x$ alloys, along with the reported theoretically calculated values [10] and reported ones for $x = 0$ [11]. The blue and light blue circles are θ_D in M phase and P phase, respectively, and the red and pink circles are γ in M phase and P phase, respectively. The four lines are guide to the eye.

4. Conclusions

Specific heat measurements were performed at low-temperatures in $Ni_{50}Mn_{50-x}In_x$ alloys for $0 \leq x \leq 25$ to investigate the concentration dependences of electronic specific heat coefficient (γ) and the Debye temperature (θ_D) over a wide concentration region. In the martensite (M) phase ($x \leq 15$), θ_D decreases linearly and γ increases with increasing In content. The change in θ_D attributed to the substitution effect by the heavy element In. The increase in γ is attributed to the change in the density of states (DOS), and the clear pseudo-gap formed around the Fermi energy (E_F) may widen with the disappearance of the strong antiferromagnetism in $L1_0$-type NiMn equiatomic alloy, whereas the pseudo-gap is somewhat maintained within the M phase.

In the ferromagnetic parent (P) phase ($x \geq 16.2$), γ increases with decreasing In content. The change in DOS, in accordance with the first-principle density-functional calculation results reported earlier, is that the minor peak in the minority band shifts closer to E_F with decreasing

In content. From the extrapolations of the composition dependences of γ and θ_D from both the phase regions, these values should change discontinuously during the martensitic phase transformation. The value of θ_D in the M phase is larger than that in the P phase, in accordance with the fact that the volume of the M phase is more compressive than that of the P phase. It is suggested that the DOS in P phase decreases during martensitic phase transformation. This behavior is explained by the theoretical calculations that the Ni-3d e_g band splits in the M phase and the pseudo-gap is formed around the E_F.

Acknowledgments. The authors thank Emeritus Takeshi Kanomata in Tohoku-Gakuin University for his useful discussion. This study was supported by Grant-in-Aids from the Japanese Society for the Promotion of Science (JSPS), Ministry of Education, Culture, Sports, Science and Technology (MEXT), Japan. Parts of this study were done at the Center for Low Temperature Science, Institute for Materials Research, Tohoku University.

Author Contributions: Rie Y Umetsu and Ryosuke Kainuma conceived and designed the experiments; Rie Y Umetsu, Xiao Xu, and Wataru Ito performed the experiments; Rie Y Umetsu analyzed the data and wrote the paper; discussion was held among all of the authors, and the manuscript was finalized under their agreements.

Conflicts of Interest: The authors declare no conflict of interest.

References

1. Sutou, Y.; Imano, Y.; Koeda, N.; Omori, T.; Kainuma, R.; Ishida, K.; Oikawa, K. Magnetic and martensitic transformations of NiMnX (X = In, Sn, Sb) ferromagnetic shape memory alloys. *Appl. Phys. Lett.* **2004**, *85*, 4358–4360. [CrossRef]

2. Kainuma, R.; Imano, Y.; Ito, W.; Sutou, Y.; Morito, H.; Okamoto, S.; Kitakami, O.; Oikawa, K.; Fujita, A.; Kanomata, T.; et al. Magnetic field-induced shape recovery by reverse phase transformation. *Nature* **2006**, *439*, 957–960. [CrossRef] [PubMed]

3. Kainuma, R.; Imano, Y.; Ito, W.; Morito, H.; Sutou, Y.; Oikawa, K.; Fujita, A.; Ishida, K.; Okamoto, S.; Kitakami, O.; et al. Metamagnetic shape memory effect in a Heusler-type Ni$_{43}$Co$_7$Mn$_{39}$Sn$_{11}$ polycrystalline alloy. *Appl. Phys. Lett.* **2006**, *88*, 192513. [CrossRef]

4. Krenke, T.; Duman, E.; Acet, M.; Wassermann, E.F.; Moya, X.; Mañosa, L.; Planes, A. Inverse magnetocaloric effect in ferromagnetic Ni–Mn–Sn alloys. *Nat. Mater.* **2005**, *4*, 450–454. [CrossRef] [PubMed]

5. Oikawa, K.; Ito, W.; Imano, Y.; Sutou, Y.; Kainuma, R.; Ishida, K.; Okamoto, S.; Kitakami, O.; Kanomata, T. Effect of magnetic field on martensitic transition of Ni$_{46}$Mn$_{41}$In$_{13}$ Heusler alloy. *Appl. Phys. Lett.* **2006**, *88*, 122507. [CrossRef]

6. Han, Z.D.; Wang, D.H.; Zhang, C.L.; Tang, S.L.; Gu, B.X.; Du, Y.W. Large magnetic entropy changes in the Ni$_{45.4}$Mn$_{41.5}$In$_{13.1}$ ferromagnetic shape memory alloy. *Appl. Phys. Lett.* **2006**, *89*, 182507. [CrossRef]

7. Koyama, K.; Okada, H.; Watanabe, K.; Kanomata, T.; Kainuma, R.; Ito, W.; Oikawa, K.; Ishida, K. Observation of large magnetoresistance of magnetic Heusler alloy Ni$_{50}$Mn$_{36}$Sn$_{14}$ in high magnetic fields. *Appl. Phys. Lett.* **2006**, *88*, 132505. [CrossRef]

8. Yu, S.Y.; Liu, Z.H.; Liu, G.D.; Chen, J.L.; Cao, Z.X.; Wu, G.H.; Zhang, B.; Zhang, X.X. Large magnetoresistance in single-crystalline Ni$_{50}$Mn$_{50-x}$In$_x$ alloys (x = 14–16) upon martensitic transformation. *Appl. Phys. Lett.* **2006**, *89*, 162503. [CrossRef]

9. Zhang, B.; Zhang, X.X.; Yu, S.Y.; Chen, J.L.; Cao, Z.X.; Wu, G.H. Giant magnetothermal conductivity in the Ni–Mn–In ferromagnetic shape memory alloys. *Appl. Phys. Lett.* **2007**, *91*, 012510. [CrossRef]

10. Zhu, S.; Ye, M.; Shirai, K.; Taniguchi, M.; Ueda, S.; Miura, Y.; Shirai, M.; Umetsu, R.Y.; Kainuma, R.; Kanomata, T.; et al. Drastic change in density of states upon martensitic phase transition for metamagnetic shape memory alloy Ni$_2$Mn$_{1+x}$In$_{1-x}$. *J. Phys. Condens. Matter* **2015**, *27*, 362201. [CrossRef] [PubMed]

11. Umetsu, R.Y.; Sakuma, A.; Fukamichi, K. Magnetic properties and electronic structures of $L1_0$-type MnTM (TM = Ir, Pt, Pd and Ni) alloy systems. *Met. Mater. Process.* **2003**, *15*, 67–94. [CrossRef]

12. Umetsu, R.Y.; Fukamichi, K.; Sakuma, A. Electrical and magnetic properties, and electronic structures of pseudo-gap-type antiferromagnetic $L1_0$-type MnPt Alloys. *Mater. Trans.* **2006**, *47*, 2–10. [CrossRef]

13. Umetsu, R.Y.; Fukamichi, K.; Sakuma, A. Effective exchange constant and electronic structure of pseudo-gap-type $L1_0$-MnPd Alloys. *J. Phys. Soc. Jpn.* **2006**, *75*, 104714. [CrossRef]

14. Ito, W.; Imano, Y.; Kainuma, R.; Sutou, Y.; Oikawa, K.; Ishida, K. Martensitic and magnetic transformation behaviors in Heusler-type NiMnIn and NiCoMnIn metamagnetic shape memory Alloys. *Metall. Mater. Trans. A* **2007**, *38*, 759–766. [CrossRef]

15. Pál, L.; Krén, E.; Kádár, G.; Szabó, P.; Tarnóczi, T. Magnetic structures and phase transformations in Mn-based CuAu-I type alloys. *J. Appl. Phys.* **1968**, *39*, 538–544. [CrossRef]

16. Aksoy, S.; Posth, O.; Acet, M.; Meckenstock, R.; Lindner, J.; Farle, M.; Wassermann, E.F. Ferromagnetic resonance in Ni-Mn based ferromagnetic Heusler alloys. *J. Phys. Conf. Ser.* **2010**, *200*, 092001. [CrossRef]

17. Priolkar, K.R. Role of local disorder in martensitic and magnetic interactions in Ni–Mn based ferromagnetic shape memory alloys. *Phys. Status Solidi B* **2014**, *251*, 2088–2096. [CrossRef]

18. Krenke, T.; Acet, M.; Wassermann, E.F.; Moya, X.; Mañosa, L.; Planes, A. Ferromagnetism in the austenitic and martensitic states of Ni-Mn-In alloys. *Phys. Rev. B* **2006**, *73*, 174413. [CrossRef]

19. Yan, H.; Zhang, Y.; Xu, N.; Senyshyn, A.; Brokmeier, H.G.; Esling, C.; Zhao, X.; Zuo, L. Crystal structure determination of incommensurate modulated martensite in Ni–Mn–In Heusler alloys. *Acta Mater.* **2015**, *88*, 375–388. [CrossRef]

20. Umetsu, R.Y.; Fujita, A.; Ito, W.; Kanomata, T.; Kainuma, R. Determination of the magnetic ground state in the martensite phase of Ni–Mn–Z (Z = In, Sn and Sb) off-stoichiometric Heusler alloys by nonlinear AC susceptibility. *J. Phys. Condens. Matter* **2011**, *23*, 326001. [CrossRef] [PubMed]

21. Khovaylo, V.V.; Kanomata, T.; Tanaka, T.; Nakashima, M.; Amako, Y.; Kainuma, R.; Umetsu, R.Y.; Morito, H.; Miki, H. Magnetic properties of $Ni_{50}Mn_{34.8}In_{15.2}$ probed by Mössbauer spectroscopy. *Phys. Rev. B* **2009**, *80*, 144409. [CrossRef]

22. Kanomata, T.; Yasuda, T.; Sasaki, S.; Nishihara, H.; Kainuma, R.; Ito, W.; Oikawa, K.; Ishida, K.; Neumann, K.-U.; Ziebeck, K.R.A. Magnetic properties on shape memory alloys $Ni_2Mn_{1+x}In_{1-x}$. *J. Magn. Magn. Mater.* **2009**, *321*, 773–776. [CrossRef]

23. Şaşıoğlu, E.; Sandratskii, L.M.; Bruno, P. First-principles calculation of the intersublattice exchange interactions and Curie temperatures of the full Heusler alloys Ni_2MnX (X = Ga, In, Sn, Sb). *Phys. Rev. B* **2004**, *70*, 024427. [CrossRef]

24. Umetsu, R.Y.; Kusakari, Y.; Kanomata, T.; Suga, K.; Sawai, Y.; Kindo, K.; Oikawa, K.; Kainuma, R.; Ishida, K. Metamagnetic behaviour under high magnetic fields in $Ni_{50}Mn_{50-x}In_X$ (X = 14.0 and 15.6) shape memory alloys. *J. Phys. Appl. Phys.* **2009**, *42*, 075003. [CrossRef]

25. Miyamoto, T.; Ito, W.; Umetsu, R.Y.; Kainuma, R.; Kanomata, T.; Ishida, K. Phase stability and magnetic properties of $Ni_{50}Mn_{50-x}In_x$ Heusler-type alloys. *Scr. Mater.* **2010**, *62*, 151–154. [CrossRef]

26. Gopal, E.S.R. *Specific Heats at Low Temperatures*; Plenum Press: New York, NY, USA, 1966.

27. Abematsu, K.; Umetsu, R.Y.; Kainuma, R.; Kanomata, T.; Watanabe, K.; Koyama, K. Structural and magnetic properties of magnetic shape memory alloy $Ni_{46}Mn_{41}In_{13}$ under magnetic fields. *Mater. Trans.* **2014**, *55*, 477–481. [CrossRef]

28. D'Souza, S.W.; Chakrabarti, A.; Barman, S.R. Magnetic interactions and electronic structure of Ni–Mn–In. *J. Electron Spectrosc. Relat. Phenom.* **2016**, *208*, 33–39. [CrossRef]

![metals logo] *metals*

MDPI

Article

Effect of Hydrogen on the Elastic and Anelastic Properties of the R Phase in Ti$_{50}$Ni$_{46.1}$Fe$_{3.9}$ Alloy

Konstantin Sapozhnikov [1,2,*], Joan Torrens-Serra [3], Eduard Cesari [3], Jan Van Humbeeck [4] and Sergey Kustov [2,3]

[1] Division of Solid State Physics, Ioffe Institute, Politekhnicheskaya 26, 194021 St. Petersburg, Russia
[2] Department of Modern Functional Materials, ITMO University, Kronverkskiy 49, 197101 St. Petersburg, Russia; sergey.kustov@uib.es
[3] Departament de Física, Universitat de les Illes Balears, Cra Valldemossa km 7.5, E 07122 Palma de Mallorca, Spain; j.torrens@uib.es (J.T.-S.); eduard.cesari@uib.cat (E.C.)
[4] Departement Materiaalkunde (MTM), Katholieke Universiteit Leuven, Kasteelpark Arenberg 44, B-3001 Leuven, Belgium; jan.vanhumbeeck@kuleuven.be
* Correspondence: k.sapozhnikov@mail.ioffe.ru; Tel.: +7-812-2927119

Received: 12 October 2017; Accepted: 7 November 2017; Published: 10 November 2017

Abstract: The linear and non-linear internal friction, effective Young's modulus, and amplitude-dependent modulus defect of a Ti$_{50}$Ni$_{46.1}$Fe$_{3.9}$ alloy have been studied after different heat treatments, affecting hydrogen content, at temperatures of 13–300 K, and frequencies near 90 kHz. It has been shown that the contamination of the alloy by hydrogen gives rise to an internal friction maximum in the R martensitic phase and a complicated pinning stage in the temperature dependence of the effective Young's modulus at temperatures corresponding to the high-temperature side of the maximum. Dehydrogenation of the H-contaminated alloy transforms the internal friction maximum into a plateau and minimizes the pinning stage. The internal friction maximum is associated with a competition of two different temperature-dependent processes affecting the hydrogen concentration in the core regions of twin boundaries. The amplitude-dependent anelasticity of the R phase is also very sensitive to hydrogen content, its temperature dependence reflects the evolution of extended hydrogen atmospheres near twin boundaries.

Keywords: shape memory alloy; acoustic properties; internal friction; lattice defects; hydrogen

---

1. Introduction

Ti-Ni-based shape memory alloys are quite attractive functional materials [1]. Anelastic properties of Ti-Ni-based alloys are widely explored, both for microstructural characterization of the alloys and for their application as high-damping materials [2]. The effect of hydrogen on elastic and anelastic properties of Ti-Ni-based alloys have attracted considerable attention (see [3] for a review), especially after Fan et al. [4,5] have shown that hydrogen is unintentionally introduced during conventional heat treatment (water quenching after high-temperature annealing in argon-filled [4] or vacuum-sealed [5] quartz tubes) by the chemical reaction of residual water in quartz tubes with Ti at high temperatures. Fan et al. [4,5] have also shown that the relaxation internal friction (IF) peak observed by many researchers at low frequencies (about 200 K at 1 Hz) is due to an interaction of twin boundaries with hydrogen. Recently, we have studied the effect of hydrogen contamination during heat treatments on the low-temperature elastic and anelastic properties of a Ni$_{50.8}$Ti$_{49.2}$ alloy at ultrasonic frequencies, for which the '200 K' relaxation IF peak cannot be observed [6]. It has been shown that the hydrogen contamination gives rise to a non-relaxation internal IF maximum, whose temperature and height depend strongly on the hydrogen content. The IF maximum was interpreted as a pseudo-peak formed due to a competition of two different temperature-dependent processes affecting the hydrogen concentration in the core regions of twin boundaries [6].

It seems important to check whether the new hydrogen-related anelastic effect, reported in [6], is inherent only in the B19′ martensitic phase, or it is common phenomenon for different martensitic phases of Ti-Ni-based alloys. The goal of the present work is to study the effect of hydrogen contamination on the elastic and anelastic properties of a Ti-Ni-based alloy with the R martensitic phase. The alloy composition was chosen in such a way that the B2↔R transformation type occurred, and the R martensitic phase could be tested over a wide temperature range. Such a situation takes place for TiNi$_{50-x}$Fe$_x$ alloys with $3 < x < 5$ [7]. Although there are data in the literature on the elastic and anelastic properties of Ti-Ni$_{50-x}$Fe$_x$ alloys [7–15], most of the data refer either to low x values, providing a narrow temperature range of the R phase, or to high ones, suppressing the martensitic transformation. Data for intermediate compositions are scarce and requires replenishment. The present paper is devoted to investigations of low-temperature elastic and anelastic properties of a Ti$_{50}$Ni$_{46.1}$Fe$_{3.9}$ alloy after two distinct heat treatments: conventional heat treatment contaminating the alloy by hydrogen and dehydrogenation treatment.

2. Materials and Methods

The Ti$_{50}$Ni$_{46.1}$Fe$_{3.9}$ alloy was prepared by induction melting of 99.99 wt. % pure components. Rod-shaped samples for acoustic and resistivity measurements were spark cut and mechanically polished. Two different heat treatments were used: (1) water quenching (WQ), (2) vacuum annealing (VA). The WQ treatment included annealing for 2 h at 1273 K in a vacuum-sealed quartz tube followed by quenching into water with breaking the tube. The VA treatment consisted in annealing for 26 h at 950 K under a residual gas pressure $P \approx 10^{-3}$ Pa. The annealing temperature was chosen to be higher than the temperature ranges of hydrogen desorption from Ni-Ti alloys reported in the literature [16–18]. Two samples were used for acoustic measurements: one after only the WQ treatment and another one subjected to a combination of the WQ and VA treatments.

The IF and effective Young's modulus, E, of the samples were measured at temperatures of 13–300 K by means of the resonant piezoelectric composite oscillator technique [19] using longitudinal oscillations at frequencies near 90 kHz. The logarithmic decrement δ, defined as $\delta = \Delta W/2W$, where ΔW is the energy dissipated per cycle and W is the maximum stored vibrational energy, was used as a measure of the IF. A computer-controlled setup [20] enabled us to measure quasi-simultaneously temperature dependences of the IF and Young's modulus at two values of oscillatory strain amplitude. At a low value of strain amplitude (10^{-6}), we monitored the linear (strain amplitude-independent) behaviour of the IF and Young's modulus. The non-linear (strain amplitude-dependent) effects were observed at a high value of strain amplitude (5×10^{-5}). The amplitude-dependent parts of the IF and Young's modulus defect, δ_h and $(\Delta E/E)_h$, were routinely derived from the differences of the IF and Young's modulus values registered at high and low strain amplitudes: $\delta_h = \delta\,(\varepsilon_m) - \delta_i$ and $(\Delta E/E)_h = (E_i - E\,(\varepsilon_m))/E_i$, where ε_m is the oscillatory strain amplitude, $\delta\,(\varepsilon_m)$, δ_i and $E\,(\varepsilon_m)$, E_i, are the values of the IF and Young's modulus in the amplitude-dependent and amplitude-independent ranges, respectively. The samples were cooled/heated in a helium atmosphere in an Oxford close-loop cryostat (Oxford Instruments, Abingdon, UK) at a temperature change rate of about 2 K/min.

The type of the martensitic transformation was verified by electrical resistance data. The four-wire alternating current impedance measurements were performed at frequency of 686 Hz and temperature change rate of 2 K/min. The real part of the impedance R was measured using a lock-in amplifier (Stanford Research Systems, Inc., Sunnyvale, CA, USA).

Hydrogen content in variously treated samples was quantified by means of the vacuum hot extraction method. An industrial hydrogen analyser AV-1 (Electronic and Beam Technologies Ltd., St. Petersburg, Russia) was used with mass-spectrometric registration of the time dependence of hydrogen flux from samples heated in vacuum [21]. The analyser was calibrated on certified hydrogen-containing samples of an aluminium alloy with the error of the certified value of hydrogen concentration of 6%. Hydrogen was extracted from samples in two successive steps at temperatures of 803 K and 1073 K under a working pressure of 10^{-4} Pa.

3. Results

Figure 1 shows the normalized resistance data for the WQ sample. It is seen that the resistance exhibits a sharp change at temperatures near 240 K with narrow temperature hysteresis of about 3 K. Such behaviour is typical for the B2↔R transformation [1,12,22].

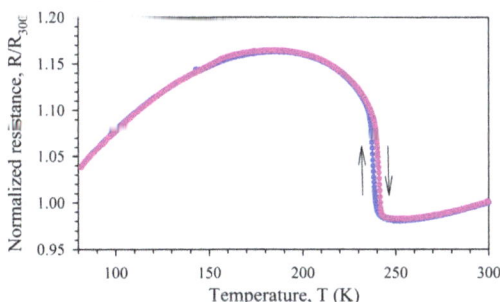

Figure 1. Temperature dependence of resistance for the water-quenched $Ti_{50}Ni_{46.1}Fe_{3.9}$ sample on cooling (blue) and heating (pink). Data are normalized to the resistance values at 300 K.

Results of the hydrogen content evaluation in differently treated samples (WQ and VA) by means of the vacuum hot extraction method are summarized in Table 1. It is seen that hydrogen content varies considerably with heat treatment. The total extracted hydrogen content in the WQ sample is about 5.7 times higher as compared to the VA sample. It is important to note that the effect of the heat treatments is much stronger for the hydrogen fraction extracted at 803 K, because weakly bound states of hydrogen prevail in the WQ sample with increased hydrogen content.

Table 1. Hydrogen extracted from differently treated samples.

Sample	Hydrogen Content, at. ppm		
	Extracted at 803 K	Extracted at 1073 K	Total
WQ	372	79	450
VA	41	38	79

The temperature spectra of the amplitude-independent IF (a) and effective Young's modulus (b) measured in a cooling-heating cycle after the WQ and VA heat treatments are depicted in Figure 2. Two IF maxima are observed after the WQ treatment: (1) a broad maximum at 90 K (on cooling) or 100 K (on heating); and (2) a sharp asymmetric maximum at temperatures of the martensitic transformation, accompanied by a minimum of the effective Young's modulus. The IF of the austenitic phase diminishes after the VA treatment. On the contrary, both the high-temperature IF maximum and the IF of the martensitic phase strongly increase, except the IF at the low-temperature side of the low-temperature IF maximum. As a result of such an effect of the VA treatment, the low-temperature IF maximum transforms into a plateau or a very smooth maximum at temperatures of 150–160 K. Temperature hysteresis of the IF is observed between 80 and 170 K after both heat treatments. $E(T)$ curves measured after the WQ and VA treatments cross at a temperature near 130 K, since $E(T)$ dependence for the WQ sample is weaker in this temperature range.

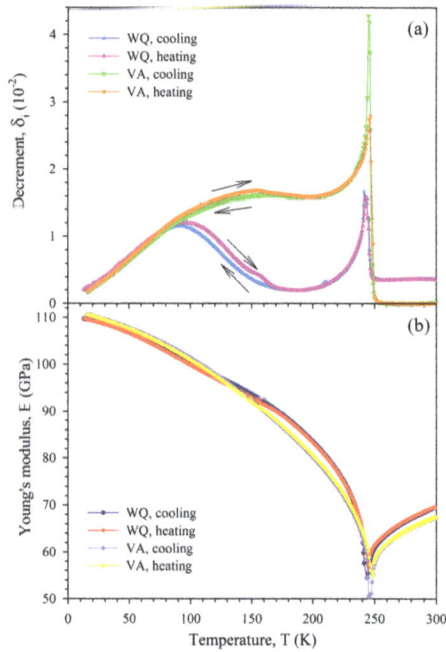

Figure 2. Temperature dependence of the decrement (**a**) and effective Young's modulus (**b**) of the $Ti_{50}Ni_{46.1}Fe_{3.9}$ sample measured in a cooling-heating cycle at strain amplitude of 10^{-6} after water quenching (WQ) or vacuum annealing (VA).

Figure 3 displays the temperature dependence of the temperature coefficient of the Young's modulus, $\alpha_E = (1/E_0) \times (dE/dT)$, calculated for all the curves shown in Figure 2b, with E_0 taken as the Young's modulus value at $T = 300$ K. One can see that the difference between E (T) curves for WQ and VA samples extends from 80 to 200 K, with lower absolute values of the temperature coefficient of the Young's modulus for the WQ sample. Two different effects can be distinguished in the heating curves: (1) smooth variations between 80 and 200 K, which are reproduced in the cooling curves with some temperature hysteresis; and (2) abrupt changes between 150 and 170 K, which are not observed in the cooling curves.

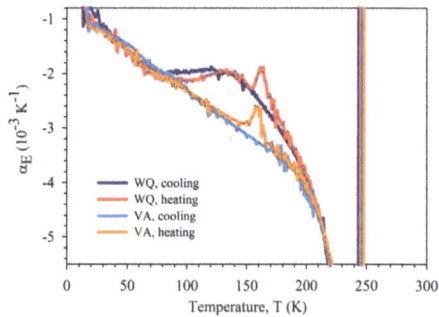

Figure 3. Temperature dependence of the temperature coefficient of the effective Young's modulus of the $Ti_{50}Ni_{46.1}Fe_{3.9}$ sample after water quenching (WQ) or vacuum annealing (VA), derived from the curves shown in Figure 2b.

Figure 4 shows δ_h (a) and $(\Delta E/E)_h$ (b) versus T dependences measured in a cooling-heating cycle for WQ and VA samples. The amplitude-dependent anelasticity (both δ_h and $(\Delta E/E)_h$) in the R phase is strongly promoted by the VA treatment as compared to the WQ treatment, except temperatures below 70 K. For $T < 70$ K the difference between VA and WQ samples is small. Similar to the δ_i (T) dependence, the δ_h and $(\Delta E/E)_h$ versus T dependences for the WQ sample exhibit a maximum, but at a much higher temperatures (140–160 K for δ_h, about 200 K for $(\Delta E/E)_h$). These maxima transform into plateaux or shoulders after the VA treatment. Temperature hysteresis is observed in both characteristics of the amplitude-dependent anelasticity between 140 K and phase transformation temperatures.

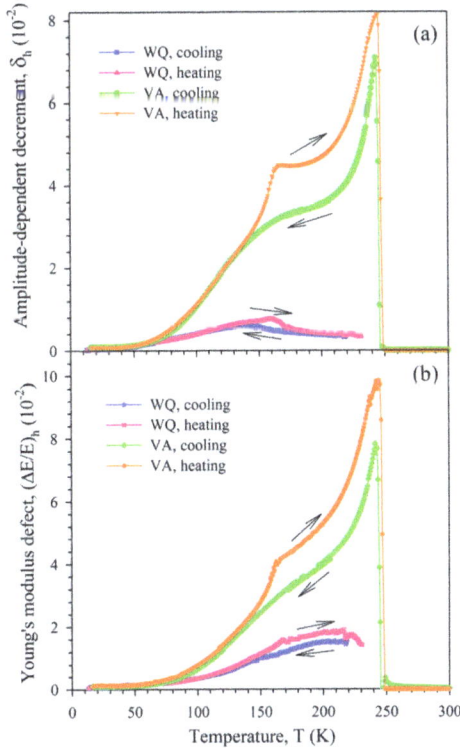

Figure 4. Temperature dependence of the amplitude-dependent components of the decrement (**a**) and Young's modulus defect (**b**) of the $Ti_{50}Ni_{46.1}Fe_{3.9}$ sample measured in a cooling-heating cycle after water quenching (WQ) or vacuum annealing (VA). The dependence is derived from the temperature spectra of the decrement and effective Young's modulus measured at strain amplitudes of 10^{-6} and 5×10^{-5}.

4. Discussion

The anelasticity of the martensitic phases of Ni-Ti-based alloys is usually attributed to the motion of twin boundaries (see, for example, [2,3]). Like other structural defects in solids creating lattice distortions, twin boundaries trap hydrogen atoms [23]. Linear and planar defects trap hydrogen both into core regions and into surrounding stress fields, but separation of these effects is not always possible [24]. Investigations of linear and non-linear components of anelasticity (IF and modulus defect), related respectively to atomic-scale and mesoscopic displacements of linear and planar defects, may enable such separation: the linear (amplitude-independent) anelasticity is sensitive to point defects situated in the core regions of dislocations/boundaries, whereas non-linear anelasticity is efficient to monitor the changes of point defect concentration in extended point defect atmospheres.

Our data evidence that elastic and anelastic properties of the R phase are very sensitive to variations of hydrogen content. After the conventional heat treatment (WQ) contaminating the alloy with hydrogen, a broad IF maximum is observed at temperatures of 90–100 K (Figure 2a). The high-temperature side of this maximum corresponds to stages in the E (T) dependence with decreasing absolute values of the temperature coefficient (Figure 3). Such behaviour of the effective elastic modulus on heating is conventionally ascribed to a pinning (annealing, recovery) stage (see, for example, [25–27]). Dehydrogenation treatment suppresses strongly the pinning stage provoking the increase of δ_i and transformation of the well-defined low-temperature δ_i (T) maximum into a plateau (or very smooth maximum at higher temperatures of 150–160 K). Similar behaviour of δ_i and E was reported for the B19′ martensitic phase of the $Ni_{50.8}Ti_{49.2}$ alloy [6]. The characteristic pattern of δ_i and E on the high-temperature side of the low-temperature IF maximum on heating/cooling was ascribed to a pinning/depinning stage due to increase/decrease of the hydrogen concentration in the core regions of twin boundaries [6]. The pinning/depinning stage was associated with redistribution of hydrogen between twin boundaries and other structural defects, such as bulk dislocations and/or grain boundaries [6]. As in the case of the $Ni_{50.8}Ti_{49.2}$ alloy [6], only traces of the pinning/depinning stage, caused by a residual hydrogen content, are observed after dehydrogenation treatment of the $Ti_{50}Ni_{46.1}Fe_{3.9}$ alloy (see VA curves in Figures 2–4). Thus, the overall pattern of the δ_i (T) spectra can be qualitatively interpreted as a superposition of the more (high hydrogen content) or less (low hydrogen content) pronounced pinning stage, superimposed on a general IF rise with increasing temperature. In [6], this general trend was associated with a decrease of hydrogen concentration in the core regions of twin boundaries caused by entropy contributions to the binding free energy of a hydrogen atom to a twin boundary. Figure 2a shows that the amplitude-independent IF does not depend on heat treatment at $T < 80$ K. This is an indication that the IF variations for $T < 80$ K is not controlled be the overall hydrogen content (at least for concentrations between 79 and 450 at. ppm).

It should be noted that two different pinning stages are distinguished in the δ_i (T) and E (T) heating curves. The extended smooth stage starting near 80 K is also found on cooling, with certain temperature hysteresis (Figures 2 and 3). This stage is strongly suppressed by dehydrogenation treatment. The abrupt pinning stage at 150–170 K has no correspondence in the cooling curves and is not affected by dehydrogenation treatment. A comparison of linear and non-linear anelastic effects may shed some light on the origin of the stages. The smooth pinning stage is observed only in δ_i (T) and E (T) curves, whereas the abrupt stage can be distinguished, in addition, in δ_h and $(\Delta E/E)_h$ versus T dependences. In Figure 4, the pinning stage is manifested in δ_h (T) curves as a decrease with increasing temperature (heating run for VA sample, cooling and heating runs for WQ sample) or as a shoulder (cooling run for VA sample). Different manifestation of the two pinning stages in the linear and non-linear anelastic properties indicates distinct spatial localization of the hydrogen diffusion paths involved in the redistribution of hydrogen between twin boundaries and other structural defects. During the initial smooth stage, pinning suppresses only short-range mobility of twin boundaries, while their non-linear dynamics on the mesoscopic scale remains unaffected. This evidences that hydrogen diffusion proceeds in the core regions of the defects and, hence, is defect-assisted. The second pinning stage suppresses both short-range and long-range mobility of twins. It is indicative that the second pinning stage is much more extended (30–40 K) in the non-linear anelastic properties as compared to the linear ones. A possible interpretation of the high-temperature pinning stage, affecting both core and extended atmosphere regions of twin boundaries, can involve bulk diffusion of hydrogen. We note here that much faster pipe diffusion of hydrogen along the dislocations, as compared to bulk diffusion, has been reported in Pd [28].

Finally, we note that the present study provides at least two indications that the diffusion mobility of hydrogen interstitials in the R phase is lower than in the B19′ phase: (1) the low-temperature IF maximum and pinning stage in the effective Young's modulus are shifted to higher temperatures in the R phase as compared to the B19′ phase; (2) temperature hysteresis of the IF on the high-temperature

Metals **2017**, *7*, 493

side of the low-temperature IF maximum is observed in the R phase, but not in the B19′ phase. Since hydrogen diffusion over the range of the low-temperature IF maximum is most probably defect-assisted, different structure of twin boundaries may contribute to this distinction between diffusion in the R and B19′ phases.

5. Conclusions

(1)   Contamination of the $Ti_{50}Ni_{46.1}Fe_{3.9}$ alloy by hydrogen gives rise to an IF maximum in the R martensitic phase and a complicated pinning stage in the temperature dependence of the effective Young's modulus at temperatures corresponding to the high-temperature side of the IF maximum.

(2)   Dehydrogenation heat treatment of the H-contaminated $Ti_{50}Ni_{46.1}Fe_{3.9}$ alloy transforms the IF maximum into a plateau or a smooth maximum caused by a residual hydrogen content.

(3)   The IF maximum is associated with a competition of two different temperature-dependent processes affecting the hydrogen concentration in the core regions of twin boundaries, similarly to the non relaxation IF maximum observed earlier in the B19′ martensitic phase of a $Ni_{50.8}Ti_{49.2}$ alloy.

(4)   Amplitude-dependent anelasticity (IF and Young's modulus defect) of the R phase is also very sensitive to hydrogen content, but its maximum values are shifted to higher temperatures as compared to those of the amplitude-independent IF. The temperature dependence of the amplitude-dependent anelasticity is associated with evolution of extended hydrogen atmospheres near twin boundaries.

(5)   Diffusion mobility of hydrogen interstitials in the R phase is lower than in the B19′ phase.

Acknowledgments: The work was supported by the Ministry of Education and Science of the Russian Federation, goszadanie No. 3.1421.2017/4.6 and by Spanish Ministerio de Economía y Competitividad, Project MAT2014-56116-C04-01-R. SK acknowledges the support by the Government of Russian Federation (grant No. 074-U01) through the ITMO Fellowship and Professorship Program.

Author Contributions: Konstantin Sapozhnikov, Joan Torrens-Serra and Sergey Kustov conceived, designed and performed the experiments; Eduard Cesari prepared the alloy; Konstantin Sapozhnikov, Joan Torrens-Serra, Eduard Cesari, Jan Van Humbeeck and Sergey Kustov discussed the results; Konstantin Sapozhnikov and Sergey Kustov wrote the paper.

Conflicts of Interest: The authors declare no conflict of interest.

References

1.   Otsuka, K.; Ren, X. Physical metallurgy of Ti-Ni-based shape memory alloys. *Prog. Mater. Sci.* **2005**, *50*, 511–678. [CrossRef]
2.   Blanter, M.S.; Golovin, I.S.; Neuhäuser, H.; Sinning, H.-R. *Internal Friction in Metallic Materials. A Handbook*; Springer: Berlin/Heidelberg, Germany, 2007.
3.   Mazzolai, G. Recent progresses in the understanding of the elastic and anelastic properties of H-free, H-doped and H-contaminated NiTi based alloys. *AIP Adva.* **2011**, *1*, 040701. [CrossRef]
4.   Fan, G.; Zhou, Y.; Otsuka, K.; Ren, X.; Nakamura, K.; Ohba, T.; Suzuki, T.; Yoshida, I.; Yin, F. Effects of frequency, composition, hydrogen and twin boundary density on the internal friction of $Ti_{50}Ni_{50-x}Cu_x$ shape memory alloys. *Acta Mater.* **2006**, *54*, 5221–5229. [CrossRef]
5.   Fan, G.; Otsuka, K.; Ren, X.; Yin, F. Twofold role of dislocations in the relaxation behavior of Ti-Ni martensite. *Acta Mater.* **2008**, *56*, 632–641. [CrossRef]
6.   Sapozhnikov, K.; Torrens-Serra, J.; Cesari, E.; Van Humbeeck, J.; Kustov, S. On the effect of hydrogen on the low-temperature elastic and anelastic properties of Ni-Ti-based alloys. *Materials* **2017**, *10*, 1174. [CrossRef] [PubMed]
7.   Zhang, J.; Wang, Y.; Ding, X.; Zhang, Z.; Zhou, Y.; Ren, X.; Wang, D.; Ji, Y.; Song, M.; Otsuka, K.; et al. Spontaneous strain glass to martensite transition in a $Ti_{50}Ni_{44.5}Fe_{5.5}$ strain glass. *Phys. Rev. B* **2011**, *84*, 214201. [CrossRef]

8.   Chui, N.Y.; Huang, Y.T. A study of internal friction, electric resistance and shape change in a Ti-Ni-Fe alloy during phase transformation. *Scr. Mater.* **1987**, *21*, 447–452. [CrossRef]

9.   Fan, G.; Zhou, Y.; Otsuka, K.; Ren, X. Ultrahigh damping in R-phase state of Ti–Ni–Fe alloy. *Appl. Phys. Lett.* **2006**, *89*, 161902. [CrossRef]

10.   Yoshida, I.; Monma, D.; Ono, T. Damping characteristics of $Ti_{50}Ni_{47}Fe_3$ alloy. *J. Alloys Compd.* **2008**, *448*, 349–354. [CrossRef]

11.   Fan, G.; Zhou, Y.; Otsuka, K.; Ren, X. Comparison of the two relaxation peaks in the $Ti_{50}Ni_{48}Fe_2$ alloy. *Mater. Sci. Eng. A* **2009**, *521–522*, 178–181. [CrossRef]

12.   Wang, D.; Zhang, Z.; Zhang, J.; Zhou, Y.; Wang, Y.; Ding, X.; Wang, Y.; Ren, X. Strain glass in Fe-doped Ti–Ni. *Acta Mater.* **2010**, *58*, 6206–6215. [CrossRef]

13.   Zuo, S.; Jin, M.; Chen, D.; Jin, X. Origin of the anelastic behavior in $Ti_{50}Ni_{44}Fe_6$ alloy. *Scr. Mater.* **2015**, *108*, 113–116. [CrossRef]

14.   Chang, S.H.; Chien, C.; Wu, S.K. Damping characteristics of the inherent and intrinsic internal friction of $Ti_{50}Ni_{50-x}Fe_x$ (x = 2, 3, and 4) shape memory alloys. *Mater. Trans.* **2016**, *57*, 351–356. [CrossRef]

15.   Zhang, J.; Xue, D.; Cai, X.; Ding, X.; Ren, X.; Sun, J. Dislocation induced strain glass in $Ti_{50}Ni_{45}Fe_5$ alloy. *Acta Mater.* **2016**, *120*, 130–137. [CrossRef]

16.   Yokoyama, K.; Tomita, M.; Sakai, J. Hydrogen embrittlement behavior induced by dynamic martensite transformation of Ni-Ti superelastic alloy. *Acta Mater.* **2009**, *57*, 1875–1885. [CrossRef]

17.   Saito, T.; Yokoyama, T.; Takasaki, A. Hydrogenation of TiNi shape memory alloy produced by mechanical alloying. *J. Alloys Compd.* **2011**, *509* (Suppl. 2), S779–S781. [CrossRef]

18.   Ribeiro, R.M.; Lemus, L.F.; dos Santos, D.S. Hydrogen absorption study of Ti-based alloys performed by melt-spinning. *Mater. Res.* **2013**, *16*, 679–682. [CrossRef]

19.   Robinson, W.H.; Edgar, A. The piezoelectric method of determining mechanical damping at frequencies of 30 to 200 kHz. *IEEE Trans. Sonics Ultrason.* **1974**, *21*, 98–105. [CrossRef]

20.   Kustov, S.; Golyandin, S.; Ichino, A.; Gremaud, G. A new design of automated piezoelectric composite oscillator technique. *Mater. Sci. Eng. A* **2006**, *442*, 532–537. [CrossRef]

21.   Polyanskiy, A.M.; Polyanskiy, V.A.; Yakovlev, Y.A. Experimental determination of parameters of multichannel hydrogen diffusion in solid probe. *Int. J. Hydrogen Energy* **2014**, *39*, 17381–17390. [CrossRef]

22.   Choi, M.S.; Fukuda, T.; Kakeshita, T. Anomalies in resistivity, magnetic susceptibility and specific heat in iron-doped Ti-Ni shape memory alloys. *Scr. Mater.* **2005**, *53*, 869–873. [CrossRef]

23.   So, K.H.; Kim, J.S.; Chun, Y.S.; Park, K.T.; Lee, Y.K.; Lee, C.S. Hydrogen delayed fracture properties and internal hydrogen behavior of a Fe-18Mn-1.5Al-0.6C TWIP steel. *ISIJ Int.* **2009**, *49*, 1952–1959. [CrossRef]

24.   Myers, S.M.; Baskes, M.I.; Birnbaum, H.K.; Corbett, J.W.; DeLeo, G.G.; Estreicher, S.K.; Haller, E.E.; Jena, P.; Johnson, N.M.; Kirchheim, R.; et al. Hydrogen interactions with defects in crystalline solids. *Rev. Mod. Phys.* **1992**, *64*, 559. [CrossRef]

25.   Thompson, D.O.; Pare, V.K. Effect of fast neutron bombardment at various temperatures upon the Young's modulus and internal friction of copper. *J. Appl. Phys.* **1960**, *31*, 528–535. [CrossRef]

26.   Magalas, L.B.; Moser, P. Internal friction in cold worked iron. *J. Phys. Coll.* **1981**, *42*, C5-97–C5-102. [CrossRef]

27.   Takamura, S.; Kobiyama, M. Dislocation pinning in Al and Ag alloys after low-temperature deformation. *Phys. Stat. Sol. A* **1986**, *95*, 165–172. [CrossRef]

28.   Kirchheim, R. Interaction of hydrogen with dislocations in palladium—I. Activity and diffusivity and their phenomenological interpretation. *Acta Metall.* **1981**, *29*, 835–843. [CrossRef]

metals

MDPI

Article

Effect of Ni-Content on the Transformation Temperatures in NiTi-20 at. % Zr High Temperature Shape Memory Alloys

Matthew Carl [1],*, Jesse D. Smith [1], Brian Van Doren [2] and Marcus L. Young [1]

[1]   Department of Materials Science and Engineering, University of North Texas, Denton, TX 76203, USA; jessesmith@my.unt.edu (J.D.S.); Marcus.Young@unt.edu (M.L.Y.)

[2]   ATI Specialty Alloys and Components, Albany, OR 97321, USA; Brian.Vandoren@ATImetals.com

*   Correspondence: matthewcarl@my.unt.edu; Tel.: +1-(940)-369-7170

Received: 18 October 2017; Accepted: 16 November 2017; Published: 21 November 2017

Abstract: The effect of Ni-content on phase transformation behavior of NiTi-20 at. % Zr high temperature shape memory alloy (HTSMA) is investigated over a small composition range, i.e., 49.8, 50.0 and 50.2 at. % Ni, by differential scanning calorimetry (DSC), high-energy synchrotron radiation X-ray diffraction (SR-XRD), scanning electron microscopy (SEM), and transmission electron microscopy (TEM). All samples show a monoclinic B19′ martensitic structure at room temperature. It is shown that even with these small variations in Ni-content, the alloy shows vastly different transformation temperatures and responds in a drastically different manner to aging treatments at 550 and 600 °C. Lastly, a discussion on H-phase composition with respect to bulk composition is presented.

Keywords: shape memory alloys; high temperature shape memory alloy (HTSMA); NiTiZr; NiTi

---

1. Introduction

Since the discovery of the shape memory effect in near equi-atomic binary Ni–Ti alloys [1], considerable research efforts have been made to control the transformation temperatures (TTs), microstructure, and shape memory properties through alloying and thermo–mechanical processing [2]. Through this research, Ni–Ti-based shape memory alloys (SMAs) have now become an important technological material for a wide array of applications [3,4], i.e., specifically medical devices [5], actuators [6], etc. Despite this eclectic and ever-expanding list of applications, there exists more opportunity to enlarge the realm of potential uses by increasing the temperatures at which the characteristic phase transformation occurs, most notably in the aerospace industry [6–9].

It has been shown that binary Ni–Ti can exhibit maximum TTs of approximately 115 °C [2,6], specifically the austenitic finish (A_f), but this can readily be increased by adding ternary additions such as Au, Pt, Pd, Hf or Zr to create high temperature SMAs (HTSMAs); however, with the exception of Hf and Zr, these additions increase production cost greatly, making them economically impractical for almost all applications, especially in the private sector where profit is the main concern. Following this logic, there still exists a need to create viable cost-effective HTSMAs. While NiTiHf HTSMAs are currently being heavily investigated due to their ease of processability and excellent shape memory characteristics, it is worth noting that Hf still costs ten times the price of Zr and is twice as dense; meaning that producing acceptable NiTiZr HTSMAs would allow for cheaper and lighter devices, two of the most important aspects in the aerospace industry. Therefore, it has been determined by the authors that the pursuit of NiTiZr HTSMAs is of both technological and intellectual importance.

The beneficial effects of Zr additions to Ni–Ti were first reported by Eckelmeyer et al. [10] in 1976 and later expanded by Adu Judom et al. [11] and Hsieh et al. [12,13] to include higher atomic percentages while still maintaining the characteristic austenite-to-martensite phase transformation, up to 25 at. % Zr when substituting for Ti and the amounts of Ti and Zr sum to above 51 at. %. Early research focused on Ti-rich compositions due to their higher TTs but soon proved to be too brittle, exhibited poor thermal stability, and formed a large number of intermetallic phases such NiTiZr Laves phase, $Ni_7(Ti,Zr)_2$, $Ni_{10}(Ti,Zr)_7$, and NiZr [13,14]; however, recent research has shifted to Ni-rich compositions due to the formation of nanoscale precipitates first recognized by Sandu et al. [15,16].

Since its indexing by Han et al. [17], hence the common name "H-phase", in a NiTiHf HTSMA, numerous groups have reported and characterized the same structure in both NiTiHf and NiTiZr HTSMAs [7,18–27]. Most notably Yang et al. [24], who performed high resolution S/TEM and atomic modeling to verify the stability of the orthorhombic structure in Ni-rich NiTiHf, but spectrum of compositions has been reported for the phase, most notably with respect to the Ti:(Zr/Hf) ratio while maintaining a Ni:(Ti,Hf/Zr)approximately equal to 1-1.2:1 [17,23,24,26–28]. By this point, NiTiZr HTSMAs have slowly been disregarded, not only due to their initial failures but also due to the rapid success of comparable Hf alloys; although Evirgen et al. [18,19] studied the microstructural and shape memory effects of slightly Ni-rich NiTiZr and showed that the addition of these nanoprecipitates of the appropriate size could not only increase TTs but also the shape memory properties and thermal stability of the alloys. However, to the authors' knowledge, there has yet to be a systematic study that examines how small changes in Ni-content of near equimolar compositions in NiTiZr HTSMAs can affect precipitation, TTs, transformation stability, and microstructure. It is the intent of this study to fill this void and to elucidate precipitation regions based on composition and aging temperature by examining three NiTi-20 at. % Zr HTSMAs with small changes in Ni-content. The three HTSMAS are subsequently characterized by differential scanning calorimetry (DSC), scanning electron microscopy (SEM), transmission electron microscopy (TEM), and high-energy synchrotron X-ray diffraction (SR-XRD).

2. Materials and Methods

Three 200 g ingots of NiTi-20 at. % Zr with Ni-content of 49.8, 50.0 and 50.2 at. % (referred to Ni-lean, equimolar and Ni-rich, respectively, throughout the paper) were arc-melted in an inert argon environment using high purity elemental Ni, Ti, and Zr at ATI Specialty Alloys and Components and supplied to the University of North Texas for further investigation. The as-cast ingots were then sectioned into approximate $7 \times 7 \times 45$ mm^3 strips and solutionized at 1000 °C for 1 h in air to remove any casting effects and subsequently mechanically polished to remove the formed oxide layer. Rolling of alloys was not attempted because all three compositions exhibited gross cracking during water quenching after solutionizing treatment. Solutionized samples were then sectioned into $7 \times 7 \times 1$ mm^3 slices, so that aging at 550, 600 and 800 °C for various times could be examined under the same starting metallurgical conditions. Aging at 550 °C was chosen since it is the generally accepted peak aging temperature for Ni-rich NiTiHf/Zr HTSMA [7,18,26]. Aging at 600 °C was chosen since this will increase the precipitation rate. Aging at 800 °C was chosen to investigate possible precipitation during hot rolling applications.

TTs were examined using a Netzsch differential scanning calorimetry (DSC) (Selb, Deutschland) 204 F1 Phoenix at a heating rate of 10 °C/min while under a helium environment to negate oxidation. Sample weights were kept between 20–30 mg and then thermally cycled from 50–400 °C. In addition, some samples were cycled 10 times to assess stability of the transformation. The austenitic and martensitic start, peak, and finish temperatures (A_s, A_p, A_f, M_s, M_p and M_f respectively) represent the onset, peak position and the end of the transformation peaks measured in the DSC experiments.

High-energy synchrotron radiation X-ray diffraction (SR-XRD,) measurements were collected at the Advanced Photon Source (APS) in Argonne National Laboratory at the sector 11-ID-C beam line in transmission mode with a sample thickness of approximately 1 mm. For ex situ experiments, Debye–Scherrer diffraction patterns (APS, Argonne, IL, USA) were taken at a beam energy of approximately 105.1 keV and rectangular beam size of 0.3×0.3 mm^2 for an exposure time of 0.5 s/frame incorporating 8 summed frames for approximately 4 s total exposure time for each measurement. Diffraction data was collected using a Perkin Elmer amorphous silicon detector, which was positioned approximately 1.8 m from the sample and calibrated using Ce_2O powder. For in situ experiments, Debye–Scherrer diffraction patterns were continuously collected as the sample was heated from 30 to 550 °C at a rate of 30 °C/min and held for 3 h to observe precipitate growth. These measurements were recorded at a beam energy of 105.1 keV with a beam size of 0.5×0.5 mm^2 for an exposure time of 0.1 s/frame incorporating 100 summed frames for an approximate total exposure time of 10 s for each diffraction pattern. A 2 mm stainless steel block was placed in front of the beam before the sample to act as an absorber to prevent overexposure of the diffraction patterns. Data analysis was performed using a combination of Fit2D [29], custom matlab code [30], Fiji [31], an imageJ package, for qualitative phase analysis, and MAUD [32], Materials Analysis Using Diffraction, for Reitveld refinement of the lattice parameters. Powder diffraction data for binary NiTi B19' (PDF number: 03-065-0145) [33] was used as a starting point for refinement of NiTiZr martensitic phase. Ti_2Ni phase structural parameters were obtained from the COD database [34] (COD ID: 2310267 [35]) and H-phase parameters were obtained from Yang et al. [21] and used for identification of XRD reflections.

Scanning electron microscopy (SEM) was performed using a Hitachi TM3030 (Tokyo, Japan). Samples were etched with a 26 mL glycerol, 6 mL concentrated HNO$_3$, and 1 mL concentrated HF solution for 5 s. Area fraction was calculated using a total image area of 0.25 mm^2 and measuring at least 3000 particles for each alloy. Transmission electron microscopy (TEM) samples were prepared using the lift-out method on an FEI Nova 200 Nanolab dual beam SEM (Hillboro, OR, USA)/focused ion beam (FIB) and thinned to an appropriate thickness for TEM imaging. High angle annular dark field scanning TEM (HAADF-STEM) images were collected using an FEI Techai G2 F20 S-Twin 200 keV field-emission S/TEM equipped with an EDS system to allow for compositional analysis. The HAADF detector specifically allows for Z-contrast imaging in STEM mode so that precipitation could easily be identified within the martensitic matrix.

3. Results

3.1. Differential Scanning Calorimetry (DSC) for Transformation Temperatures

Figure 1a–c shows the DSC data over two thermal cycles for the Ni-lean, Equimolar, and Ni-rich samples, respectively. In the solutionized condition, the Ni-lean alloy exhibits the highest TTs, followed by the equimolar alloy and then the Ni-rich alloy. After aging at 550 and 600 °C, all three alloys show different responses with respect to changes in TTs. For the Ni-lean case, both the austenitic and martensitic phase transformation peaks continually fall to lower temperatures and slightly broaden with increasing aging times. The Ni-rich alloy exhibits the opposite trend with respect to TTs position. As aging times increase, TTs also increase while a peak broadening is also observed. The equimolar alloy is observed to be less sensitive to aging effects with only minor changes in the DSC curves despite what aging treatment is employed. The one exception to this is the 550 °C for 1 h (lowest temperature and shortest time); a two-stage transformation is observed at this heat treatment.

Figure 2a–c shows the transformation peak positions, specifically A_f and M_s, from DSC measurements over ten thermal cycles for the solutionized, 550 °C for 3 h, 600 °C for 1 h, and 600 °C for 6 h conditions for the Ni-lean, equimolar, and Ni-rich alloy, respectively. For the rest of this article, we will define a stable transformation as a temperature drop less than 0.5 °C between consecutive cycles for both A_f and M_s. Using this criterion and examining the solutionized conditions, it is observed that the Ni-lean and equimolar exhibit a stable phase transformation after only three heating cycles, while the Ni-rich achieves stability after eight thermal cycles. Aging the Ni-lean and equimolar samples does not affect the overall stability of the transformation with all three aging treatments showing a transformation temperature stabilization after 3–4 heating cycles. For the Ni-rich sample, all three heat treatments again reach transformation stabilization after eight heating cycles, similar to the solutionized sample, but the overall transformation temperature drop for the Af, i.e., the transformation temperature at cycle one minus cycle ten, is approximately 15 °C less for all three heat treatments,

39.5 °C for the solutionized and approximately 25 °C for the heat-treated samples. Overall, the Ni-lean and the equimolar samples show much better stability in TTs than the Ni-rich sample no matter which heat treatment schedule is performed.

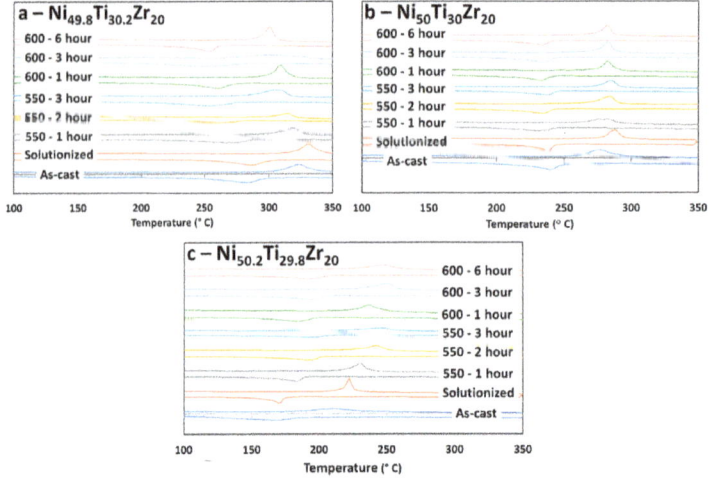

Figure 1. Offset Raw DSC data for the (**a**) Ni-lean, (**b**) equimolar and (**c**) Ni-rich alloys showing transformation temperatures (TTs) as a function of heat treatment.

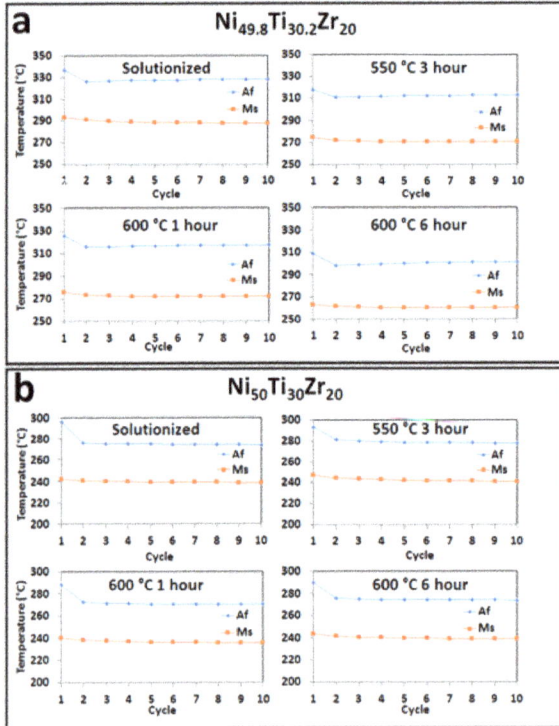

Figure 2. *Cont.*

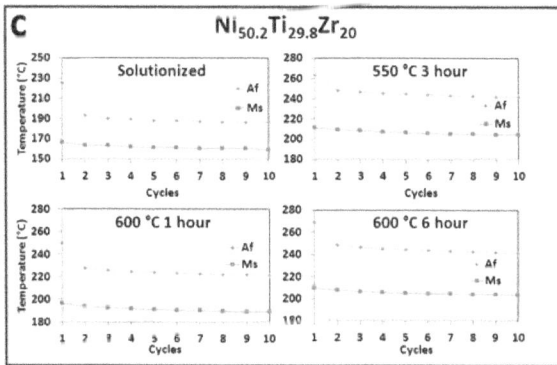

Figure 2. TT position acquired from DSC thermal cycling data for the (a) Ni-lean, (b) equimolar and (c) Ni-rich alloys over ten heating and cooling cycles. Only A_f and M_s are shown for clarity, however both A_s and M_f show a similar trend in all cases.

3.2. Microstructural and High Energy Synchrotron Radiation X-ray Diffraction (SR-XRD) Analysis

Figure 3a–c shows SEM images of the Ni-lean, equimolar, and Ni-rich alloy, respectively, in the solutionized condition. The micrographs indicate a mostly martensitic structure with the presence of $(Ti,Zr)_2Ni$ or $(Ti,Zr)_4Ox$ type precipitates in all the samples, black phases in the micrographs. These will be discussed as $(Ti,Zr)_2Ni$ precipitates for the remainder of this article. The area fraction of the $(Ti,Zr)_2Ni$ precipitates increases with decreasing Ni-content and the size of the precipitates also increases slightly with the Ni-lean showing a higher distribution of large particles. Other conditions, are not shown due to a lack of change associated with the micrographs at SEM resolution and therefore will no longer be discussed.

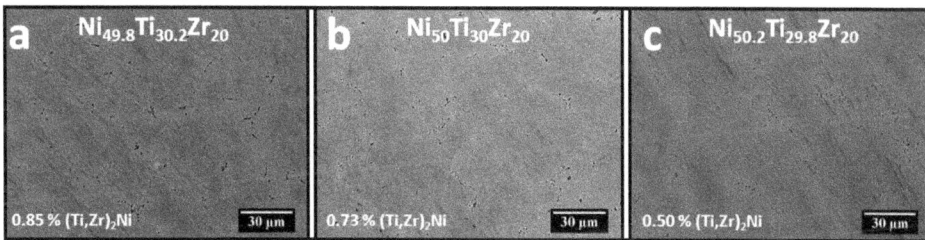

Figure 3. Backscattered electron (BSE) micrographs of the solutionized conditions for the (a) Ni-lean, (b) equimolar, and (c) Ni-rich alloys. All alloys show a similar martensitic structure with increasing amounts of the $(Ti,Zr)_2Ni$ precipitates (black phase) scattered within the martensitic matrix as (Ti,Zr) amount is increased. Calculated % area fraction is given in the bottom left of each micrograph.

Figure 4a–c shows both the full Debye–Scherrer diffraction patterns and the corresponding 1-D integrated diffraction pattern from ex situ experiments for the solutionized conditions of the Ni-lean, equimolar and Ni-rich alloy, respectively. The calculated lattice parameters using a Reitveld refinement fitting method are also given. All three alloys show similar results that indicate a fully monoclinic B19' structure and residual texture, i.e., oriented diffraction spots, associated with the casting process. In addition, all of the diffraction patterns appear "spotty" due to the small beam size in relation to the grain size although enough grains are being irradiated to confirm that the alloys exhibit polycrystalline behavior. Based on SEM images, it is estimated that approximately 2500 grains, martensitic lathes, are within the irradiated volume. Lastly, the only notable second phase present is a $(Ti,Zr)_2Ni$ peak centered at 2.26 Å, which is observable in all the samples but is highest in intensity in the Ni-lean case, Figure 4a. No residual austenite is detected in any of the solutionized conditions.

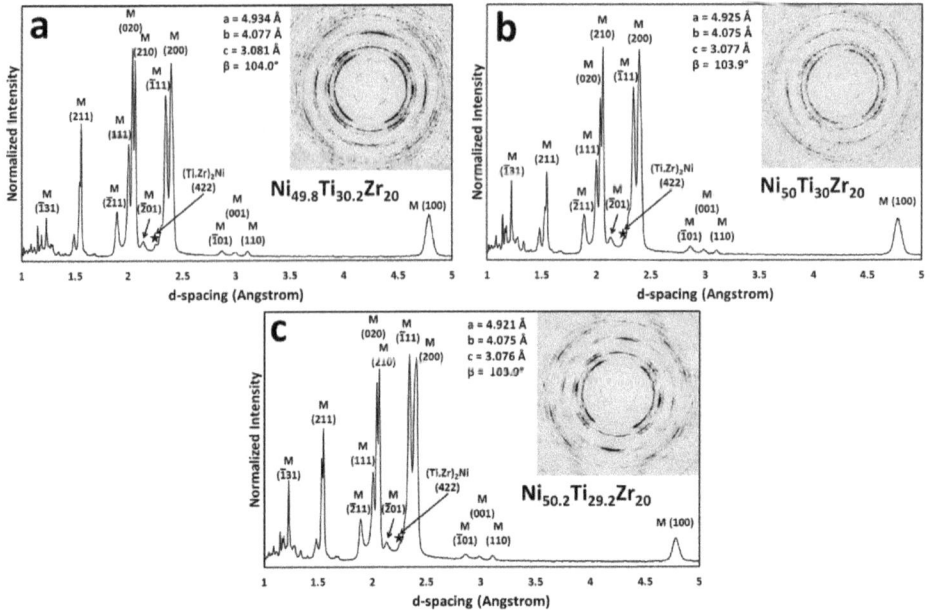

Figure 4. SR-XRD diffraction data from ex situ experiments including both the full Debye–Scherrer diffraction pattern and the integrated normalized 1-D diffraction pattern collected for the (**a**) Ni-lean, (**b**) equimolar, and (**c**) Ni-rich in the solutionized condition. Confirms the B19′ martensitic structure present in all three alloys at room temperature. Selected martensite peaks indexed for clarity and calculated lattice parameters obtained by the Reitveld method also included in each graph.

Figure 5a–c shows integrated 1-D diffraction patterns from ex situ experiments, which have been normalized and offset, for the 550 °C for 3 h, 600 °C for 6 h, and 800 °C for 3 h heat treatment conditions for the Ni-lean, equimolar and Ni-rich case. Looking at the Ni-lean case, Figure 5a, it is observed that a sharp peak is apparent between the martensitic fingers at approximately 2.19 Å and is indexed to be residual austenite residing in the martensite matrix. This peak is most prevalent in the 550 °C for 3 h and least visible in the 600 °C for 6 h condition, which is most likely due to the small interaction volume associated with the micro-beam used to collect the diffraction data in addition to the low intensity, spotty nature of the residual austenite throughout the matrix. The main point of interest however is tracking the (Ti,Zr)$_2$Ni precipitate phase, e.g., small peak at 2.25 Å. From the diffraction patterns taken during ex situ experiments, it is not readily obvious that any meaningful change in amount of precipitates can be observed. Figure 5b shows the diffraction data for the equimolar alloy. No observable phase growth is observed at any condition when the alloy is treated at 550 °C or 600 °C for any amount of time. However, once the temperature is raised to 800 °C for 3 h, the same peak for the residual austenite at 2.19 Å is observed. In Figure 5c, the Ni-rich 1-D diffraction data, a different broad peak is observed, centered at 2.21 Å, in the diffraction data when the alloy is treated at 550 °C or 600 °C. This peak has been indexed as the nano-strengthening H-phase seen in other NiTiZr and NiTiHf HTSMAs compositions [7,17,18,23,26,28] using the basis positions from Yang et al. [24]. Once the temperature increases to 800 °C, the presence of the broad H-phase peak is no longer observed, but rather the sharp residual austenite peak at 2.19 Å is observed instead.

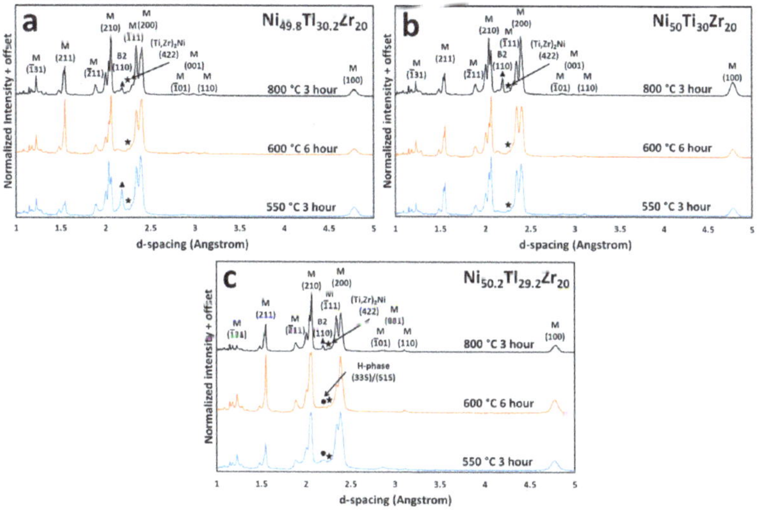

Figure 5. SR-XRD integrated normalized 1-D diffraction patterns from ex situ experiments for select heat treatment conditions for the (**a**) Ni-lean, (**b**) equimolar and (**c**) Ni-rich condition illustrating different precipitation growth. The appearance of residual austenite is evident in some of the conditions (peak at 2.19 Å indicated with a triangle). Growth of the $(Ti,Zr)_2Ni$ precipitates is not obvious in any of the samples (peak at 2.25 Å indicated by a star), while H-phase growth is observed in the Ni-rich alloy by the appearance of a broad peak (peak at 2.21 Å indicated by a circle)

Figure 6a–c shows HAADF-STEM images for the 600 °C 6 h heat treatment for the Ni-lean, equimolar, and Ni-rich alloys, respectively. For the Ni-lean case, a low contrast image is observed with no major precipitation on the sub-micron scale. The inset in Figure 6a shows small localized grain boundary precipitation on the martensitic lathes of the nanoscale H-phase and confirmed by EDS measurements, although this was the only location on the entire TEM liftout where these precipitates were observed. The equimolar alloy shows a few $(Ti,Zr)_2Ni$ precipitates but no major amount of precipitation; however, small pockets of H-phase along the martensitic lathes are observed. These are evident by the bright white spots in the micrographs due to the higher atomic concentration of Zr with respect to the matrix. The Ni-rich alloy exhibits an appreciable concentration of the lenticular H-phase precipitates on the order 130 ± 50 nm in length and 35 ± 8 nm in width homogenously dispersed throughout the martensitic matrix. In addition, two large $(Ti,Zr)_2Ni$ precipitates are also observed.

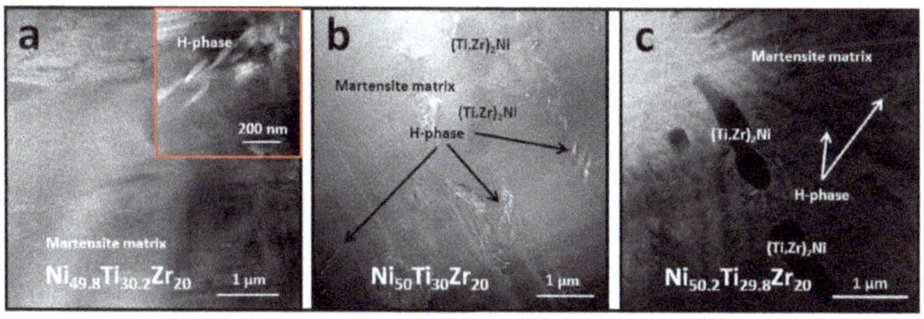

Figure 6. Representative HAADF-STEM images for the (**a**) Ni-lean, (**b**) equimolar and (**c**) Ni-rich alloys after a 600 °C 6 h heat treatment. The bright white lenticular phase are the H-phase due to its higher Zr content with respect to the matrix and the large dark phase seen in (**b**,**c**) correspond to $(Ti,Zr)_2Ni$ precipitates. Inset in (**a**) shows the only small pocket of H-phase precipitates within the Ni-lean martensitic matrix. In both (**b**,**c**) it is observed that $(Ti,Zr)_2Ni$ and H-phase precipitates are in the material, with more H-phase growth in the (**c**) Ni-rich alloy.

Figure 7 presents the measured average composition for the observed H-phase precipitates over the nominal Ni-content. It should be noted that only three H-phase precipitates were observed in the entire Ni-lean TEM liftout and only approximately 15–20 were found in the equimolar TEM liftout. The measured composition of the H-phase gives an approximate at. % Ni:(Ti,Zr) ratio of 1:1 for all three alloys but the relative amount of Ti and Zr appears to deviate when Ni-content is above 50 at. % with a 10 at. % increase in Zr-content for the measured precipitates in the Ni-rich alloy.

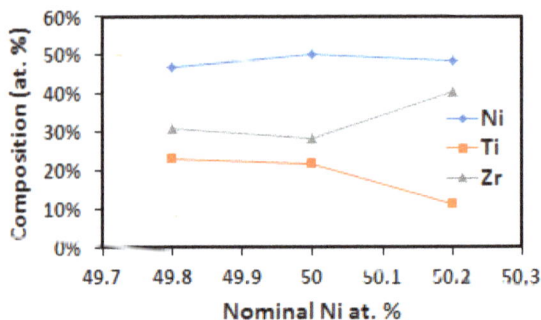

Figure 7. Average measured composition of the observed H-phase precipitates over nominal at. Ni content obtained with from EDS measurements.

Figure 8a–c shows the integrated normalized 1-D diffraction patterns from in situ SR-XRD experiments for the Ni-lean, equimolar, and Ni-rich alloys, respectively, at 550 °C for 0, 1, 2, and 3 h of the isothermal hold in the d-spacing range of 2.00–2.75 Å. XRD measurements collected during in situ experiments were performed with a slightly larger beam size, with respect to the diffraction data collected during ex situ experiments, to improve grain statistics. In addition, in situ measurements allow for observation of the phase evolution at a single location on the samples, and therefore remove any area dependent errors that can be produced using a micro-beam for ex situ measurements. In all cases, a small amount of the Ti_2O and hexagonal Ti_2ZrO are observed. Also, a small reflection originating from TiC is observed and likely was formed during casting of the alloys. Al_2O_3 peaks are also observed, but these are due to the insulation in the furnace used during the experiment. For the Ni-lean alloy, it is evident from the relative intensity of the austenite (110) and the $(Ti,Zr)_2Ni$ (422), that there exist a larger fraction of the $(Ti,Zr)_2Ni$ precipitates within the irradiated volume upon initially reaching 550 °C (0 h diffraction pattern) than is observed in either the equimolar or Ni-rich alloy. It is observed that the relative intensity of the austenite (110) and the $(Ti,Zr)_2Ni$ (422) is changing with time due to the decrease in intensity of the austenite (110) peak because of orientation changes within the B2 phase within the interaction volume, while the intensity of $(Ti,Zr)_2Ni$ (422) peak is relatively the same throughout the duration. As time is increased, the intensity of the other two $(Ti,Zr)_2Ni$ reflections shown in Figure 8 grow, that is the (511) at 2.16 Å and the (331) at 2.57 Å, throughout the duration of the experiment. This observation can be explained as the $(Ti,Zr)_2Ni$ phase does not act like a homogenous powder phase as it is in a low volume fraction. In the equimolar alloy, there exist an initial amount of the $(Ti,Zr)_2Ni$ precipitates but no major growth of the $(Ti,Zr)_2Ni$ phase is observed within the 3 h of the experiment. The Ni-rich alloy also shows an initial amount of $(Ti,Zr)_2Ni$ precipitates but with a much lower intensity with respect to both the Ni-lean and equimolar alloys. As the sample is heated, little change is observed with regards to phase precipitation.

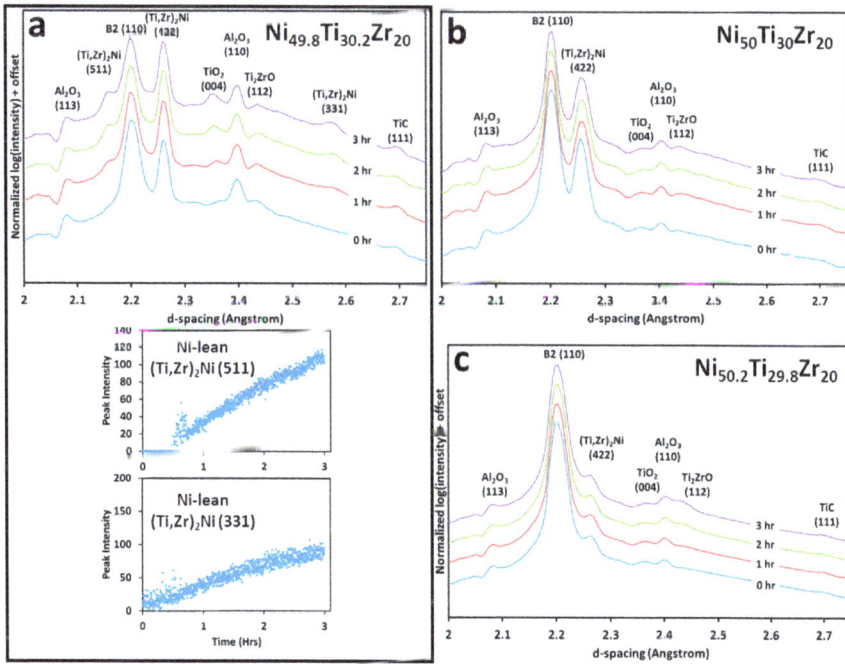

Figure 8. Integrated normalized 1-D diffraction patterns from the in situ SR-XRD experiments for an isothermal hold at 550 °C at 0, 1, 2, and 3 h over the d-spacing range 2–2.75 Å for the (**a**) Ni-lean with insets of the peak growth of the (331) and (511) for the (Ti,Zr)$_2$Ni phase, (**b**) equimolar, and (**c**) Ni-rich condition illustrating different precipitation and oxide growth. Al$_2$O$_3$ reflections correspond to the insulation in the furnace.

Figure 9a–c shows the integrated normalized 1-D diffraction patterns from the in situ SR-XRD experiments for the Ni-lean, equimolar, and Ni-rich alloys, respectively, at 550 °C for 0, 1, 2 and 3 h of the isothermal hold in the d-spacing range 3.4–3.8 Å with an inset of the 0 and 3 h stacked to emphasize the change. In this d-spacing range, it is possible to independently observe the growth of (206)/(220) H-phase reflections, centered at 3.57 and 3.61 Å, respectively. It is clear, that no measureable growth is observed in either the Ni-lean or equimolar but a small broad hump becomes visible as aging proceeds in the Ni-rich alloy indicating the growth of the H-phase precipitates. The peak is observed to be very low in intensity and extremely broad in the aging condition due to the volume fraction and size scale associated with its formation.

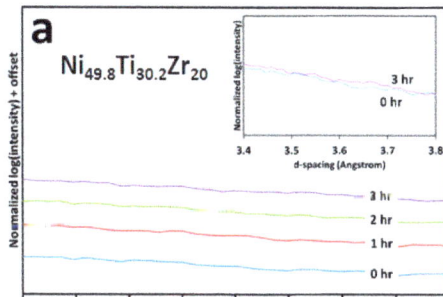

Figure 9. *Cont.*

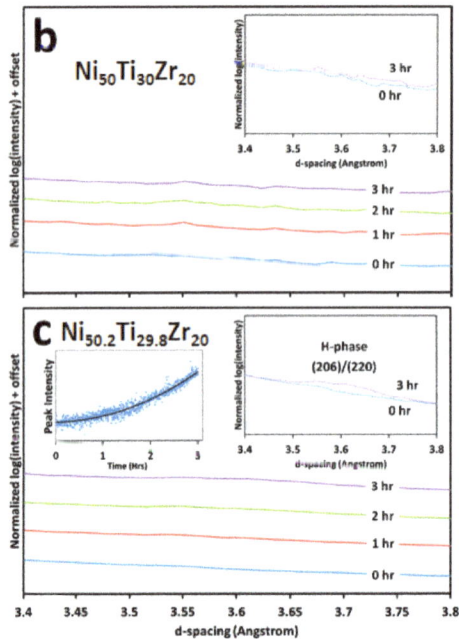

Figure 9. Integrated normalized 1-D diffraction patterns from the in situ SR-XRD experiment for an isothermal hold at 550 °C at 0, 1, 2 and 3 h over the d-spacing range 3.4–3.8 Å for the (**a**) Ni-lean, (**b**) equimolar, and (**c**) Ni-rich condition. The only measureable phase growth is due to the formation of the nanoscale H-phase precipitates. The insets allow for better visualization of the subtle change caused by the formation of the precipitate phase and in (**c**) the growth of the peak intensity over time is shown.

4. Discussion

4.1. Transformation Temperatures and Precipitation with Respect to Ni-Content

Figure 8a shows a plot of the transformation temperatures of the solutionized condition vs. nominal Ni at. %. It is important to note the sharp drop in TTs from the Ni-lean, 49.8 at. % Ni, to the Ni-rich, 50.2 at. % Ni, composition. There is an approximate 110 °C change in both the austenitic and martensitic TTs. When compared with previous research for binary NiTi of similar compositions ($Ni_{49.8}Ti_{50.2}$ and $Ni_{50.2}Ti_{49.8}$), only an approximate 36 °C change [36] is seen over this composition range. This confirms that strict composition control during production is even more important in NiTiZr HTSMAs than in conventional binary NiTi SMAs due to the inclusion of the Zr atoms substituting on the Ti-sites in the B19′ and B2 structures. All the alloys investigated in this study respond very differently when aging at 550 and 600 °C. For brevity, only the peak positions for the two characteristic phase transformations, A_p and M_p, will be discussed but other TT metrics, i.e., A_s, A_f, M_s and M_f follow the same trends.

Starting with the Ni-lean case, TTs drop as the alloy is aged at 550 and 600 °C. At 550 °C, a continual drop in both the A_p and M_p is observed from 1–3 h with a maximum drop of 25 °C and 31 °C for the A_p and M_p, respectively, after aging for 3 h. When the temperature is increased to 600 °C, the drop in TTs almost completely levels off after a 3 h heat treatment with a drop of approximately 30 and 33 °C for the A_p and M_p, respectively. The drop in TTs can be explained by the formation of the $(Ti,Zr)_2Ni$ precipitates, as confirmed by the SR-XRD patterns from the in situ experiment presented in Figure 8a, which upon their formation or coarsening will cause the depletion of Ti from the martensitic matrix, therefore, creating a slightly higher Ni-content in the B19′ structure, which is directly proven to drop TTs in Figure 10a. It can be concluded that the precipitation/coarsening rate of $(Ti,Zr)_2Ni$ precipitates is faster at 600 °C than at 550 °C, although no in situ data was collected at 600 °C, since the TTs drop at 600 °C for 1 h and 550 °C for 3 h is approximately the same, 25 °C for the A_p and 30 °C for the M_p, In addition, precipitation of the $(Ti,Zr)_2Ni$ phase is observed at 800 °C, thereby confirm its growth stability in the range of 550–800 °C.

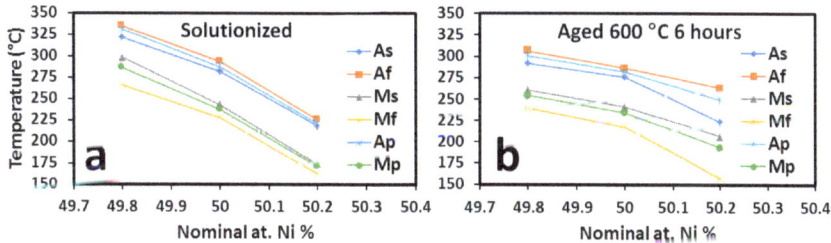

Figure 10. Transformation temperatures for the (**a**) solutionized and (**b**) aged at 600 °C for 6 h over the nominal at. Ni content. As the non-equimolar alloys are aged, the matrix composition moves closer to equimolar values therefore shifting transformation temperatures to overall closer values.

Aging the equimolar alloy show little change, with only slight drops, e.g., 5 and 4 °C for the A_p and M_p, respectively, and this behavior is confirmed by the in situ experiment in Figure 8b, in the TTs peak position observed when heat treating at 600 °C for 6 h. The exception to this is the 550 °C for 1 h, in which a multi-step transformation is observed. This is generally attributed to grain boundary precipitation or an intermediate R-phase transformation in binary NiTi alloys [37], and this is most likely the cause in this instance as H-phase grain boundary precipitation has been directly observed in HAADF-STEM image for the 600 °C for 6 h heat treatment, Figure 6b. Once the fine precipitates grow larger, the effect is no longer observed since the precipitate density along the grain boundaries has decreased due to coarsening [38]. The volume fraction of these H-phase precipitates is so small, and therefore no major changes in TTs peak position are observed. In addition, there is likely some co-precipitation of the $(Ti,Zr)_2Ni$ precipitates throughout the alloy due to local changes in composition throughout the bulk, thereby leaving the B19′ matrix composition relatively unaffected.

The Ni-rich alloy responds by increasing TTs upon heat treating at 550 and 600 °C. The maximum temperature increases observed are 28 and 22 °C for the A_p and M_p, respectively, due to the 600 °C 6 h treatment. The reason for this temperature increase can be attributed to the formation of the Ni-rich nano-scale H-phase observed in the SR-XRD data collected during both ex situ and in situ experiments (Figures 5c and 9c respectively) and the HAADF-STEM micrograph (Figure 6c) which cause a Ni depletion in the B19′ martensitic matrix. In addition, the homogenous distribution of the nano-scale H-phase, like Ni_4Ti_3 in binary NiTi, creates strain fields in the martensitic matrix that also aid in increasing transformation temperatures [2,7,18,19]; however, this amount of change in TTs is already achieved between 2–3 h at 600 °C, so it is likely that further heat treatments are causing coarsening of the precipitates rather an increase in phase amount. Aging at 550 °C has the same effect, but an obviously slower precipitation rate as the TTs observed at 550 °C for 3 h never fully reach the maximum observed when aging at 600 °C. This observation leads to the conclusion that the equilibrium volume fraction has yet to be reached and longer aging times at 550 °C can produce slightly higher TTs than what is reported in this article.

It should be noted that the ex situ experimental data (Figure 5) did not lead to solid experimental evidence with regards to the precipitation paths and the shifts in TTs. This can be attributed to the small interaction volume, the changing of the observation point, the low volume fraction change of the precipitates, and the high amount of reflections present due to the martensitic structure. All these factors together make it difficult to see the minor changes happening within the martensitic matrix due to the aging conditions. This result speaks to the limitations of experimental techniques to determine relative change in such compositionally dependent alloys, even with high resolution SR-XRD. There is still value in presenting the data, as not only does it show limitations, but also provides a link to the DSC experimental data and the aging conditions. The main conclusion, however, comes from the in situ experimental data (Figures 8 and 9) that Ni-lean alloys can shift TTs to lower temperature by aging due to the formation of (Ti, Zr)$_2$Ni (causing depletion of Ti from the matrix) and Ni-rich alloys can shift TTs to higher temperature by aging due to the formation of H-phase (causing depletion of Ni from the matrix).

Figure 10b presents the samples aged at 600 °C for 6 h over the nominal Ni at. %. Examining the peak positions of both the austenitic and martensitic transformation, it is observed that the transformation temperatures are much closer than in the previously solutionized condition. In this condition, there is now only 50 and 65 °C for the A$_p$ and M$_p$, respectively, because of the now closer B19' matrix compositions. It is evident that as the alloys are heat treated they are slowly moving toward an equimolar composition in the martensite, and therefore, TTs are moving closer together with respect to the solutionized condition.

Examining the thermal stability of the phase transformation with respect to Ni-content (Figure 2a–c), the Ni-lean and equimolar samples exhibit a much more stable transformation regardless of the heat treatment. This can be explained since the precipitation reactions have a much smaller effect at the cycling temperature. Cycling was performed up to 500 °C, allowing for the Ni-rich sample to potentially undergo changes in the H-phase size and amount. This is not the case in the Ni-lean sample as the formation of (Ti,Zr)$_2$Ni precipitates are generally already large (greater than 1 micron), and therefore do not cause changes in the TTs due to the dispersion of nano-precipitates which induce a strain field in the martensite matrix, and the precipitation rate of the (Ti,Zr)$_2$Ni precipitates is slow at this temperature; however, in the Ni-rich alloy, cycling to 500 °C can cause meaningful changes in the microstructure of the material, and therefore, the material is different from the previous cycle in regards to amount and internal strain produced by the nano-precipitation of H-phase. The equimolar alloy shows little sensitivity to aging at 550 and 600 °C, therefore, the phase transformations are already stable after only 2–4 thermal cycles.

4.2. Aging at 550 °C and 600 °C vs. Annealing at 800 °C

It has already been previously discussed that the precipitation reactions and sensitivity at 550 and 600 °C are very different in the Ni-lean, equimolar, and Ni-rich compositions even over the small composition range examined in this study; however, it is worth discussing that as temperature is increased by solutionizing or heat treating, the alloys begin to behave more similarly. Looking specifically at the 800 °C SR-XRD results presented in Figure 5a–c, it is observed that all three alloys show very similar diffraction patterns. At this temperature, the Ni-rich alloy shows the same sharp peak associated with the formation of residual austenite and does not show the broad peak associated with the growth of the H-phase. The lack of precipitation of the H-phase in the 800 °C 3 h condition can also be deduced by assuming that transformation temperatures would either (1) show little change or (2) drop slightly because of the relaxation of residual strain due to casting. Although the data is not presented in this work, DSC was performed on the Ni-rich 800 °C 3 h sample, A$_p$ = 220 °C and M$_p$ = 150 °C, and it was confirmed that TTs slightly decreased from the solutionized condition, A$_p$ = 221 °C and M$_p$ = 170 °C. Therefore, it is concluded that the formation of the nano-scale H-phase can be avoided, or already formed nano-scale H-phase can be removed by annealing at temperatures greater than 800 °C for Ni-rich NiTiZr HTSMAs.

4.3. Composition of the H-Phase

Figure 6a–c shows that there is an appreciable volume fraction of lenticular precipitates homogenously dispersed in the matrix of the Ni-rich alloy that is not apparent in either the Ni-lean or equimolar composition. EDS analysis confirms that the composition of the precipitate is consistent with the H-phase reported by other researchers [17,23,24,26–28], i.e., a slightly Ni-rich precipitate, as expected. As noted earlier, the formation of H-phase precipitates was also observed in both the Ni-lean and equimolar alloys, but only in small clusters centered along the grain boundaries of the martensitic lathes. As expected, these clusters were more prevalent in the equimolar alloy, due to the slightly higher Ni-content, but still were not in a significant amount. However, EDS analysis was performed on the precipitates for each alloy (Figure 7), and the results suggest that the H-phase has a variable (Ti,Zr) composition based on the alloy composition. This variable Zr content change would adjust the Zr content in the matrix as the precipitates form, but it has been shown that changes in Zr content only affect TTs by 0.17 °C/0.1 at. % Zr [11]; therefore, this effect is muted by the changing Ni content which as shown in this article results in an approximate 27.5 °C/0.1 at. % Ni in the range studied. Hornbuckle et al. [28] investigated the possibility of compositionally dependent H-phase based on heat treatment time and found no statistical difference but stopped short of examining alloy composition with respect to H-phase composition. However, reported results of H-phase composition have differed vastly since its discovery leading to the hypothesis that H-phase could have a composition range depending on the alloy studied [17,23,24,26–28]. Our results appear to support this position, but more rigorous methods (i.e., atom probe tomography) are needed to completely verify this hypothesis

5. Conclusions

Three different NiTi-20 at. % Zr alloys with varying Ni-content were created and characterized using DSC, SR-XRD, SEM, and TEM. Based on analysis of the data, the following conclusions were made:

1. Small changes in at. % Ni have a dramatic effect on the TTs of the NiTi-20 at. % Zr system, even more so than in binary NiTi. From 49.8–50.2 at. % Ni, a drop of the 110 °C of the A_f is observed while in binary the same composition range only changes the A_f by approximately 36 °C, meaning composition control is even more crucial in the NiTi-20 at. % Zr system.

2. TTs can be tuned for a given application through aging treatments. The Ni-rich alloy, 50.2 at. %, can be shifted to higher transformation temperatures due to the formation of nanoscale H-phase precipitates, while the Ni-lean alloy can be shifted to lower transformation temperatures due to the formation of $(Ti,Zr)_2Ni$ precipitates. In both Ni-lean and Ni-rich cases, precipitation results in moving the martensite matrix composition toward the equimolar composition. The equimolar alloy, however, shows little response to aging.

3. The Ni-rich alloy forms fine nano-scale H-phase precipitates when aging at lower temperatures, i.e., 550 and 600 °C, but these precipitates no longer form when temperature is increased, i.e., 800 °C. Although equimolar and Ni-lean alloys do not show an appreciable amount of H-phase, some local pockets exist along martensitic grain boundaries.

4. H-phase is shown to shift compositionally with changes in Ni-content. Specifically by increasing the Zr content in the precipitate from 30 at. % to 40 at. % Zr in the equimolar and Ni-rich conditions, respectively. This behavior helps explain the wide range of H-phase compositions reported in the literature in both NiTi-Hf and -Zr alloys.

Acknowledgments: The authors acknowledge Nathan Ley for collecting ex situ SR-XRD measurements. The authors acknowledge the Materials Research Facility (MRF) at the University of North Texas for access to electron microscopes used in the characterization. This research used resources of the Advanced Photon Source, a U.S. Department of Energy (DOE) Office of Science User Facility operated for the DOE Office of Science by Argonne National Laboratory under Contract No. DE-AC02-06CH11357. The authors would also like to thank Yang Ren at APS sector 11 for help with collecting diffraction patterns.

Author Contributions: M.C. performed the experiments, analyzed the data, and wrote/edited manuscript; J.D.S. performed TEM imaging of samples; B.V.D. provided samples and funding; M.L.Y. designed experiments, provide experimental feedback, and wrote/edited manuscript.

Conflicts of Interest: The authors declare no conflict of interest.

References

1. Buehler, W.; Wiley, R. Tini-ductile intermetallic compound. *Trans. Am. Soc. Met.* **1962**, *55*, 269–276.
2. Otsuka, K.; Ren, X. Physical metallurgy of Ti–Ni-based shape memory alloys. *Prog. Mater. Sci.* **2005**, *50*, 511–678. [CrossRef]
3. Mohd Jani, J.; Leary, M.; Subic, A.; Gibson, M.A. A review of shape memory alloy research, applications and opportunities. *Mater. Des. (1980–2015)* **2014**, *56*, 1078–1113. [CrossRef]
4. Van Humbeeck, J. Non-medical applications of shape memory alloys. *Mater. Sci. Eng. A* **1999**, *273–275*, 134–148. [CrossRef]
5. Pelton, A.; Duerig, T.; Stöckel, D. A guide to shape memory and superelasticity in nitinol medical devices. *Minim. Invasive Ther. Allied Technol.* **2004**, *13*, 218–221. [CrossRef] [PubMed]
6. Benafan, O.; Brown, J.; Calkins, F.; Kumar, P.; Stebner, A.; Turner, T.; Vaidyanathan, R.; Webster, J.; Young, M Shape memory alloy actuator design: Casmart collaborative best practices and case studies. *Int. J. Mech. Mater. Des.* **2014**, *10*, 1–42. [CrossRef]
7. Benafan, O.; Noebe, R.; Padula, S., II; Vaidyanathan, R. Microstructural response during isothermal and isobaric loading of a precipitation-strengthened Ni-29.7Ti-20Hf high-temperature shape memory alloy. *Metall. Mater. Trans. A* **2012**, *43*, 4539–1552. [CrossRef]
8. Hartl, D.J.; Lagoudas, D.C. Aerospace applications of shape memory alloys. *Proc. Inst. Mech. Eng. Part G J. Aerosp. Eng.* **2007**, *221*, 535–552. [CrossRef]
9. Calkins, F.; Mabe, J.; Ruggeri, R. Overview of boeing's shape memory alloy based morphing aerostructures. In Proceedings of the SMASIS08: ASME Conference on Smart Materials, Adaptive Structures and Intelligent Systems, Ellicott City, MD, USA, 28–30 October 2008; pp. 1–11.
10. Eckelmeyer, K. The effect of alloying on the shape memory phenomenon in nitinol. *Scr. Metall.* **1976**, *10*, 667–672. [CrossRef]
11. David, N.A.I.; Thoma, P.E.; Kao, M.Y.; Angst, D.R. High Transformation Temperature Shape Memory Alloy. U.S. Patents 5114504, 19 May 1992.
12. Hsieh, S.; Wu, S. A study on lattice parameters of martensite in $Ti_{50.5-x}Ni_{49.5}Zr_x$ shape memory alloys. *J. Alloys Compd.* **1998**, *270*, 237–241. [CrossRef]
13. Hsieh, S.F.; Wu, S.K. Room-temperature phases observed in $Ti_{53-x}Ni_{47}Zr_x$ high-temperature shape memory alloys. *J. Alloys Compd.* **1998**, *266*, 276–282. [CrossRef]
14. Pu, Z.J.; Tseng, H.-K.; Wu, K.-H. Martensite transformation and shape memory effect of NiTi-Zr high-temperature shape memory alloys. *Proc. SPIE* **1995**, *2441*, 171–178.
15. Sandu, A.M.; Tsuchiya, K.; Yamamoto, S.; Todaka, Y.; Umemoto, M. Influence of isothermal ageing on mechanical behaviour in Ni-rich Ti–Zr–Ni shape memory alloy. *Scr. Mater.* **2006**, *55*, 1079–1082. [CrossRef]
16. Sandu, A.M.; Tsuchiya, K.; Tabuchi, M.; Yamamoto, S.; Todaka, Y.; Umemoto, M. Microstructural evolution during isothermal aging in Ni-rich Ti-Zr-Ni shape memory alloys. *Mater. Trans.* **2007**, *48*, 432–438. [CrossRef]
17. Han, X.D.; Wang, R.; Zhang, Z.; Yang, D.Z. A new precipitate phase in a tinihf high temperature shape memory alloy. *Acta Mater.* **1998**, *46*, 273–281. [CrossRef]
18. Evirgen, A.; Karaman, I.; Noebe, R.; Santamarta, R.; Pons, J. Effect of precipitation on the microstructure and the shape memory response of the $Ni_{50.3}Ti_{29.7}Zr_{20}$ high temperature shape memory alloy. *Scr. Mater.* **2013**, *69*, 354–357. [CrossRef]
19. Evirgen, A.; Karaman, I.; Santamarta, R.; Pons, J.; Hayrettin, C.; Noebe, R.D. Relationship between crystallographic compatibility and thermal hysteresis in Ni-rich NiTiHf and NiTiZr high temperature shape memory alloys. *Acta Mater.* **2016**, *121*, 374–383. [CrossRef]
20. Karaca, H.; Acar, E.; Tobe, H.; Saghaian, S. NiTiHf-based shape memory alloys. *Mater. Sci. Technol.* **2014**, *30*, 1530–1544. [CrossRef]
21. Meng, X.; Cai, W.; Zheng, Y.; Zhao, L. Phase transformation and precipitation in aged Ti–Ni–Hf high-temperature shape memory alloys. *Mater. Sci. Eng. A* **2006**, *438*, 666–670. [CrossRef]
22. Pérez-Sierra, A.M.; Pons, J.; Santamarta, R.; Karaman, I.; Noebe, R.D. Stability of a Ni-rich Ni-Ti-Zr high temperature shape memory alloy upon low temperature aging and thermal cycling. *Scr. Mater.* **2016**, *124*, 47–50. [CrossRef]
23. Prasher, M.; Sen, D. Influence of aging on phase transformation and microstructure of $Ni_{50.3}Ti_{29.7}Zr_{20}$ high temperature shape memory alloy. *J. Alloys Compd.* **2014**, *615*, 469–474. [CrossRef]

24. Yang, F.; Coughlin, D.R.; Phillips, P.J.; Yang, L.; Devaraj, A.; Kovarik, L.; Noebe, R.D.; Mills, M.J. Structure analysis of a precipitate phase in an Ni-rich high-temperature NiTiHf shape memory alloy. *Acta Mater.* **2013**, *61*, 3335–3346. [CrossRef]

25. Stebner, A.P.; Bigelow, G.S.; Yang, J.; Shukla, D.P.; Saghaian, S.M.; Rogers, R.; Garg, A.; Karaca, H.E.; Chumlyakov, Y.; Bhattacharya, K.; et al. Transformation strains and temperatures of a Nickel–Titanium–Hafnium high temperature shape memory alloy. *Acta Mater.* **2014**, *76*, 40–53. [CrossRef]

26. Coughlin, D.R.; Phillips, P.J.; Bigelow, G.S.; Garg, A.; Noebe, R.D.; Mills, M.J. Characterization of the microstructure and mechanical properties of a 50.3Ni–29.7Ti–20Hf shape memory alloy. *Scr. Mater.* **2012**, *67*, 112–115. [CrossRef]

27. Saghaian, S.M.; Karaca, H.E.; Tobe, H.; Pons, J.; Santamarta, R.; Chumlyakov, Y.I.; Noebe, R.D. Effects of Ni content on the shape memory properties and microstructure of Ni-rich NiTi-20Hf alloys. *Smart Mater. Struct.* **2016**, *25*, 095029. [CrossRef]

28. Hornbuckle, B.C.; Sasaki, T.T.; Bigelow, G.S.; Noebe, R.D.; Weaver, M.L.; Thompson, G.B. Structure–property relationships in a precipitation strengthened Ni–29.7Ti–20Hf (at%) shape memory alloy. *Mater. Sci. Eng. A* **2015**, *637*, 63–69. [CrossRef]

29. Hammersley, A. *Fit2d: An Introduction and Overview*; European Synchrotron Radiation Facility Internal Report ESRF9/HA02T; ESRF: Grenoble, France, 1997; Volume 68.

30. Young, M.; Almer, J.; Daymond, M.; Haeffner, D.; Dunand, D. Load partitioning between ferrite and cementite during elasto-plastic deformation of an ultrahigh-carbon steel. *Acta Mater.* **2007**, *55*, 1999–2011. [CrossRef]

31. Schindelin, J.; Arganda-Carreras, I.; Frise, E.; Kaynig, V.; Longair, M.; Pietzsch, T.; Preibisch, S.; Rueden, C.; Saalfeld, S.; Schmid, B. Fiji: An open-source platform for biological-image analysis. *Nat. Methods* **2012**, *9*, 676–682. [CrossRef] [PubMed]

32. Lutterotti, L.; Matthies, S.; Wenk, H. Maud: A friendly java program for material analysis using diffraction. *IUCr Newslett. CPD* **1999**, *21*, 14–15.

33. Kudoh, Y.; Tokonami, M.; Miyazaki, S.; Otsuka, K. Crystal structure of the martensite in Ti-49.2 at. % Ni alloy analyzed by the single crystal x-ray diffraction method. *Acta Metall.* **1985**, *33*, 2049–2056. [CrossRef]

34. Gražulis, S.; Daškevič, A.; Merkys, A.; Chateigner, D.; Lutterotti, L.; Quiros, M.; Serebryanaya, N.R.; Moeck, P.; Downs, R.T.; Le Bail, A. Crystallography open database (cod): An open-access collection of crystal structures and platform for world-wide collaboration. *Nucleic Acids Res.* **2011**, *40*, D420–D427. [CrossRef] [PubMed]

35. Yurko, G.; Barton, J.; Parr, J. The crystal structure of Ti2Ni (a correction). *Acta Crystallogr.* **1962**, *15*, 1309. [CrossRef]

36. Frenzel, J.; George, E.P.; Dlouhy, A.; Somsen, C.; Wagner, M.F.X.; Eggeler, G. Influence of Ni on martensitic phase transformations in NiTi shape memory alloys. *Acta Mater.* **2010**, *58*, 3444–3458. [CrossRef]

37. Allafi, J.K.; Ren, X.; Eggeler, G. The mechanism of multistage martensitic transformations in aged Ni-rich NiTi shape memory alloys. *Acta Mater.* **2002**, *50*, 793–803. [CrossRef]

38. Moshref-Javadi, M.; Seyedein, S.H.; Salehi, M.T.; Aboutalebi, M.R. Age-induced multi-stage transformation in a Ni-rich NiTiHf alloy. *Acta Mater.* **2013**, *61*, 2583–2594. [CrossRef]

MDPI AG

St. Alban-Anlage 66

4052 Basel, Switzerland

Tel. +41 61 683 77 34

Fax +41 61 302 89 18

http://www.mdpi.com

Metals Editorial Office

E-mail: metals@mdpi.com

http://www.mdpi.com/journal/metals

www.ingramcontent.com/pod-product-compliance
Lightning Source LLC
Chambersburg PA
CBHW051854210326
41597CB00033B/5893